# Farm Management in Peasant Agriculture

**Also of Interest**

*Farming Systems in the Nigerian Savanna: Research and Strategies for Development,* David W. Norman, Emmy B. Simmons, and Henry M. Hays

† *Farming Systems Research and Development: Guidelines for Developing Countries,* W. W. Shaner, P. F. Philipp, and W. R. Schmehl

*Readings in Farming Systems Research and Development,* edited by W. W. Shaner, P. F. Philipp, and W. R. Schmehl

*Multiple Cropping and Tropical Farming Systems,* Willem C. Beets

† Available in hardcover and paperback

# Farm Management in Peasant Agriculture

Michael Collinson

Routledge
Taylor & Francis Group
LONDON AND NEW YORK

First published 1983 by Westview Press

Published 2018 by Routledge
52 Vanderbilt Avenue, New York, NY 10017
2 Park Square, Milton Park, Abingdon, Oxon OX14 4RN

*Routledge is an imprint of the Taylor & Francis Group, an informa business*



Library of Congress Catalog Card Number: 83-50090
ISBN 0-86531-558-2

ISBN 13: 978-0-367-01977-8 (hbk)

## ACKNOWLEDGMENTS

This book is the result of a period of work in various roles in African agriculture, and my thanks are due to a wide range of people who have molded my views between 1959 and 1971.

Despite periods of more formal study over these years there is no doubt that the farmers of Sukumaland, Tanzania, played the largest role in instilling at least some understanding of the complexities of traditional African agricultural systems. I owe them my biggest debt. With them, I remember the enumerators who made up the core of the survey team between 1961 and 1966: Lawrence Komanya, Lawrence Mzee, Lawrence Vita, Bartholomew Nyalada, Benedict Primi, and Ernest Methusaleh. Their tolerance for "modest" living conditions in the field was significantly greater than my own, they taught me many valuable lessons, and we learned others together.

Last, I am grateful for the financial support of the Overseas Development Administration, which allowed me to write up the work, and to the Department of Agricultural Economics, the University of Reading. Within the department I am particularly grateful to Doug Thornton for his help in sifting some of the wood from the trees in what is a fairly wide-ranging study.

## CONTENTS

Chapter | Page

| Chapter | | Page |
|---|---|---|
| | **PART III: PLANNING EXTENSION STRATEGY AND CONTENT** | |

| Chapter | Page |
|---|---|

# LIST OF TABLES

| Table | | Page |
|---|---|---|

| Table | | Page |
|---|---|---|

| Table | | Page |
|---|---|---|

| Table | | Page |
|---|---|---|

| Table | | Page |
|---|---|---|

## LIST OF FIGURES

## LIST OF ACTIVITY CODES

### EXISTING ENTERPRISES

| | |
|---|---|
| EML | Early maize, legumes with sweet potatoes. |
| ECML | Early maize, cassava, legumes with sweet potatoes. |
| EP | Early sweet potatoes. |
| R | Rice. |
| LCML | Late maize, cassava, legumes, and sweet potatoes. |
| LML | Late maize, legumes, and sweet potatoes. |
| LP | Late sweet potatoes. |
| OC | Old cassava (its second year in the ground). |
| COT | Cotton. |

### IMPROVED ENTERPRISES: CASH CROPS

| | |
|---|---|
| COTA | Cotton. Variable time of planting: Dec. 7 to Jan. 22. Optimal plant population and timely weeding. |
| COTB | Cotton. Time of planting: Jan. 1. Optimal plant population and timely weeding. |
| COTC | Cotton. Optimal time of planting: Dec. 7. Optimal plant population and timely weeding. |
| COTD | Cotton. Time of planting: Jan. 22. Optimal plant population and timely weeding. |
| COTE | Cotton. As COTA, plus fertilizer: 112 lbs. 21 percent sulphate of ammonia topdressed and 112 lbs. double superphosphate. |
| COTF | Cotton. As COTB, plus fertilizer as for COTE. |

| | | |
|---|---|---|
| COTG | Cotton. | As COTC, plus fertilizer as for COTE. |
| COTH | Cotton. | As COTD, plus fertilizer as for COTE. |
| COTI | Cotton. | As COTA, plus insecticides: Six applications of one lb. active ingredient of 75 percent DDT wettable powder. First application tenth week after germination and subsequently at ten-day intervals. |
| COTJ | Cotton. | As COTB, plus insecticide as for COTI. |
| COTK | Cotton. | As COTC, plus insecticide as for COTI. |
| COTL | Cotton. | As COTD, plus insecticide as for COTI. |
| COTM | Cotton. | As COTA, plus fertilizer as for COTF, plus insecticide as for COTI. |
| COTN | Cotton. | As COTB, plus fertilizer as for COTF, plus insecticide as for COTI. |
| COTO | Cotton. | As COTC, plus fertilizer as for COTF, plus insecticide as for COTI. |
| COTP | Cotton. | As COTD, plus fertilizer as for COTF, plus insecticide as for COTI. |

## IMPROVED ENTERPRISES: FOOD CROPS

| | | |
|---|---|---|
| MAA | Maize. | Variable time of planting: Nov. 30 to Feb. 28. Optimal plant population and timely weeding. |
| MAB | Maize. | Variable time of planting: Nov. 30 to Jan. 15. Optimal plant population and timely weeding. |
| MAC | Maize. | Time of planting: Dec. 15. Optimal plant population and timely weeding. |
| MAD | Maize. | Time of planting: Jan. 15. Optimal plant population and timely weeding. |
| MAE | Maize. | Time of planting: Feb. 7. Optimal plant population and timely weeding. |

| | | |
|---|---|---|
| MAF | Maize. | As MAA, plus fertilizer: 224 lbs. 21 percent sulphate of ammonia, topdressed, plus insecticide: 51 lbs. 5 percent DDT dust shaken into plant funnels as required. |
| MAG | Maize. | As MAB, plus fertilizer and insecticide as for MAF. |
| MAH | Maize. | As MAC, plus fertilizer and insecticide as for MAF. |
| MAI | Maize. | As MAD, plus fertilizer and insecticide as for MAF. |
| MAJ | Maize. | As MAE, plus fertilizer and insecticide as for MAF. |
| GNA | Groundnuts. | Variable time of planting: Dec. 15 to Jan. 25. Optimal plant population and timely weeding. |
| GNB | Groundnuts. | Time of planting: Dec. 30. Optimal plant population and timely weeding. |
| GNC | Groundnuts. | Optimal time of planting: Dec. 7. Optimal plant population and timely weeding. |
| GND | Groundnuts. | Time of planting: Jan. 15. Optimal plant population and timely weeding. |
| GNE | Groundnuts. | Time of planting: Feb. 7. Optimal plant population and timely weeding. |
| MGN | Maize and Groundnuts. | Time of planting: Dec. 15. Optimal plant population and timely weeding. |
| CASSA | Cassava. | Time of planting: Nov. 25. Optimal plant population and timely weeding. |
| CASSB | Cassava. | Time of planting: Mar. 15. Optimal plant population and timely weeding. |
| SPOTA | Sweet Potatoes. | Time of planting: Nov. 15. Present plant population and weeding. |
| SPOTB | Sweet Potatoes. | Time of planting: Jan. 15. Present plant population and weeding. |

# PREFACE

This book was originally published in 1972; university colleagues who were using it to teach farm management for developing African countries urged me to have it republished. The only additions are this brief preface to point out parts of the book I believe to be useful for students, teachers, and researchers, and a brief postscript, summarizing developments over the last ten years in the orientation and approach of farm management as it applies to peasant farming.

Part I of the book introduces an approach (given in Chapter 6) for farm management economists to apply their discipline, in a cost-effective way, to the development of peasant agriculture. The approach is a departure from traditional Western approaches to the application of farm management. Due to the small scale of farms in developing countries and the dearth of farm management economists, the discipline must contribute at the system level, rather than the individual farm level. To this end, the approach evaluates the potential impact of extension-program content on the goals, managerial task, and resource productivity embodied in the existing systems of small farmers.

Parts II and III of the book detail the methods for implementing this system-level approach for farm management. Part II is valuable for its discussion of methods of investigating farm management in peasant agriculture. Chapter 7 covers farm classification as a prerequisite to investigation and analyzes the accuracy/cost compromise at the heart of all survey design. Chapter 8 sets out a method of building representative farm models that seek some control of aggregation bias. Chapters 9–13 give a detailed description and comparison of data-collection methods for key attributes in investigation. Chapter 14 draws conclusions for survey organization and design, using different data-collection methods.

Part III, as a whole, is an example of how evaluation criteria important to farmers are used in interpreting the results of modeling. The planning sequence set out here is based on the concept that small farmers move away from their existing systems in relatively small steps consistent with their goals, low-cash incomes, low-risk preferences, and the dominant pressures on their available land and labor resources. The aim in planning is a sequence of steps for the development of the system that draws from available research results, is acceptable to farmers, and is consistent with foreseeable changes in sector economics. Each step represents the content of an extension program for farmers operating the system. Chapter 16 reviews problems in using research results for planning extension work.

I feel the book has aged well and believe this new edition will be useful to students, teachers, researchers, and farm economists in applying the principles of farm management to peasant agriculture in developing countries.

*M. P. Collinson*
Nairobi, Kenya
Spring 1983

# Farm Management in Peasant Agriculture

CHAPTER

# 1

# INTRODUCTION

This book describes the way in which farm management economics can make an effective contribution to the development of traditional African agriculture. The importance of agriculture for national economic growth is reflected by the proliferation of literature and of aid programs for agricultural development. A contribution from management economics has been called for in the literature, but there has been no comprehensive application of the discipline in development programs. Part I outlines some reasons for this failure. Parts II and III detail the conditions of rurally based African economies within which the discipline must operate and set out a sequence of methods of investigation and planning for an effective contribution under these conditions.

A wide range of roles has been attributed to agriculture in economic development: as a source of food for the expanding urban population and as a captive market for expanding industrial production. It is seen as a source of both labor and capital for industrialization, with export crops in particular earning foreign exchange to pay for vital capital goods. More recently, in the African context at least, it has been seen as a sponge for the growing numbers of urban unemployed. Not all these roles are compatible; combining them requires a balance of government policy and action often difficult to achieve: for example, agriculture as a source of industrial capital and a market for industrial products has contradictory implications for tax and fiscal policy. The point to be stressed here is that all, or indeed any, of these roles demand increasing productivity in agriculture.

Nowhere is the importance of increasing agricultural productivity greater than in the rurally dominated economies of Africa. Table 1

TABLE 1

Agricultural and Nonagricultural Working Populations, Selected African Countries

| Country | Percent Economically Active Agriculture | Other |
|---|---|---|
| Ethiopia | 90 | 10 |
| Guinea | 87 | 13 |
| Nigeria | 78 | 22 |
| Cameroon | 91 | 9 |
| Ghana | 70 | 30 |
| Ivory Coast | 91 | 9 |
| Kenya | 80 | 20 |
| Tanzania | 88 | 12 |

Source: U.N. Economic Commission for Africa Estimates, in M.J. Herskovits and M. Horowitz, eds., Economic Transition in Africa (Evanston, Illinois: Northwestern University Press, 1964), with Tanzania added.

illustrates the extent of rural domination in selected African countries and presents estimates of the proportions of the economically active population engaged in agriculture and other livelihoods.

Throughout the continent over 100 million people are dependent on traditional forms of agriculture, a number which must increase. Several writers have described the arithmetic of structural development, emphasizing that the high rates of population growth, 2-3 percent per annum, aggravate the difficulty of shifting from an overwhelming dependence on agriculture to a balance among primary production, manufacturing industry, and services. The trend can be illustrated by the example of Tanzania over the twenty years 1948-67, using the proportion of the population centered in towns as a measure of structural change. (See Table 2.)

TABLE 2

The Growth of Rural- and Urban-Based Population in Tanzania, 1948-67

| Area | | 1948 | 1957 | 1967 |
|---|---|---|---|---|
| Rural | Thousands | 7,494.6 | 8,647.9 | 11,470.2 |
| | Percent | 96.8 | 95.2 | 93.8 |
| Urban | Thousands | 250.0 | 436.1 | 761.1 |
| | Percent | 3.2 | 4.8 | 6.2 |
| Total Population | Thousands | 7,744.6 | 9,084.0 | 12,231.3 |
| Annual Growth Rate | Percent | 1.8 | 3.0 | 2.4 |

Source: Tanzania Central Statistical Bureau, Tanzania: Population Changes 1948-67 (Dar es Salaam, 1968).

H. M. Southworth and B. F. Johnston provide an example with 80 percent of the country's labor now in agriculture and with the total labor force increasing at a rate of 2 percent per year—somewhat below prevailing rates in many countries—and the nonagricultural force at 3 percent per year. They show that 67 percent of the total labor force will still be in agriculture in fifty years, the first downturn in the absolute size of the agriculture labor force appearing after 125 years.[1] The problem is clearly not a transitory one. The search for national economic growth is destined to retain its agricultural roots for a long time.

The establishment of national government at independence brought increased emphasis on organized effort for development. At the same time, during the late 1950's and early 1960's a few farm economists began to appear in the agricultural departments and faculties of what had been British Africa. During the 1960's there were appeals in the literature for a contribution from farm economists to the development effort. J. C. De Wilde, in his survey of tropical agricultural development in Africa, stressed the need for farm management studies to "diagnose the problems confronting the farmer and devise more economic ways of using available resources."[2] The reasons why the discipline made little impression in this period are readily brought out by a brief review of the history of farm economics in East Africa during the 1960's.

Of the three East African countries Kenya was the exception at the beginning of the 1960's and remains so today. Kenya had an agricultural economist in the treasury in the 1930's and the 1950's; and although farm surveys started in the mid-1950's, they concentrated wholly on the large-scale European farms, an orientation from which they have not yet fully recovered in terms of objectives and methodology. Agricultural economic data was originally collected by lecturers at Egerton College, Njoro, for teaching purposes. In 1960 this was formalized by the establishment of the Farm Economics Survey Unit, initially located within the Ministry of Agriculture but transferred in 1963 to the Statistics Division of the Ministry of Economic Planning, where it remains. The unit turned toward African smallholder farming, although little progress was made until commitments to programs on large-scale farms were completed in 1962.[3] Some twenty-three farm economic reports were issued by 1964 and contained descriptive and comparative information on both large and small farms. After 1964 the unit became preoccupied with data collection and accumulated information at a rate well beyond its capacity to process it. The number of farmers covered, mainly in the resettled areas of the former white highlands, rose to 2,616 in 1965-66. When the resettlement program was completed in 1968, the unit was again reorganised. Coverage was broadened to all small-scale farming, although the number of sample units was reduced. In the 1969-70 financial year the unit was to cover 1,700 farms[4] and was by far the largest and most securely established agricultural economic data collection program in the three countries.

In Uganda and Tanzania the picture is very different. In the late 1950's senior agricultural administrators observed the growing emphasis on economics in agriculture. These observations were stimulated through professional contact at home in the United Kingdom and by the rapid growth of multilateral aid. USAID experts were recruited to set up an agricultural economic unit within each Ministry of Agriculture. British counterpart staff were recruited to help. Such a unit was established in Uganda, and in the early 1960's several agricultural economists were recruited. A modest survey program was started in liaison with the Department of Rural Economy and Extension at Makerere College. The program was frustrated by the rapid turnover of contract staff and by the diversion of senior economists to other work. Descriptive reports were written by individual fieldworkers and issued by the Ministry of Agriculture in 1966 and 1967.[5]

No unit was established in Tanganyika; the two agricultural economists worked under the Director of Research in the Ministry of Agriculture and were stationed at research centers. Surveys were carried out from the centers and issued by the Ministry of Agriculture between 1961 and 1965. The reports described the farming of particular areas, and later ones evaluated the impact of available technical recommendations on the productivity of the farming systems.[6]

The initial emphasis on the settler minority in Kenya allowed a transplanting of the then current British farm management approach, based on the comparative analysis of individual and standard or premium performances. Although the unit refocused its efforts toward smallholder farming after 1962, there was little change in the approach followed; collection techniques had to be modified to cover a largely illiterate population, but there was no rethinking of the approach through all its stages. General uses for the collected data proliferated. The sponsors of the resettlement program were eager to monitor its progress, and development planning needs reached a climax in 1964-65 with the preparation of the Second Five-Year Plan. Once this cycle of collection and general use was firmly established in government recurrent expenditure, it calcified and prevented the development of a farm management approach tailored to the needs of Kenya's smallholders. It is significant that J. D. MacArthur neither discusses objectives in the study of small farms nor mentions the usefulness of farm management in extension.[7]

The Uganda and Tanzania pictures are, in some respects, even bleaker, for Kenya is the one country where farm management has an "establishment," albeit in a limited role. In Uganda the farm management program disintegrated with a loss of senior staff in 1966. Even earlier their farm management effort had been dissipated by a wide range of other commitments, mainly to macro planning. In a paper presented to the Second East African Agricultural Economists Conference in 1965, J. Cleave sets out the intended uses of the survey data being collected, under the Small Farm Data Scheme, as a part of the First Five-Year Plan (1961-66).[8] He clearly sees applications for the data in planning extension but also reveals a methodological predisposition that perhaps stems from experience of the comparative approach, used in Britain, in which the performance of premium farmers is used as a base for improving that of others. This conviction created two methodological problems: covering adequate numbers of improved farmers by sampling procedures and relating inputs and outputs for single plots to show the productivity increases of improved practices. These two problems focused Ugandan efforts on the methodological detail in investigation, at the cost of a balance between investigation and utilization in planning or extension.

The Tanzanian story is similar, although the discipline never found the ministerial niche it did in the other two countries. Agricultural economists are still allocated to research centers—in itself a good practice—but the weakness was and remains the lack of a senior economist to coordinate and direct the fieldworkers. New staff who arrive are left very much to their own devices, and the whole methodological routine tends to be examined anew. My later work in interpolating improvements into the traditional farm system to evaluate profitability and acceptability has been credited as "a well integrated approach to peasant farm improvement . . . [forming] . . . the only example to date of an attempt to improve general extension advice through Farm Management techniques."[9] Nevertheless, efforts in Tanzania fell short of a comprehensive farm management approach because of the weakness of the contact with the extension services. Even the link between technical research and extension was tenuous, and the willingness of a technically based extension service to absorb economic recommendations was understandably limited. This isolation was aggravated by the lack of a senior officer in the Ministry of Agriculture who could influence extension planning.

This brief analysis of farm management in East Africa in the 1960's has so far been limited to government efforts. A number of academic bodies sponsored ad hoc farm studies in all three territories: prominent were the work of the Institute of Foreign Agriculture, Munich, as part of its Africa Studies series, directed by H. Ruthenburg, and J. Heyer's study at Machakos, Kenya.[10] Contributions were made by the Economic Research Bureau, Dar es Salaam; the Department of Rural Economy and Extension, Makerere; and the Institute of Development Studies, Nairobi. The studies fall into two broad categories: demonstrations of the application of sophisticated techniques to traditional agriculture and farm studies aimed at verifying hypotheses of development theory and postulating possible strategy.

The universities, with relatively senior personnel, must carry a share of the blame for failing to explore the potential of farm management economics in both extension and development planning and for failing to orient their own researchers to a problem-solving approach of direct value to government development efforts. In mitigation it must be added that these are difficult roles complicated by the fact that the universities, as new institutions, need to create their own niches in the national infrastructure. In order to attract talent they must allow latitude for individuals to follow their own interests. Most professional workers are stimulated by professional recognition and motivation may be lost if they are sidetracked into a more immediately useful role. Short-term contract commitments add to the difficulty of identifying with urgent local problems. Nevertheless, senior writers have been critical of this phenomenon in professional motivation. E. O. Heady has written: "So marked has the trend towards stylishness been, at least in the past, that the question implicitly posed was sometimes: 'Here is a new tool, where can I use it.' But in the quest to solve applied social problems we expect somewhat the opposite: 'Here is a problem, where is an efficient tool for solving it', . . ."[11] E. R. Kiehl has been particularly outspoken against Ph.D. programs which engender personal aggrandizement in the exploitation of new-found theoretical skills: "Somehow some have . . . embarked on a personal program of emphasizing their newly acquired competence."[12] Enough has been said to indicate some professional shortcomings in addition to those difficulties of the new environment and the established order, both of which require penetration to allow a contribution from a new discipline.

The reasons for the failure of farm management to contribute to agricultural improvement in the 1960's in East Africa are fairly clear. Kenya is unique in pattern but has a great deal in common with Uganda and Tanzania in cause. Two factors were circumstantial. All three countries gained their independence in the early 1960's—with wide ramifications. There was political disillusion with peasant agriculture and colonial improvement policies. Machinery and large-scale production methods were identified with progress in what is now referred to as the transformation approach. There was a lack of sympathy with investigational work, not least with agricultural research itself, which was often designated unproductive by politicians eager for results. Institutionally, newly qualified local professionals were elevated to senior posts without much opportunity to supplement their training with field experience.

At the same time the era of development plans came to East Africa, absorbing all available expertise. Planning started at the national level and worked down to programs for the local implementation of policies thought desirable for the economy. In Kenya the established data collection system was diverted to macro planning uses; and in both Kenya and Uganda economists were caught up in a planning surge that needed all the data, experience, and manpower it could find. Farm management economics was preempted in a quest for limited government funds by the macro planners.

Two other factors—the technical tradition of the government agricultural services and professional failure in orientation—were more fundamental. In the Colonial Agricultural Service the improvement of agriculture was synonymous with increased production through better yields per acre, in the British tradition. High yield was the criterion for both research and extension efforts. Indeed, the very methodology of experimental design grew up in the same tradition of land scarcity and does not easily relate output to other resources in the production process. In addition to this inherited orientation there was an inherited insularity, particularly in research. C. P. McMeekan has commented critically on the preoccupation of British workers with pure research, a bias apparent in the expatriate research programs in East Africa and made worse for both research and extension by the isolation of the European minority from the farming population.[13] Language barriers, a wide educational gap, and, as B. Brock put it: ". . . the social pressure of the elite enclave . . . created difficulties for those who wanted to get to know the farmer."[14] A teacher-pupil atmosphere and an attitude that the new farming must be built up ab initio ignored the fact that the existing system met the problems of the national and economic environments, however modestly, and that these problems could not be opted out of by any solution, and it prevented a dialogue to elicit the felt needs of the farmer.

Parallel to this the emergent status of farm management economics in Britain itself was at once a reason for a lack of impact on the technical orientation of colonial agricultural departments and for professional disarray in the face of the new needs and conditions of Africa. The discipline had hardly found its feet in agricultural extension in Britain in the 1950's. Only during that decade did the blind faith in increased production for increased profits, the aftermath of World War II, begin to waver in the face of overseas competition and rising costs. Peart, writing in 1969, sees the 1950's as the decade which "eventually saw the Farm Management approach widely accepted amongst advisers and the progressive sections of the farming community . . . saw its position reinforced [in the 1960's] by a changing emphasis from historical analysis to forward planning."[15] But he stresses that this change is by no means complete and makes it a major need for the 1970's. Thus, despite the Provincial Agricultural Economics Service and the National Agricultural Advisory Service in Britain and fifteen years of farm management liaison officers between the two, agricultural extension has not yet fully absorbed forward-looking planning techniques and its farm management advice remains rooted in the historical approach of comparative analysis. It is small wonder that junior members of the profession and technically qualified agricultural administrators failed to identify its place in agricultural development in East Africa during the 1960's.

We have already touched on a second professional failing: the emphasis on technique- rather than problem-oriented research effort, stemming particularly from the nature of graduate program organization in the United States. V. W. Ruttan has added a more fundamental criticism: He sees farm management economics research in the United States as traditionally aimed at aiding decision making by individual farm operations.[16] Rightly so, because the agricultural economy is characterized by relatively large farms and limited public intervention in markets and in farm source supply. He holds that this same orientation is characteristic of American farm management experts in the developing countries, which have few professional services, limited extension personnel, and often very active public intervention in both markets and resource supply. He feels it is clearly a misuse of professional talent to treat the farm operator, or even the individual extension worker, as the appropriate client for microeconomic research. This is the philosophy dominating the approach to be detailed here.

Breaking down the technical domination and isolation was a formidable task in the 1960's and remains so today. The new agricultural administrators have very often been cast from the same mold. Several factors suggest a new opportunity for a contribution from farm management economics:

1. There is now a core of experience of applied farm management economics in East Africa to aid the adaptation of the discipline to the needs and conditions of this type of economy.

2. The macro planners are now part of the establishment infrastructure of these economies. Their experience during the 1960's has created an awareness that plans are missing a link with the production unit, particularly the smallholder. Such a link, although it presents methodological problems, is well based in theory and growing in favor.

3. Aid organizations, particularly the IBRD, are conscious that traditional agriculture holds the key to broad-based development in

the rural sector and that an economic approach at the farm level is required. However, it is a sphere which they remain reluctant to enter because of its complexities.

The continuing isolation of the technically dominated agricultural services and the need for a micro base for effective planning create a role for farm management in bridging the gap between farmer and extension service, on the one hand, and between farmer and development planner, on the other. Many of the agricultural departments remain formidable barriers to participation, but the influence of the macro planners and the aid organizations seems a useful vehicle for renewed assault by the discipline, this time with its objectives and approach more clearly defined.

We have noted the opportunity for a contribution from a farm management economics with its objectives and approach clearly defined. The initial task is this clear definition, and an aid to it is the identification of the types of interaction with the economy which mold the way the discipline must operate.

The objectives of farm management economics are universal; indeed, they define the discipline. Heady acknowledges both micro and macro roles: "To guide individual farmers in the best use of their resources in a manner compatible with the welfare of society. . . . To provide fundamental analyses of the efficiency of farm resource combinations which can serve as a basis for bettering the public administration of resources where agricultural policy or institutions which condition production efficiency are concerned."[17] In both roles he emphasizes the importance of the link between farm and economy: for the micro function in shifting farm resources in harmony with national economic objectives, and for the macro function in giving a more accurate picture of the likely effects of public policy on agricultural production.

The breadth of Heady's definition leaves us a wide array of objectives. L. M. Eisgruber, among others, has criticized modern agricultural information systems, specifically the lack ". . . of a consensus on the primary objective and on the relative weights to be assigned to other objectives."[18] The lack of orientation in early farm management efforts in East Africa was due in part to badly defined objectives. This book describes an approach to fulfilling the micro function of improving extension services. A final section draws some pointers from the discussion for the macro role for the discipline in developing agriculture; these are corollaries of conclusions reached on the most effective way to contribute to extension.

The role of the discipline is also universal. Wherever it is applied, a diagnosis of economic problems on the farm is followed by the evaluation of possible solutions. This is as relevant to traditional as to advanced agriculture—often more so, with the gulf of language and attitude between peasant farmers and trained agricultural administrators.

It is the approach which is the variable aspect. While it has a universal format in investigation, planning, and extension, the links between the phases and their organization and institutionalization depend on the conditions of the particular type of economy. Ruttan's criticism of the inappropriateness of the American approach for developing agriculture is focused on the individual production unit. The approach of the Farm Economic Survey Unit in Kenya is rooted in the comparative analysis of British farm management. Both fail to appreciate the need to adapt to local conditions. The evolution of the whole-farm planning approach, used mainly by private consultants in Britain, is an example of a development of method within the profession which altered the balance between phases. It encouraged a new approach to farm economic advice for farms operating under certain conditions and is a useful example of adaptation to changing circumstances within a single economy.

Comparative analysis was rooted in the investigational phase. In Britain, where it originated with the early 1950's, farm survey data were used by the Provincial Agricultural Economics Service to draw up standards for defined types of farming. The standards were used by the National Agricultural Advisory Services to diagnose weaknesses in the performance of individual farmers requesting advice. Partial budgets were used to evaluate the effects of management improvements. Planning and extension phases were indistinguishable, both being carried out by the advisory worker on the farm. For the whole-farm planning approach the nationwide farm survey is superfluous. The resource position is set out for the individual farm and the production opportunities defined. Records from the farm itself provide the coefficients for the planning model. The production pattern is optimized, improvement being sought mainly from a more profitable combination of enterprises. With the comparative approach, the cost of the large sampling investigations virtually limited farm management extension work to government organizations; the whole-farm planning approach removes this burden. Private consultants can now carry out all three phases on the individual farm. The cost remains fairly high, limiting the technique to large-scale units on which even a modest increment in efficiency can meet the expenses involved. The trend to large-scale production units, together with the evolution of whole-farm planning techniques, has allowed a new approach to the application of farm management economics.

Part I examines the conditions of developing economies to identify how they will influence the approach to be adopted by the discipline. It covers three groups of factors. The first group consists of characteristics of the farmers themselves, particularly their objectives in farming and their level of education. The second is made up of characteristics of the agricultural sector: the producer-market balance, the scale of the typical production unit, and community organization. The third, the structure of the economy itself, emphasizes the rural-urban balance, the infrastructure framework (covering both institutions and qualified manpower as slow-changing constraints), and government policy.

There are both direct and indirect influences on the three phases of the approach. Direct influences mold the organisation of each phase—for example, the fact that extension service contact workers are not qualified to handle farm management techniques is of consequence for the institutional location of the planning phase. It implies a division between planning and extension, unlike current approaches in advanced agriculture. Both direct and indirect influences may limit the usefulness of certain techniques dominant in advanced agriculture and mold the approach in this way. Some direct influences are fairly obvious: the level of literacy and the lack of penetration of the postal service preclude the use of postal survey techniques. Indirect influences may be less obvious. Farm management economics is an amalgam of sciences dependent on a wide range of social and natural sciences for its data and techniques. The amalgam may alter, certainly in emphasis, under the conditions of developing economies. The close coordination of production and consumption on the individual farm, without reference to the market, increases the importance of consumption patterns in resource allocation. At the same time, the close-knit community organization, with reciprocal responsibilities between member and group, creates a role for sociology. In advanced agriculture farmers themselves request advice. In developing economies governments are the initiators, seeking to pull their vast rural populations into the twentieth century. They need to make innovation more attractive to farmers. Adoption and diffusion theory has a more positive role in planning and extension, particularly in broadening the criteria for evaluating possible farm improvements. Farm management economics needs to become a much wider amalgam, almost multidisciplinary, for developing agriculture.

In the same way, the basis of some theoretical components may be inappropriate. Economic techniques have evolved under the conditions of the market. It has been apparent to many workers that sheer survival dominates decision making in smallholder agriculture—operating, as it does, close to the subsistence level. Under such circumstances optimizing differs from profit maximizing, but the particular qualifications required present as yet unsolved problems of theory and technique. Nevertheless, straightforward profit motivations are certainly inadequate and more realistic approximations can be made.

Part I details the range of direct and indirect influences of the three groups of factors, outlining their impact on the investigation, planning, and extension phases of the approach.

Although Tanzanian and East African examples are predominant, the usefulness of this book is bounded not geographically but by the conditions to be discussed in Part I. Certainly parts of the agricultural sector in East Africa are unsuitable for treatment by the approach to be described. Indeed, a farm management department in a Ministry of Agriculture needs a range of approaches to meet various types of farm organization. Enclave developments such as plantations or settler farmers on alienated land might absorb the cost of individual whole-farm planning. Especially progressive and entrepreneurial small farmers, perhaps wholly oriented to the market, might need management advice on an individual basis. Further, the discipline must of course subordinate itself to government policy, and this may influence the approach it uses. The thesis here is that smallholder agriculture is improved within its existing structure as speeded evolution. In view of the popularity of transformation programs in postindependence Africa, associated particularly with settlement schemes, this must be a controversial point. Improvement is justified in Chapter 5. When government has a policy of transformation, the schemes lend themselves to sophisticated planning techniques and an alternative approach will be more appropriate.

Within the general context of rurally based African economies three conditions bound the valid application of the approach to be outlined, the first two defining the traditional agricultural sector and the third the policy alternative:

1. Smallholder agriculture is primarily dependent on family labor and usually limited in scale by seasonal labor needs. Typically it is organized in tribal communities.

2. There must be a system of agriculture with significant subsistence production but with market opportunities.

3. There must be a government committed to agricultural development by improving smallholder productivity within its traditional context.

The crux of the problem is smallholder communities in which improved productivity requires the reorientation of resources from subsistence to production for the market. In discussing the farmer and the values he places on his food supplies, the first chapters of Part I highlights the clash between individual and national priorities, which later chapters seek to resolve.

## NOTES

1. H. M. Southworth and B. F. Johnston, Agricultural Development and Economic Growth (Ithaca, N.Y.: Cornell University Press, 1967).

2. J. C. De Wilde, Experiences with Agricultural Development in Tropical Africa (Baltimore: Johns Hopkins Press, 1967).

3. J. D. MacArthur, "The Economic Study of African Small Farms—Some Kenya Experiences," Journal of Agricultural Economics, XIX, 2 (May, 1968).

4. R. H. Clough, "Recent Experience with Farm Management Surveys in Kenya," East African Agricultural Economists Conference paper (1970).

5. D. Pudsey, "A Pilot Study of 12 Farms in Toro" (Kampala: Uganda Dept. of Agriculture, 1966); "An Economic Survey of the Farming of the Wet Long Grass Area of Toro" (Kampala: Uganda Dept. of Agriculture, 1967); "The Economics of Outgrower Tea Production in Toro, Western Uganda" (Kampala: Uganda Dept. of Agriculture, 1967). (All mimeographed.)

6. M. P. Collinson, "Bukumbi Area," Farm Economic Survey no. 1 (Dar es Salaam: Tanzania Dept. of Agriculture, 1961); "Usmao Area," Farm Economic Survey no. 2 (Dar es Salaam: Tanzania Dept. of Agriculture, 1962); "Maswa Area," Farm Economic Survey no. 3 (Dar es Salaam: Tanzania Dept. of Agriculture, 1963); "Lwenge Area," Farm Economic Survey no. 4 (Dar es Salaam: Tanzania Dept. of Agriculture, 1964). (All mimeographed.)

7. MacArthur, op. cit.

8. J. Cleave, "The Collection, Analysis and Use of Farm Management Data in Uganda," East African Agricultural Economists Conference paper (1965).

9. M. Hall, "A Review of Farm Management Research in East Africa," East African Agricultural Economists Conference paper (1970).

10. H. Ruthenberg, ed., Smallholder Agriculture and Development in Tanzania, "Africa Studies," XXIV (Munich: IFO, 1968); J. Heyer, "Seasonal Labour Inputs in Peasant Agriculture," East African Agricultural Economists Conference paper (1965).

11. E. O. Heady, "The Agricultural Economist and His Tools," Proceedings of the International Conference of Agricultural Economists (1961).

12. E. R. Kiehl, "A Critical Appraisal of the State of Agricultural Economics in the Mid-Sixties," Journal of Farm Economics, XLIII (1961).

13. C. P. McMeekan, "Co-ordinating Economic and Technical Research," Proceedings of the International Association of Agricultural Economists (1964).

14. B. Brock, "The Sociology of the Innovator," East African Agricultural Economists Conference paper (1969).

15. B. Peart, "Farm Management Advisory Work," Agriculture (U.K.), LXXVI, 9 (1969).

16. V. W. Ruttan, "Issues in the Evolution of Production Economics," Journal of Farm Economics, XLIX (1967).

17. E. O. Heady, "Objectives of Farm Management Research," Journal of Farm Economics, XXX (1948).

18. L. M. Eisgruber, "Micro and Macro Potential of Agricultural Information Systems," Journal of Farm Economics, XLIX (1967).

# PART I

# CONDITIONS IN AFRICAN PEASANT AGRICULTURE AFFECTING THE APPLICATION OF FARM MANAGEMENT TECHNIQUES

# CHAPTER 2

# THE FARMER: HIS MOTIVATIONAL BALANCE AND FARM ORGANIZATION

In terms of government-sponsored agricultural innovation, the farmers are consumers. As with any market, penetration depends on the nature of effective demand and attractive advertising and packaging of the product offered. We examine here the motivations and priorities of peasant farmers as a basis for devising a product that meets their needs and for mounting an effective sales campaign.

Simplifying a borrowed psychological concept of primary and secondary drives helps to categorize motivations important in the economic process. Primary drives produce their effects through the actions of inherited bodily mechanisms; they are "primary" in the sense of being first essentials to life. The withholding of air, food, and water is a stimulus which activates such drives. Behavioral characteristics dependent on these drives stem from motivations for survival; an assured food supply and the personal security needed to allow the productive activity to guarantee it are the two important here. The emergence of secondary drives depends on learning and can be identified with achievement motivations, of which profit maximization is a central tenet of Western economics. While survival motivations are confined to the satisfaction of primary drives, profit maximization can satisfy both bodily needs and the urge for achievement.

Development can be seen as a continuum from hunting and gathering to full specialization in an exchange economy. The balance of motivational forces changes along this continuum. Before productive forces are coordinated through the market, allowing the advantages of specialization, survival dominates the balance. Where the farmer has begun to produce for the market, achievement urges are emerging. J. Mellor has pointed out:

> Substantial decline in the marginal utility of income once subsistence is met is only of academic interest in high income societies because high labour productivity places essentially all decision making at utility positions above subsistence. In contrast, in a low income society the marginal product of labour may be low enough to make this position on the utility surface a crucial decision making area.[1]

Diagnosis of the balance of motivational forces in productive activity is a prerequisite for effective problem identification and solution.

This book is concerned with African peasant agriculture, which fits Mellor's diagnosis well. We will examine the two aspects of survival motivations—personal security and an assured food supply—in more detail.

In tribal society, personal security is manifest in community organization and perhaps represents the earliest step in welfare economics. It is largely of historical interest in Africa, where the rule of law introduced by the colonial powers removed the need for much of the custom designed to protect the individual. Several writers have noted the vestiges of the protective role in social organization. Referring to Sukumaland in northwestern Tanzania, an area which will be used for examples throughout this study, H. Cory quotes from a report written in the 1930's:

> It is easy to forget that as recently as 60 years ago the distribution of population was controlled to a large extent by consideration of security. Most independent chiefdoms were isolated units and people . . . were unable to move into uninhabited lands because of internecine warfare. The traditional Sukuma village . . . was roughly circular in shape and protected by Euphorbia hedges. Within these fortifications lived the whole village community with their stock, and the present organisation of collective labour is a result of conditions in which it would have been dangerous to hoe alone.[2]

Cory himself characterizes the right to graze community cattle on the fields of individuals once the crop had been taken as being due to the limitation on grazing facilities under the secure control of the group. L. A. Fallers attributes the sex differentiation in work tasks in Buganda to the need for the men to be available for the Kabaka's army.[3] Similarly, the traditions of load carrying by women, common throughout Africa, allowed the men to be free to use their weapons against would-be thieves or predators. Many of these customs linger on within the traditional community, often having become closely linked with the need for security of food supply, which still remains.

J. L. Joy has listed the food supply priorities in the day-to-day existence of the hunter and the gatherer in the following order: quantity, nutritional adequacy, and preferred taste.[4] The hunter used his mobility to assure his supply of food, moving on as local sources were exhausted. We have noted the continuum from hunting and gathering to the exchange economy. A second, parallel continuum is the transition from shifting to sedentary agriculture. With semi-permanent and permanent settlement, mobility is sacrificed and technique must be used to guarantee the supply of foods, thus requiring foresight in planning production. As Joy points out, only where feasible alternatives can meet the quantity and quality requirements will taste arise as a factor in choosing between them.

The quantity of food, its nutritional quality, its reliability of supply, and preferred taste are four types of motivation dominating decision making and resource allocation in traditional African agriculture. Frequent market dealings in a large number of inputs and products characterizes an organizationally sophisticated production unit with the market as a common denominator for its transactions. Traditional agriculture has little sophistication in this sense; its complexity arises from the lack of a common denominator. Balancing the four subsistence motivations to optimize his satisfactions is the central decision-making task for the traditional farmer. However, the individual farmer is not required to take all those decisions personally, for the balance of subsistence motivations is built into the farming system traditional to the community. Two types of decision remain under the control of the farmer. For each season he will weigh his requirements, which depend upon the number of people he must feed and his expectations, guided by his knowledge of his own land, the capacity of his labor force, the stocks on hand from the last harvest, and reserve crops he has in the ground. The balancing will give a cropping pattern which coordinates preferred and reliable foods and meets his individual risk preferences.

Part of his expectations are variable yields which cannot be forecast for the particular season. The farmer operates so close to subsistence that a failure in supply may jeopardize the very survival of his family. Insurance is built in partly by his own risk preferences, partly by the traditional husbandry practices used, and partly by a second field of decision, in day-to-day management, allowing the farmer flexibility to meet seasonal contingencies. If an early planted staple fails, he may replant with a more drought-resistant species or variety, or he may replant with a faster-maturing variety. If he feels none will be successful, he may increase the area of his famine reserve crop, thinking ahead to times of potential shortage. Under conditions of heavy rainfall he may have to concentrate on crops which will flourish on the lighter hill soil; under drought, to use crops which will flourish in valley bottom areas. These types of alternatives, and the decisions that they imply, are those which distinguish good and indifferent farmers in traditional agriculture.

A balance of quantity, quality, preference, and reliability is manifest in the traditional practices of the system. These are susceptible to evolution over time and to trial and error by the individual farmer. The more important practices are outlined briefly.

## INTERCROPPING

Intercropping is a widely used traditional practice, particularly as an insurance technique. Crops which are more resistant to drought or local pests are planted together with preferred foods to ensure that some return is gleaned from the effort put into the preparation of the land, often the limiting operation in the traditional system. Sukumaland provides an interesting example of double insurance with maize and sorghum mixtures. As recently as the 1940's sorghum was the preferred grain in many parts of Sukumaland, but the unbearded variety grown was particularly susceptible to bird damage.[5]

As the bird problem grew, maize was interplanted with the sorghum, for although less drought-resistant, it was not attacked by the birds. Tastes gradually changed and the balance altered; now sorghum is sometimes intercropped with maize as an insurance against rain failure, especially in the drought-prone parts of the area to the east and south.

Although insurance is acknowledged to be probably the most important reason for intercropping, R. N. Parker has classified other reasons which play a part in particular farming systems.[6]

1. Increased use of environmental factors, under which he includes three subcategories: (a) the complementary use of light in space, where crops planted together have different growth habits and each can get enough sunlight; (b) the complementary use of light in time, where one crop can reach a stage at which it has minimal

light needs and another can then be interplanted to use the light now available; (c) the complementary use of both water and nutrients in space—groundnuts and maize are the commonly quoted example—usually characterized by differential root depth and spread.

2. Reduction of adverse conditions in the ecosystem, under which are subcategories covering the environment—mainly shelter effects, conservation of soil moisture, increase of soil and air temperatures, and decrease of soil erosion—and disease—he quotes N. S. Simmons, who states that in heterogeneous populations physiological specialization is minimized and the variation acts as a buffer against pathogen spread.[7] Reduction of weed level is supported by work by A. C. Evans, who showed that certain intercrops did not have any worse effect than weeds and were, of course, themselves useful plants.[8]

3. Physical protection of the soil is essential because large drops of water form on the leaves and with tall crops, particularly trees, great damage is caused to the soil because the water running off rapidly achieves its terminal velocity. Low leaf cover breaks the force of the water, thus protecting the soil.

## STAGGERED PLANTING

H. Conklin emphasizes that an understanding of timing is crucial to an understanding of swidden intercropping and that staggered plantings allow for a large and varied crop inventory:

> Exact timing of agricultural activity is unknown to the Hanunoo, relative timing is however of great importance. This is accomplished mainly by the sensitive observation of phenologically determined changes in the wild flora, changes of wind and the stars. Agricultural time is reckoned forward and backward from the critical operation of grain planting.[9]

Several aspects of staggered planting havc already been touched on and are listed in summary here.

1. Crops with different growth habits that are used in the same dishes, will be planted at different times to ensure complementary maturity.

2. Small plantings of key staples are made at different times so that periods of water stress will find the plantings at various

stages of growth. While some plantings may suffer, those with relatively limited needs at that time are more likely to survive.

3. Small plantings at different times allow a continuous flow of a food preferred at a particular stage of maturity.

4. Staggered planting is often the result of meeting seasonal contingencies as they arise.

5. Staggered planting may be a means of flattening out seasonal labor peaks. Although yields may be reduced by delayed planting, total production may be increased where the off-peak resources have a low opportunity cost.

In examining a particular farming system it will be important to evaluate the reasons for intercropping and for staggered planting. Improving the system will depend on which reasons are dominant. The emphasis so often placed in extension recommendations on pure stands and optimal planting times can be an important difference between technical and economic optima.

## OTHER INSURANCE MEASURES

Particular crops are often grown specially for insurance. Cassava is a case in point with the Sukuma. It is not a preferred starch staple, but its high productivity in terms of bulk and its capacity for storage in situ in the ground for up to three years make it an ideal famine reserve crop. The aim of the farmer will be to have cassava available and mature in future periods of expected food crisis. Cassava requires little work after being weeded at the beginning of its second season. In Sukumaland the labor required for planting is minimized by intercropping and by planting the cassava at a time when there are few other demands on the labor. This late planting means the sacrifice of optimal yields, but it still gives five-ten times the bulk of grain crops per unit of labor used. When cassava supplies are low after a famine season, reestablishment may take first priority when the rains arrive. The amount of mature cassava likely to be available will influence decisions on the urgency of staple grain supply and contingency plantings of late grains in a poor season. It will thus be important in the allocation of resources to other crops.

H. A. Luning describes the deliberate fragmentation of staple crop fields in northern Nigeria in order to offset microclimatic variations and localize pests and disease attacks.[10] Cattle hoarding is also a measure of insurance, and again the Sukuma are a good example. The animals themselves offer a good supply of food, and they can also be sold for cash with subsequent command over the foodstuffs for sale. In Sukumaland there were standard measures of staples as the equivalent of different types of stock, although the emphasis has now moved to exchange through the medium of cash.

The combination of inherited practices and short-term farmer decisions allows the continuing satisfaction of subsistence priorities. Work in Sukumaland offers interesting comments on the success of the traditional system in stabilising family food supply. Though the data are by no means complete, they are quoted in evidence that the system works. Studies carried out in three similar areas of Sukumaland in 1962, 1963, and 1964 show a level of staple grain supply much more stable than the amount of critical rainfall (November-February totals). During an overlapping period, improved food growing practices were tested on a Trial Farm Unit rated at a very high level of management. All the indices are based on the three year average equal to 100.

The variation in grain supplies per adult equivalent was very much greater on the trial farm. A failure to forecast yield levels was compounded by a failure to get the planned acreage on the ground. The trial farm as a single unit was subject to microclimatic as well as interseasonal variation, although the figures demonstrate that stability was achieved within the communities surveyed. Community organization itself has also had reciprocal responsibilities between the household and the group to alleviate individual food shortages. P. Hill and the Forteses have given fascinating examples of food aid between families to balance supplies within the group.[11] Whereas intercropping and staggered planting are important practices in the systems covered by survey, recommended times of planting were fixed for the pure stands grown on the trial farm. The coefficient of variation of rainfall for the period November-February for the area is 37 percent, compared with 53 percent for December and 55 percent for January—further evidence of the benefits to be gained from flexibility in planting times, as in traditional practice.

The traditional farming pattern is thus a complex balance between practices which have evolved to meet subsistence priorities, the natural resources peculiar to the area, and the family resources available to the individual farmer. Before the advent of the exchange economy there is no reason to believe the system would be static, and indeed the idea of a continuum from shifting to semipermanent cultivation contradicts this. Technical know-how would have increased

TABLE 3

Comparison of Traditional and Recommended Food Supply Techniques for Reliability

| | Farm Surveys | | | | Trial Farm | | | |
|---|---|---|---|---|---|---|---|---|
| | Per Adult Equivalent | | Rain | | Per Adult Equivalent | | Rain | |
| Year | Pounds | Index | Mm. | Index | Pounds | Index | Mm. | Index |
| 1962 | 392 | 109 | 504 | 130 | — | — | — | — |
| 1963 | 368 | 102 | 306 | 78 | 385 | 74 | 424 | 116 |
| 1964 | 324 | 90 | 370 | 94 | 215 | 42 | 318 | 87 |
| 1965 | — | — | | | 750 | 145 | 353 | 96 |

Source: Compiled by the author.

in the earlier stages, and merely knowing the places to expect food for gathering at particular seasons represents the start of accumulation. Farmers would have evolved practices which satisfied their priorities more effectively. The priorities themselves would have altered as new, preferred consumption opportunities arose. The rate of progress would be lower without the exchange facilities of the market, perhaps the most effective catalyst for system evolution; but changes would diffuse slowly through the community, just as contrived changes on the part of government are expected to spread. The comparison of typical farm units for 1945 and 1961 in central Sukumaland shows how rapid change and diffusion can be in both market and subsistence sectors of the system.

The staple grains, except for rice, are intercropped, usually with legumes and roots but sometimes with each other. While the most striking change is the increase in cotton from 14 to 45 percent of the cropped area, the switch to higher productivity grains is equally dramatic. The economics of subsistence production is a measure of whether producers are keeping up with the possibilities offered by their production environment. The Sukuma case above is an example of dynamism in peasant farming systems. Other workers have noted the same type of response to changed opportunities. P. T. Bauer's study of the Malayan rubber industry demonstrates the rationale of the peasant producer in producing rubber for the market and his persistence, despite government efforts to stimulate rice production and despite extremely variable rubber prices.[12] Data collected on two samples of tobacco farmers in the same area—one sample growing aromatic tobacco, a marginally viable crop, and the other Virginia, a highly profitable one—demonstrate a varied impact on subsistence production.

The impact of Virginia tobacco, a high-value cash crop eight times as productive as aromatic, demonstrates the response of small farmers to real opportunities. Some 19 percent of the Virginia growers dropped food production completely. Although the subsistence pattern of grain, legumes, and roots remained the same in both areas, it was considerably simplified among Virginia growers and yields of maize and groundnuts were intensified. In addition to the main crops aromatic farmers averaged 2.95 other types of food, compared with .85 types on Virginia farms.

A new phenomenon is introduced with the exchange economy—the possibility of input purchases. The willingness of the farmer to incur costs is limited by his expected income level, characteristically low for peasant farmers operating just above the subsistence level.

TABLE 4

Acreage Data, Central Sukumaland, 1945 and 1961

| Staple Grain Crop | Acreage 1945 | 1961 |
|---|---|---|
| Bullrush Millet | 4.09 | 1.17 |
| Maize | .29 | 1.40 |
| Sorghum | .34 | – |
| Rice | .10 | .23 |
| Cotton | 1.01 | 2.73 |

Sources: M. P. Collinson, "Bukumbi Area," Farm Economic Survey no. 1 (Dar es Salaam: Tanzania Dept. of Agriculture, 1961) (mimeographed); N. Rounce, "Crop Acreage Survey of Sukumaland" (1945) (MS).

The outlay required for innovations should be acceptable in relation to the level and variability of existing farm income. Table 6 illustrates the point, showing increasing levels of purchased inputs used on a trial farm. Clearly, extension advisers offering the innovation packages implemented in 1963-64 or 1964-65 to an average farmer would be fighting a lost cause, since the outlay required would be greater than his usual income level.

Farm costs within peasant agriculture are usually associated with the hire of labor or machinery services to expand the scale of the system. This is particularly true for seasonal systems, in which initial increments of hired resources at seasonal peaks often have very high marginal value products. It is equally true that the return to casual resources falls off rapidly as secondary peaks meet the limit of available family labor, when scale is again increased and extra hired resources are required. Each system will have its own balance, depending on the local cost/return relationships for hired resources. The return to purchased inputs associated with innovation cannot always compete with the returns to increases in scale. Credit facilities can expand the ceiling on farmer's capital outlay, for as is illustrated in Table 6, it reduces the demand on current income. However, repayment must come out of future income; and unless the

## TABLE 5

### Subsistence Production Among Smallholders in the Same Area Growing Different Cash Crops

| | Aromatic | | | Virginia | | |
|---|---|---|---|---|---|---|
| Crop | Percent Growing | Acres | Yield (lbs./acre) | Percent Growing | Acres | Yield (lbs./acre) |
| Maize | 100 | 2.27 | 358 | 59 | 1.28 | 826 |
| Groundnuts | 98 | 1.83 | 126 | 53 | 1.22 | 391 |
| Rice | 100 | .55 | 1458 | 43 | .35 | 1531 |
| Sorghum | 52 | 1.77 | 74 | 18 | .98 | 363 |
| Sunflower | 72 | 1.30 | 230 | 4 | — | — |
| Cassava | 52 | 1.36 | — | 43 | 1.12 | — |

Notes: Acreages are the average over the numbers growing the crop.

The areas are allocated to all the intercrops in a mixture.

Sources: M. P. Collinson, "Aromatic Tobacco and Virginia Tobacco: A Comparative Survey of Two Tobaccos on Family Farms in the Tabora Region of Western Tanzania," paper presented to the East Africa Agricultural Economists Conference (1970).

## TABLE 6

### Increasing Costs on a Trial Farm Unit

| | Average Local Farm | Best Local Farm | Trial Farm 1962-63 | Trial Farm 1963-64 | Trial Farm 1964-65 |
|---|---|---|---|---|---|
| Adjusted Net Cash Income | 678 | 1,283 | 1,900 | 2,630 | 1,470 |
| Adjusted Working Capital Requirements | 170 | 365 | 390 | 1,180 | 1,810 |
| Working Capital as Percent of Average Local Net Cash Income | 25 | 54 | 58 | 174 | 268 |
| Credit Provision Locally Available to Farmers | – | – | 277 | 373 | 515 |
| Cash Outlay Needed as Percent of Average Local Net Cash Income | – | – | 17 | 120 | 192 |

Notes: To allow a comparison on the same scale, trial farm data have been converted to the basis of available labor on the average local farm.

No sampled local farmers were using credit facilities, which became available for fertilizer in 1962-63 and for insecticides in 1964-65 on a limited scale.

Source: M. P. Collinson, "Experience with a Trial Management Farm. 1962-65," East African Journal of Rural Development. II. 2. Table 8 (1969).

farmer has a very definite idea of the increases likely to accrue from the use of credit, his risk preferences rapidly impose a ceiling on his willingness to incur debts. We shall see from the planning example, however, that even this small gap created by selective credit facilities may be an all-important incentive to purchase inputs in systems where the marginal value product of scale increments is high. This is of special importance in systems within a falling fertility spiral, which will be aggravated by the increased use of land.

Where the opportunities do arise, peasant farmers are rapid to respond. But in the absence of organized markets and cash crop opportunities, farmers act rationally in producing foods for their families. Where opportunities are not overwhelmingly obvious, the evaluation of subsistence activities is complicated by the diversity of motivations to be satisfied.

The frustration of government-directed, market-oriented development efforts by the dominance of nonmarket priorities has generated views of peasant irrationality and has brought disillusion with small farm improvements as an instrument for development. Yet R. Goodfellow pointed out the anomaly as early as 1939:

> . . . if we were to recommend to a Bantu people to abandon some of their organised activities . . . and devote the time saved to increased agricultural production, we might find ourselves in the absurd position of advising them to reduce their wants and increase their production. . . . this is absurd because the values that control production vest in the expressed wants.[13]

Nevertheless, views of irrationality have been reinforced from time to time by the attention of social scientists to the apparent paradox of excessive leisure preferences in a situation of dire poverty. In Africa the leisure-preference dilemma has centered on underemployment of labor resources. Studies repeatedly demonstrate a low proportion of total available labor used in farming but ignore three crucial aspects: seasonality, the extent of nonfarm tasks, and nutritional adequacy. These are discussed briefly in the hope of removing further barriers to the use of existing agriculture as a vehicle for development.

The proportion of available labor used in smallholder agriculture varies with the degree of seasonality and, almost inevitably, a far higher proportion is utilized at seasonal peaks. That there is marked underemployment due to seasonality is clear from almost any labor profile drawn up from African circumstances. It is least marked with certain perennial crops—tea and coffee are good examples—and most marked with highly labor-intensive operations in annuals with a short-growing season. Some examples for annuals are shown in Table 7.

The examples are taken from typical profiles derived from survey data; peaks will be damped by aggregation and the situation on the individual farms frequently shows oversupply in terms of the standards used. As R. H. Fox and A. Moody have pointed out, even within the short season the incidence of rainfall will reduce the availability of labor, keeping the workers out of the fields.[14] Many other workers have demonstrated the phenomenon of seasonality and, while underemployment is marked, in the absence of alternative employment opportunities it cannot be attributed to farmer motivation.

Similarly, many researchers have pointed out the importance of other work that the traditional farm family must perform. J. Heyer's study in Machakos was perhaps the first in East Africa to quantify the importance of noncrop operations in absorbing labor.[15] Using a standard forty-eight-hour week, her small sample of fourteen farmers used 37 percent of available time over the year on crop work and a further 26 percent on nonspecific work directly associated with agriculture. Other work included beer brewing, crop processing, marketing, craft work, cattle herding, and contract services. D. Pudsey has made a detailed breakdown of the use of daylight hours by different sex/age groups for both agricultural and other work.[16]

The table demonstrates the wide variety of off-farm activities and the division between different age groups and sexes. The low average number of hours worked on the farm per day is supplemented by an average of 5.0 hours per day spent on other activities. The total accounts for 7.8 hours in a twelve-hour day. Pudsey was working in a system where seasonality was not marked and periods of intensive activity were not dramatic, tending to be met by reducing nonroutine off-farm activities. Not all tasks can be dovetailed with crop work, however. The very high proportion of the time of senior women taken up by domestic chores and social obligations, such as funerals and weddings, cannot be postponed. Illness of a senior member of the family may preoccupy the whole labor force. M. P. Collinson reports the effect of the farmer falling ill on a trial farm unit operated by a local family: "On 19th December he finalised his cotton planting for the season with a total of 3.21 acres, still with adequate time to complete his remaining cultivations on schedule. On Friday 21st he went down with malaria for a full week, the whole

TABLE 7

Examples of Seasonality in Labor Supply

| Farming System | Percent Labor Use Annually | Percent Labor Use Seasonally |
|---|---|---|
| Aromatic Tobacco | 32 | 82 |
| Virginia Tobacco | 44 | 86 |
| Cotton (hand-cultivated) | 51 | 82 |

Sources: M. P. Collinson "Lwenge Area," Farm Economic Survey no. 4 (Dar es Salaam: Tanzania Dept. of Agriculture, 1964) (mimeographed); "Aromatic and Virginia Tobacco: A Comparative Survey of Two Tobaccos on Family Farms in the Tabora Region of Western Tanzania," paper presented to the East Africa Agricultural Economists Conference (1970).

family stopped work to care for him."[17] This incident reduced the labor supply in the critical month of the season by 27 percent, and the potential farm income was reduced proportionally.

The effects of poor nutrition have often been ignored in assessing labor availability. Where the food intake is low in calories, people will not have the same capacity for prolonged physical effort. Important in compounding the effects of a low protein diet is the hunger gap before the new harvest. In some areas this will clash with the period of most demanding physical effort. A. T. Richards notes that although the Bemba men are proud of their ability to lop off the high branches for fuel for the preparation of the chitemene fields—a pride generated by social recognition for a key activity for farming success—they will not start the work until primed by food from the new harvest.[18] Similarly, J. Heyer notes that after prolonged weeding, a physical maximum appears to have been reached and the local endemic diseases seem to break out again.[19]

R. H. Fox, doing medical research in Gambia, achieved what can only be described as remarkable documentation of the phenomenon of nutritional limitations on the use of available labor.[20] He plotted the path of energy intake and expenditure in a community dependent primarily on rice and groundnuts produced on the farm and related

TABLE 8

Allocation of a Twelve-Hour Day by Sex/Age Group%

| | Farmer Male House-holder | Off-Farm Male House-holder | Female House-holder | Wife | Other Men (15+) | Other Women (15+) | Boys (10-15) | Girls (10-15) | Total |
|---|---|---|---|---|---|---|---|---|---|
| Number | 15 | 9 | 6 | 22 | 10 | 20 | 11 | 11 | |
| Neighbors, Visitors | 10.9 | 2.8 | 5.1 | 5.2 | 5.7 | 19.6 | 2.9 | 1.7 | 5.5 |
| Off-Farm Work | 1.8 | 63.0 | — | 2.0 | 32.0 | 3.4 | — | — | 9.9 |
| Trade, Market | 5.0 | 2.1 | 1.9 | .7 | 11.0 | .3 | .2 | .2 | 2.3 |
| At Dispensary | 4.6 | 1.7 | 3.1 | 2.9 | 1.1 | 1.6 | 1.5 | .5 | 2.2 |
| Building Work | 2.3 | 1.2 | .1 | — | 2.8 | .1 | .2 | — | .8 |
| Household Chores | 1.9 | 1.2 | 18.4 | 20.4 | 2.7 | 12.1 | 2.3 | 9.5 | 9.6 |
| School | — | — | — | — | 8.2 | 2.3 | 23.0 | 24.4 | 6.2 |
| Other Nonfarm | 3.3 | .4 | 3.9 | .8 | 2.8 | 1.1 | 3.3 | 1.7 | 1.9 |
| Total Nonfarm | 30.0 | 72.4 | 32.5 | 32.0 | 66.3 | 40.3 | 33.8 | 38.7 | 41.0 |
| Hours per Day | 3.5 | 8.7 | 3.9 | 3.8 | 7.9 | 5.0 | 4.0 | 4.7 | 5.0 |
| Total Hours per Day | 7.1 | 9.6 | 7.6 | 7.5 | 8.6 | 7.6 | 7.0 | 7.3 | 7.8 |
| Grand Total Work | 60.2 | 80.0 | 63.5 | 62.7 | 72.0 | 63.4 | 58.7 | 61.0 | 64.0 |

Note: In deriving these percentages a twelve-hour day for 300 days is assumed. Pudsey assumed 307 days available for the year.

Source: D. Pudsey, "A Pilot Study of 12 Farms in Toro" (Kampala: Uganda Dept. of Agriculture, 1966) (mimeographed).

the balance to the operational calendar of the farming system and the availability of home-grown foods. Table 9 summarizes his energy calendar in calories per head per day.

The physical requirements of the labor involved in cultivating groundnuts caused energy deficits, particularly among the men, in May, June, and July, that time of the season which decided the potential size of harvest. Yet the hours worked per day by the farmers while ridging were lower than those worked at other times of the season on less critical operations. R. H. Fox's conclusion was "that the intensity of effort required in their work rapidly exhausted their available energy causing an early onset of fatigue and thus limiting the total time they were able to work."[21] Table 10 compares the length of day worked and energy consumed while ridging groundnuts with the figures for other operations.

Brought together, seasonality, a multiplicity of off-farm chores, and nutritional adequacy explain an apparent distaste for field work. Ignorance of the complex of objectives in peasant farming practice and the neglect of these aspects of traditional agriculture livelihood—unimportant at modern levels of living—has allowed misinterpretation of the smallholder character, which experience suggests can be a valuable attribute in agricultural development.

The colonial era brought significant changes to the community organization and often to agricultural practice. Alternative work in mining or estate agriculture attracted men away from home, reducing the family's capacity for subsistence production. A. T. Richards discusses this in relation to the Bemba, whose supply of the main staple, millet, was directly related to the availability of male labor at the season for clearing gardens and cutting wood.[22] The reported tendency for millet supplies to run out before the new harvest, even in a normal year, may be due to the absence of a portion of the male labor force at the critical season. Education takes the children away to school. P. de Schlippe and B. L. Batwell quote the example of lande, a type of sorghum grown by the Nyangwara of the southern Sudan, particularly important because its time of maturity reduced the gap between the last harvest of the old season and the first of the new one.[23] It has virtually disappeared. Quelea attack lande very severely because it matures before the grass seeds on which they normally feed. Many children are needed to chase away the birds, but now they are at school and the hungry months are extended. M. Read quotes the Paramount Chief of the Ngoni in Nyasaland as he expresses regret at the reduction in the variety of foods enjoyed by his people: "Formerly there was no other work than taking care of

## TABLE 9

Changing Energy Balance over the Agricultural Calendar

| Period | Month | Energy Intake | Energy Equivalent of Weight Change | Available Energy | Energy Expended | Available Energy Less Energy Expended | |
|---|---|---|---|---|---|---|---|
| I | 2 | 2,149 | | 1,997 | 1,601 | +396 | +396 |
| | | | -152 | | | | |
| II | 3 | | | 1,499 | 1,482 | +17 | |
| | 4 | 1,575 | | 1,575 | 1,473 | +102 | +7 |
| | 5 | | | 1,575 | 1,673 | -98 | |
| | | | 0 | | | | |
| III | 6 | | | 1,740 | 1,934 | -194 | |
| | 7 | 1,740 | | 1,776 | 1,964 | -188 | -116 |
| | 8 | | | 1,811 | 1,777 | +34 | |
| | | | +71 | | | | |
| IV | 9 | | | 1,694 | 1,709 | -15 | |
| | 10 | 1,623 | | 1,694 | 1,694 | 0 | +10 |
| | 11 | | | 1,531 | 1,486 | +45 | |
| V | 12 | | -92 | 1,868 | 1,797 | +71 | |
| | | 1,960 | | | | | +95 |
| | 1 | | | 1,868 | 1,750 | +118 | |

Source: R. H. Fox, "Studies of the Energy Intake and Expenditure Balance Among African Farmers in the Gambia," Ph.D. thesis (London: Medical Research Council, 1953).

TABLE 10

Relationship Between Length of Day Worked
and Energy Requirements of Certain Operations

| Crop | Operation | Total Time in Fields (mins.) | Percent of Time in Work | Energy Expenditure |
|---|---|---|---|---|
| Groundnuts Males | Ridging | 318 | 53 | 8.0 |
| Groundnuts Males | Lifting | 410 | 70 | 5.0 |
| | Windrowing | 365 | 61 | 3.2 |
| Rice | Weeding | 341 | 72 | 3.0 |
| | Pulling grass | 297 | 73 | 3.4 |
| Females | Transplanting | 358 | 75 | 3.8 |

Notes: Energy expenditure is measured in calories per kilogram of body weight per hour.

Operations have been selected for the length of time spent in the field. Other operations at slack periods do, of course, show shorter days.

Source: R. H. Fox, "Studies of the Energy Intake and Expenditure Balance Among African Farmers in the Gambia," Ph.D. thesis (London: Medical Research Council, 1953).

their own affairs. When the Europeans came, they came with other work, adding to the work of the people such as tax and work to receive cloth. When they were busy with such things they forgot the work of their ancestors."[24] Read makes the important point that cultural contact has destroyed the traditional channels of agricultural instruction: "People [the Ngoni] have drifted away from the scientific traditional practice and are confused and disorganised. In this form they are not receptive to advice on improvements."[25] De Schlippe and Batwell echo the same point: "In extreme cases unconsidered reform may cause social disintegration and an inability to exploit the native environment."[26]

From the point of view of long-term development, the most important impact was made by the spread of preventive and curative medicine. We have already noted its effect in increasing the rate of population growth. Unhindered by the motivational complexities of agricultural change, medical innovation has often spread throughout even the rural areas of developing economies. This has brought a new dimension to the problem of raising agricultural productivity. We have identified a balance between two continua: from hunting and gathering to the exchange economy, and from shifting cultivation to sedentary agriculture. The ease of medical innovation, by increasing population growth and thereby density, has accelerated the movement toward sedentary agriculture without a concomitant improvement in technical know-how. The total land area available to many farming systems is being reduced; the consequence is a falling fertility spiral, since traditional techniques of fertility maintenance depend on a fallow/arable sequence, and increasing numbers using the same land area causes the sequence to break down. Short-term relief may be found in using more labor to cultivate more land, a solution which aggravates the long-term problem, or in discontinuing crops which exhaust the land rapidly. Ultimately, unless the balance between the continua can be restored by innovation, absolute shortage of food will strike a new balance between the land and its population by an increase in the mortality rate. This relatively recent phenomenon adds urgency to the quest for increased agricultural productivity and implies that some improvement in techniques is necessary even to maintain the present position.

Motivations centered on food supply dominate priorities in the allocation of resources for the productive activity in smallholder farming. Given the stage of development of the exchange economy in rural African economies, particularly the inadequacy of retail food outlets, peasant behavior is rational. This is the most important characteristic of developing economies, with fundamental influences on investigation, planning, extension, and phases in the application of farm management economics.

To elicit increased efficiency in the use of resources in agriculture as a basis for national economic development, government must meet the farmer halfway. Change will be acceptable only if it is at least consistent with satisfaction of the farmer's own priorities. The urgency for recognition of this in agricultural development strategy is heightened by the accelerated rate of population growth and the threat of a falling spiral of fertility as more mouths are fed from the same total land area. As the spiral intensifies, a higher rate of technical change is required to give increments in productivity as some degree of innovation is absorbed in merely maintaining per capita production.

With the farming system directed to wider ends than marketed production, the scope of investigation increases. The system cannot be described properly in the known terms of the market; "farm economic survey" in some ways becomes a misnomer, for investigation must identify other aspects contributing to the balance between resources and production. There are three aspects in addition to those usually associated with economic analysis, each stemming from the preoccupation of the peasant with survival.

1. The food economy of the system. Investigation must establish the range of foods produced and how they are combined in consumption. It must describe their roles as preferred, expediency, or insurance crops, indicating their relative importance to the farmer. Finally, investigation should show how each influences decision making and resource allocation in terms of the quantity and timing of labor required.

2. Reciprocal obligations between household and community. Sociological inquiry will be important to show the types and effects of obligation on the community member. Where obligations involve resources, perhaps the contribution of labor to village projects and the extent and timing of commitments will be important. Other community characteristics will influence the choice of investigational technique. For example, the way in which work is allocated in the household will decide whether the farmer himself can reliably respond for the whole labor force, or whether each member should be questioned.

3. A description of farm practice. As we have seen, inherited practices are themselves a balance of food supply motivations. A description of these practices, which include customary husbandry

methods, as well as ecological guides to timing and rotation, will help in evaluating the decision complex facing the farmer.

The need to reconcile the objectives of both farmers and government in the changes to be encouraged multiplies the criteria required in evaluation. Profitability will remain the criterion of the market and of government interest, but it must be used in conjunction with acceptibility: a measure of the impact of the change on the way the farmer currently allocates his resources to meet his food requirements. The mechanical process of planning is upset by the difficulty of establishing a common denominator for the two criteria, so that the final emphasis in the planning process must rest with a subjective balancing of profitability and acceptability. This places a premium on experience and effective investigation, rather than on planning technique. At the same time, the debt ceiling of small farmers will place an important constraint on the rate of change. As with many limitations on the development process, it is a constraint which can be altered only by the increase in incomes resulting from the development process.

The implications for extension are self-evident. De Schlippe and Batwell have commented caustically but truly on the typical extension officer:

> The Agricultural Officer is not aware of the complexity of the customary system of agriculture, not aware of the social linkages through which the cash crops he introduces have a deep repercussion on food crop production and therefore on diet and social activities. He is unable to interpret the people's passive resistance to his reforms in terms of these cultural linkages and consequently explains it as conservatism, laziness or stupidity on the part of the cultivators. Cash crop production under such conditions remains alien to the culture of the people and is generally abandoned as soon as administrative supervision relaxes. So long as the supervision lasts, moreover, the cash income of the cultivators is often produced at the expense of cultural values which are not apparent to the administrator and may result in an impoverished diet, reduced security of nutrition, soil degradation, etc. In extreme cases unconsidered reforms may cause social disintegration and inability to exploit the native environment.[27]

The need in the system may be a crop with a particular growth pattern rather than one with a high yield. The content of extension must be appropriate. In addition, communications between staff and farmers in terms of farm practices and management difficulties creates the type of rapport and confidence which in turn generate the high morale so long missing from extension programs dogged by irrelevant criteria and teacher-pupil attitudes.

## NOTES

1. J. Mellor, "Family Labour in Agricultural Development," Journal of Farm Economics, XLV (1963).

2. H. Cory, Sukuma Law and Custom (London: Oxford University Press, 1954).

3. L. A. Fallers, The King's Men (London: Oxford University Press, 1964).

4. J. L. Joy, "The Economics of Food Production" Journal of Modern African Studies, XXVI.

5. H. Doggett, "Bird Damage in Sorghum," East African Agricultural and Forestry Journal, XVI, 1 (1950).

6. R. N. Parker, "Intercropping," B.Sc. dissertation (University of Reading, 1969).

7. N. S. Simmons, "Plant Interactions and Crop Yields," New Scientist, XVIII (1963).

8. A. C. Evans, "Studies of Intercropping. I," East African Agricultural and Forestry Journal, XXVI (1960).

9. H. C. Conklin, "Hanunóo Agriculture: A Report on An Integral System of Shifting Cultivation in the Phillipines" (Rome: FAO, 1957). (Pamphlet.)

10. H. A. Luning, "Patterns of Choice Behaviour on Peasant Farms in Northern Nigeria," Netherlands Journal of Agricultural Science, XV (1967).

11. P. Hill, "The Myth of the Amorphous Peasantry—a Northern Nigerian Case Study," Nigerian Journal of Social and Economic Studies, X, 2 (1963); M. and S. L. Fortes, "Food in the Domestic Economy of the Tallensi," Africa, IX.

12. P. T. Bauer, *The Rubber Industry* (London: 1958).

13. D. M. Goodfellow, *Principles of Economic Sociology* (London: 1939).

14. R. H. Fox, "Studies of the Energy Intake and Expenditure Balance Among African Farmers in the Gambia," Ph.D. thesis (London: Medical Research Council, 1953); A. Moody, "A Report on a Farm Economic Survey of Tea Smallholders in Bukoa District, Tanzania," East African Agricultural Economists Conference paper (1970).

15. J. Heyer, "Seasonal Labour Inputs in Peasant Agriculture," East African Agricultural Economists Conference paper (1965).

16. D. Pudsey, "A Pilot Study of 12 Farms in Toro" (Kampala: Uganda Dept. of Agriculture, 1966). (Mimeographed.)

17. M. P. Collinson, "Experience with a Trial Management Farm in Tanzania, 1962-65," *East African Journal of Rural Development*, II, 2 (1969).

18. A. T. Richards, *Land, Labour and Diet in Northern Rhodesia* (Oxford: Oxford University Press, 1939).

19. Heyer, *op. cit*.

20. Fox, *op. cit*.

21. *Ibid*.

22. Richards, *op. cit*.

23. P. de Schlippe and B. L. Batwell, "Preliminary Study of the Nyangwara System of Agriculture in the Southern Sudan," *Africa*, XXV, 4 (1955).

24. M. Read, "Native Standards of Living and African Cultural Change," *Africa*, XI, 3 (supp.) (1938).

25. *Ibid*.

26. De Schlippe and Batwell, *op. cit*.

27. *Ibid*.

# CHAPTER 3 THE TRADITIONAL AGRICULTURAL SECTOR

Two characteristics of traditional agriculture in African economics are important in guiding the application of farm management economics: the structure of the sector and, within it, the size of the farms.

## THE STRUCTURE OF THE TRADITIONAL AGRICULTURAL SECTOR

Historically, tribal organization has been the basis of sector structure. Lately the exchange economy has had considerable influence, certainly on the agriculture but perhaps less on the community base underlying it.

Complexity in advanced agriculture is represented by a diversity of transactions with the market of both input and output, and is governed by the market and technical alternatives open to the farmer. Both market and technical opportunities are limited in traditional agriculture.

Market opportunities depend on two factors: export possibilities and the market for agricultural produce within the economy. One of the objectives of the colonial powers was to secure raw material for their processing industries. This brought export opportunities to smallholders, and much of the development of the Colonial era was in building the marketing and transportation infrastructure to move out these exports. Such opportunities were limited in number, and each was located where natural conditions were most favorable. The major outlet for agricultural produce in most advanced economies is the food requirements of the nonagricultural population. But some African economies have 90 percent of their populations dependent on agriculture, whereas many advanced economies have less than 10 percent dependent. This differential implies market opportunities over fifty times as great facing the individual farmer in advanced economies and, in terms of absolute value, recognizing the need for international exchange, over 250 times as great. The present rural/urban balance must persist in developing economies because of the high rates of population growth. As with specialization and exchange within the rural community itself, increases in internal market opportunities are constrained by the rate of development and will be slow. Figure 1 uses von Thunen circles radiating from main population centers in the United Kingdom and Tanzania to compare the penetration of non-agricultural market opportunities in the two economies. The representation is wholly schematic, since no scale could be found to reconcile these two extremes. T. Schultz has elaborated the parallel between location and development, and the penetration of opportunities has been attributed as a cause of development under a wide variety of economic circumstances.[1]

The development of internal markets superimposes a structure of production opportunities on the possibilities dictated by conditions of climate and soil. Economic forces modify the natural advantages of particular areas. This complicates both the management of individual farms and the comparison between farms, bringing a new plane of variables into the balance. Where the network of opportunities is highly complex, as in an advanced economy with a high proportion of the population as a market for produce, farming patterns may be less related to natural than locational factors. Farm classification becomes difficult because it is more dependent on economic than natural factors.

Where internal markets are undeveloped, the opposite is true. Production alternatives remain limited only by climate and soil. Tribal communities, with their need for protection, are characteristically insular, a barrier to the buildup of internal exchanges. Within the tribal area, inherited preferences and practices are common throughout the community and form the basis of the agricultural system. Control of land resources is vested in tribal authority, and decisions as to use are made centrally. Individuals operate within this commonality bounded by the reciprocal obligations needed to give the surety required for productive activity. Throughout tribal areas the pattern of farming is homogeneous. Unlike the massive complexities in classification created by intermingling market opportunities, cropping patterns fall into geographically discrete areas within the framework of conditions of climate and soil. R. W. M. Johnson, working in Rhodesia, and H. A. Luning, working in Nigeria, have noted

FIGURE 1

Schematic Comparison of the Penetration of Markets
for Agricultural Produce in Tanzania and the United Kingdom

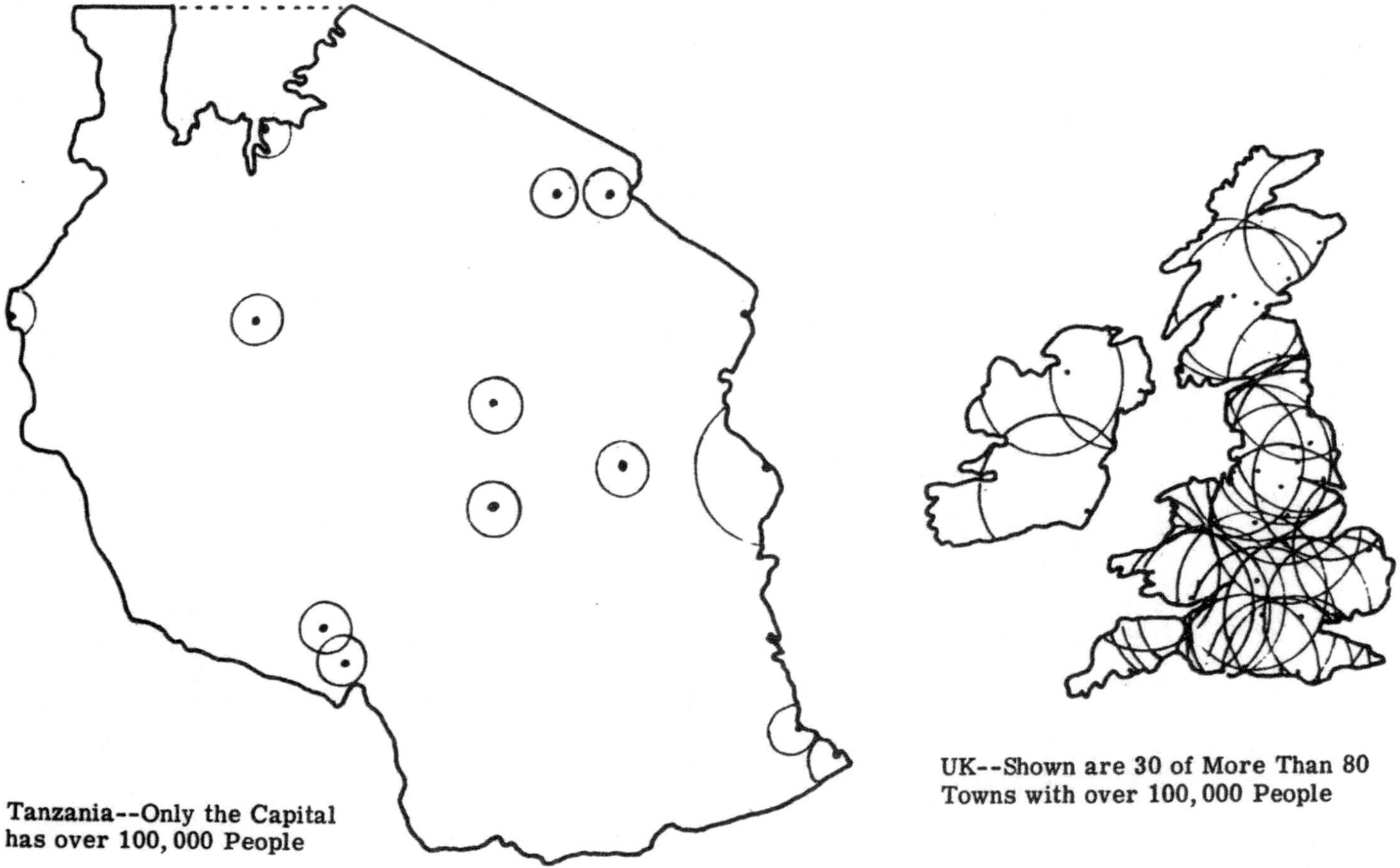

Tanzania--Only the Capital has over 100,000 People

UK--Shown are 30 of More Than 80 Towns with over 100,000 People

that this homogeneity is characteristic of peasant agriculture in Africa.[2]

Similarly, both factors—the lack of penetration by the exchange economy and tribal insularity—have limited the technology of African peasant agriculture.

As economic development has drawn labor out of agriculture, capital has filled the vacuum and made the sector an important market for the chemical and machinery industries. To remain dynamic themselves, these industries have sponsored technical progress in agriculture. The resulting alternative methods of production have raised a third tier, further complicating the natural conditions and proliferation of market opportunities forming the first two and bringing a new set of problems to farm classification. Economies of scale allow big farms to exploit a market opportunity closed to small ones, and the appropriate development paths for large and small units diverge.

Without the dynamism of industrialization cutting back its labor force, there is little of this interaction in peasant agriculture. Tribal insularity has generated its own technology, which is handed down through the generations and throughout the community. Methods and asset structure, as well as cropping patterns, tend to be homogeneous over geographically discrete tribal areas. Output levels have a linear relationship with the quantity of resources used, particularly family labor, thus removing the need for subclassifications based on means of production. As E. Clayton has noted, capital is not a significant variable in peasant agriculture and economies of scale are rare.[3] Exceptions are in areas with a degree of market penetration, associated with a cash crop for export and with the small cash surpluses available, are necessarily on a modest level.

## INFLUENCES ON THE APPLICATION OF FARM MANAGEMENT ECONOMICS

The homogeneity of peasant agriculture in geographically identifiable areas bounded by natural conditions and tribal boundaries has major consequences for the methodology of both investigation and planning. It also offers the key to extension strategy and organization within such areas.

The important attribute of the structure is the ease of farm classification, since grouping by cropping pattern and method of production is possible by discrete geographical area. The vast majority of farms in any one area face the same production opportunities, dictated by natural conditions and very limited market possibilities and weighted by community custom, and they use the same technique evolved within the tribal organization. Such areas are a logical focus for the whole approach sequence of investigation, planning, and extension. Within them, farmers face the same problems and have the same resource endowments on which to base what will be essentially the same solution.

The classification of the traditional agricultural sectcr into type of farming areas will be an efficient first step in investigation. The areas defined will form a population for first-stage sampling, which removes the three sources of between-farm variation that hamper cost-effective investigation and planning in advanced agriculture: differences in natural conditions, differences in market opportunities, and differences in methods of production.

Sampling within each area will cover two sets of attributes—descriptive information general to the community of the area and attributes susceptible to other sources of interfarm variation. Within the same natural conditions and the area of the tribal community it is important to measure the distributions of parameters affected by five other sources of variation: microclimatic differences, the quantity of available resources, individual preferences, managerial ability, and motivations. The last three deserve comment. Within community custom, individual preferences weight the balance of crops grown, particularly between dietary substitutes. The scope for variation will depend on the productivity of the system. Where food supplies are fairly readily secured, individual preferences will give large variation in food crop acreages. Managerial and motivational differentials have the same sort of influence on the pattern of cropping and production, the one arising out of relative ability and the other out of relative inclination. Where food supply remains the dominant priority, poor decisions on acreages for the season and poor manipulation of the traditional insurance practices will increase the resources required for food production and reduce the productivity of labor use. Poor motivation will reduce the total effort applied, giving high returns to labor use where management is good but a low return to available resources. These sources of variation are particularly important where a cash crop is grown with residual labor resources. Poor management and lack of motivation will reduce its acreage and production and increase variation in market activity within the area population.

Measurement of the effects of all five sources of variation gives well-described input and output coefficients for use in planning at the area level.

In the wake of this simplification in investigation comes the simplification that the representative farm technique is a useful tool in planning. As C. S. Barnard notes, it is a tool which has long been dogged by the unique and particular structure of farms in British agriculture.[4] We have shown the opposite to be true for traditional agriculture and will take up the representative farm technique as the fulcrum of the farm management approach in traditional agriculture in some detail later in the book.

## THE SIZE OF THE PRODUCTION UNIT

There is a good deal of discussion in the literature on the merits and demerits of different measures of size, and it is true that any single criterion fails to cover both resource endowment and productivity together. For our purposes, the comparison here between the size of units in advanced and peasant agriculture is so extreme, and so universally accepted, that a fairly brief description is given. It is the implications which are important.

By coincidence the number of production units in Sukuma agriculture, covering the largest tribal area in Tanzania, is of the same order of magnitude as the number in British agriculture. Each has about 300,000 farms. There the coincidence ends. Output of £100 gross for the typical Sukuma farm is dwarfed by an average of perhaps £6,000 for the British unit. Perhaps the most significant difference is the level of capital invested. A Sukuma farmer would rarely have more than 16 percent of his average net cash income invested (excluding his stockholding). The corresponding figure in British agriculture would be about 1,500 percent. The gap is wide enough to absorb errors on either side.

The average Sukuma farmer crops some eight acres each season, an area not atypical of African peasant agriculture as a whole. With family labor capacity limiting the scale of activity, differences in farm size between areas depend on the intensity of the labor requirements of the crops grown under varying natural conditions. Farm sizes increase where seasonal casual labor or machinery services are available. Secondary labor peaks, from operations it is uneconomic to mechanize, soon impose new limits for most crops.

There is a good deal of argument about the cost of applying whole-farm planning techniques to individual farm units, even in advanced agriculture. W. R. Schroder, for example, has produced evidence to show that pig farms must be over 500 acres before planning costs are covered by increments in returns.[5] The problem of how to measure returns is itself difficult. Suffice it to say that whole-farm planning techniques are available only to the large-scale units, even in advanced agriculture, and their wider application has been inhibited mainly by cost considerations. The same considerations are even more prohibitive for the small units characteristic of traditional agriculture, partly on absolute cost/benefit grounds and more particularly with reference to the opportunity costs of the qualified manpower concerned. The use of the limited numbers of agriculturally qualified manpower is examined in Chapter 5 in discussing government policy choices. Increases in output realized by program planning on a trial farm unit, on which management was directed by qualified staff, show the return to whole-farm technique. As is shown in Table 11, there were encouraging increments in productivity. However, the absolute increases in income levels are small, particularly when related to the cost of the manpower resources required in planning.

Family labor productivity was increased 180 percent above the average for local farmers. The absolute return was about £110 to the investment in planning and supervision. Even this level would not be practical on real farms, where wholly imposed management would be clearly unacceptable. For example, the cost of inputs averaged 360 percent higher than the usual outlay of the local farmer. Whole-farm planning must accept the farmers' risk ceilings as a constraint on the degree of change. Development of this magnitude would be realized only over the medium term, and returns would have to be set against accumulating costs.

Overall, the lack of complexity in the penetration of present agriculture by the exchange economy and the insularity of tribal development give geographically discrete types of farming areas, with homogeneous systems, problems, and solutions. At the same time the solution must be communicated to farmers at a low cost because the increments in output, though potentially high in percentage terms, are small. There are two major consequences for the approach in applying farm management economics:

1. Farm management economics is too sophisticated and expensive to have a direct role in extension. Emphasis shifts from extension to investigation and planning in order to select the changes that relatively unsophisticated extension services should encourage.

2. The whole approach is redirected away from the the farmer, the focus in advanced agriculture, to the area, giving the possibility of a cost-effective contribution.

TABLE 11

Increments in Labor Productivity from Whole-Farm Planning

| Source | | Net Income per Family Man/Day Available (shillings) |
|---|---|---|
| Local Farmers | Average | 1.34 |
| | Best | 2.27 |
| Trial Farm | 1962-63 | 2.98 |
| | 1963-64 | 4.63 |
| | 1964-65 | 3.80 |
| | Average | 3.80 |

Source: M. P. Collinson, "Experience with a Trial Management Farm in Tanzania, 1962-65," East African Journal of Rural Development, II, 2 (1969).

Finally, with the same solution relevant to the majority of farmers in an area, no initiative is needed on the part of contact staff in diagnosing and planning on the individual farm. The replicability of advice implies a function which can be performed with a modest level of training.

NOTES

1. T. Schultz, "The Theory of the Firm and Farm Management Economics," Journal of Farm Economics, XXI (1939).

2. R. W. M. Johnson, "The African Village Economy," Farm Economist, XI, 9 (1968). H. A. Luning, "Patterns of Choice Behaviour on Peasant Farms in Northern Nigeria," Netherlands Journal of Agricultural Science, XV (1967).

3. E. Clayton, "Opportunity Costs and Decision Making in Peasant Agriculture," Netherlands Journal of Agricultural Science, XVI (1968).

4. C. S. Barnard, "Farm Models, Management Objectives and the Bounded Planning Environment," Journal of the Agricultural Economics Society, XV (1963).

5. W. R. Schroder, "The Application of Systems Models to the Analysis of Swine Management Problems," Dissertation Abstracts (Ann Arbor), XXIX, 3 (1968).

CHAPTER

# 4

## THE ECONOMY:
## I. INFRASTRUCTURAL CONDITIONS

Absorptive capacity is a concept widely used in discussing development planning as a whole and the contribution of foreign aid in particular. The concept has grown out of experiences of project failure despite large inflows of capital. The key to creating sound development opportunities, using them effectively, and attracting capital flows is a stock of skilled manpower. Priorities therefore center on infrastructural investment in education and training. The first half of this chapter discusses the limit on the supply of qualified manpower for use in agriculture; the second, the sources and stock of improved agricultural technology.

All governments have recognized the need for intervention in sectors where the social return to capital is markedly higher than the commercial return, which is itself too low to attract private investment. With the profusion of small production units where no single firm is able to carry the overheads of research and development, agriculture has become one of these sectors. For both strategic and welfare reasons, governments of advanced economies intervene both to stimulate a flow of new technology and to protect the sector. In developing economies the role of agriculture as the basis of national economic growth creates a more positive need for intervention by government. The level of intervention is limited by the need to generate funds for public investment and to expand incomes over the mass of population, themselves dependent on agricultural livelihood. The British economy, with less than 3 percent of the working force in agriculture, provides direct support of over £1,000 per farm, including a research and extension element of £250 each year. This absorbs only 3-4 percent of total government expenditure. Tanzania, with 88 percent of her working force in agriculture, provides direct support of about £7 per farm, which requires 12 percent of total government expenditure.

Prejudging the policy issue of improvement or transformation discussed in the next chapter, we have defined the essentials of extension strategy within an improved approach as replicability and low costs per farm. The existing hierarchical structure, a residual from colonial administration, seems well adapted to these needs. Further, the relative closeness of modestly qualified contact staff assists rapport with the farmers. F. E. Emery and O. A. Oeser have noted that too wide a differential in education creates problems of communication.[1] Typically, graduates are used as local organizers, with a supervisory grade between them and the contact workers in touch with the farmers. The field supervisors are diplomats or promoted contact staff. The contact workers have eight years of education plus an in-service training course leading to a certificate in agriculture. Since independence, as primary and secondary education have expanded, there has been emphasis on upgrading the contact cadre to this certificate level. Despite the coverage that these three levels give, penetration of the smallholder population remains limited, and typically there are over 1,000 farmers to each contact worker. With such high ratios, the selection of farmers for supervision and promoting diffusion of changes through the community are necessary adjuncts of extension strategy.

It is important to look at the possibilities of increasing the intensity of coverage and of changing this structure to meet the needs of alternative strategies for development. Intensifying coverage of the existing structure implies an increased output of manpower at each level of the hierarchy. Changing the structure to favor a project or transformation strategy requires a shift of emphasis in training and education to producing higher-grade staff.

The overriding limitation is the allocation of public expenditure to education. The output of manpower for agriculture could be increased by higher expenditure, either through an increase in revenue or through a reallocation from other sectors. Such an increase could also be used to change the balance of output among the three grades. Alternatively, the structure could be altered by an internal reallocation of expenditure among primary, secondary, and higher education.

Absolute increases in public revenue are dependent on economic growth itself, though improvements in tax administration may squeeze an extra margin from existing levels of national income. Dramatic increases are precluded by the need to promote growth in agriculture, itself the major source of revenue. Within total revenue flexibility is also limited. Economic growth implies a balanced advance on several fronts; education, communications, and agriculture are all interdependent, and unilateral development of particular sectors is an anomaly. In Tanzania these sectors, together with health, defense, and administration, absorb 75 percent of the recurrent and development budget. Given the predominance of recurrent expenditure and the nature of government establishment, shifts between sectors will be marginal except as a result of policy reformulation. In the same way, only revised policy will stimulate internal reallocation within expenditure on education. Drawing the balance of primary, secondary, and higher education is a delicate task, an area in which ideals of universal primary education clash with the needs of the economy.

H. Mynt notes: "Whereas in the last resort a wrong investment in material capital can be scrapped when it proves too expensive to salvage, wrong pieces of human capital . . . tend to be self perpetuating and have the habit of not merely distorting but actually of disrupting the social infrastructure."[2] Correct gearing is dependent on the strategies adopted for manpower utilization in all sectors. The pool of skilled manpower will grow as resources permit and as historical gearing can be developed. The rate of adjustment is slow. Increasing the rate of output of manpower at any level depends on the prior increase in output of both qualified teachers and capital facilities for that level, itself financed from available revenue and in competition with established priorities.

Tanzania has been committed to a rapid expansion of extension workers in agriculture, as well as an improvement in the quality of its contact staff. Its education policy reflects the compromises which have had to be struck to meet these and other increasing requirements. Three goals were set out in 1963:

1. Full self-sufficiency at all skill levels in the economy by 1980.

2. A basic (primary) education for every Tanzanian child as soon as financial resources permit, presently planned to be achieved in 1989.

3. Provision of additional and further education only to the extent justified by the manpower requirements of the economy.

The government sees primary and secondary education as a consumer good but has been practical with the time scale, looking ahead twenty-six years from 1963 to universal primary education and rejecting for the present the right to secondary education, on grounds of national priorities. Secondary and higher education have been planned to meet the needs of the economy. The aim in agricultural manpower planning is an annual output of eighty graduates, 215 diplomats, and 600 certificate workers for the whole economy, including the extension services, by 1980. As a basis for it the Faculty of Agriculture was established at the University of Tanzania in 1970. The additional institutions to produce the diploma- and certificate-level staff are still in the planning stages. The difficulty in projecting these levels into extension organization is to know the proportion moving into extension as opposed to other areas of the economy. Data on staff representation in the regions for 1968 has been reconciled with the total from the manpower survey used as a basis for the second five-year plan. Table 12 gives estimated cadre strengths, based on expected supplies from available training facilities and the assumed levels of in-take into the extension services. It relates these numbers to the changing farmer population to give estimates of the increasing intensity of coverage.

If these manpower targets are met, extension coverage will increase dramatically over the period to 1990. However, even with this rapid expansion it is significant that the ratio of contact workers to farmers does not drop below 1:1,000 until the end of the 1970's. Even with 1:500 by 1990, selection of farmers for improvement will still be important for effective extension and the diffusion of new technology. Despite the need for positive intervention by governments in rurally dominated developing economies, penetration cannot approach the intensity of advanced economies, which divert revenues from the vast industrial sector. The British advisory services, for example, have a graduate or diplomat for every 200 farmers in addition to the large subsidy element on most products. In practice, once the educational gearing of the second Tanzanian five-year plan has been set in motion, it is unlikely that the structure of manpower output, in terms of switching funds for teachers and facilities between cadres, could be altered significantly before the 1980's. Revised strategies of agricultural development requiring a predominance of graduates and diplomats would reduce the potential penetration of the service and entail a high level of wastage in terms of trained certificate personnel and misdirected infrastructure investment.

One result of delaying universal primary education is to delay the spread of literacy. Adult literacy increases slowly, depending on the spread of primary education and increasing coverage as the population grows older. UNESCO has given the rate for East Africa as 9.8 percent in 1950 and 11.8 percent in 1960.[3] In Tanzania, where universal primary education is targeted for 1989, and where in 1969, 47 percent of the appropriate age group were newly admitted to school, literacy will remain limited to junior family members in the majority of homes for the next twenty years.

## TABLE 12

### Estimates of the Increasing Intensity of Extension Coverage, Tanzania, 1968-90

| | Graduates | | Diplomats | | Certificate | | Estimated | Farmers | Farmers |
|---|---|---|---|---|---|---|---|---|---|
| Year | No. | Assumed Intake (percent) | No. | Assumed Intake (percent) | No. | Assumed Intake (percent) | Farm Pop. (thousands) | per Grad/ Dipl. | per Certificate Worker |
| 1968 | 39 | 33 | 212 | 66 | 1,084 | 75 | 2,500 | 10,000 | 2,300 |
| 1974 | 140 | 45 | 520 | 70 | 1,950 | 80 | 2,900 | 4,400 | 1,500 |
| 1980 | 380 | 50 | 1,300 | 75 | 4,550 | 85 | 3,500 | 2,100 | 770 |
| 1990 | 780 | 50 | 2,900 | 75 | 8,900 | 85 | 4,700 | 1,300 | 530 |

Notes: Diplomats for 1968 includes Assistant Field Officers, grade I (AFO I's), senior contact staff capable of a supervisory function.

"Assumed intake" for 1968 is establishment data related to cadre totals for the whole economy from the 1968-69 survey. Veterinary staff are not included.

Source: Compiled by the author.

Guiding factors already identified for extension center are the homogeneity of problems over large numbers of farmers, allowing replicability in advice given by the service; and the need for a low-cost service because of a rate of change in practices constrained by farmers' risk preferences, giving a low return to extension on individual farm units. The existing extension structure seems well adapted to meeting these conditions, with the modestly qualified personnel as contact staff at the bottom of a hierarchy giving at least a threefold increase in the penetration of the rural community. Furthermore, though flexible over time, the existing structure will only adjust slowly. Significant changes reach back into the infrastructure of the economy and imply long-term planning to alter the gearing of educational institutions, the results of such changes having a gestation period of seventen years. Project strategies, though they could immediately draw on staff with managerial capacity from the extension services—and Tanzania did this in her efforts at transformation through village settlement schemes in the early 1960's—would be limited until a redeployment of the balance of output of qualified manpower became effective.

The effect of the growth of literacy on the correct approach for farm management is difficult to assess. Researchers have noted a relationship between education and innovativeness.[4] Other evidence suggests that the written word plays little part in extension, which remains dominated by the farm visit.[5] It seems unlikely that extension strategy will be affected by the spread of literacy.

Illiteracy and lack of education certainly limit the methods available for investigation. Data cannot be collected by mail or by the use of farm records. Some researchers have gotten around this by asking schoolchildren to record information on their family holdings.[6] This tends to bias the sampling procedure and precludes a direct relationship with the head of the household, a factor jeopardizing continued cooperation. The prospect of long-term illiteracy means that relatively high-cost investigational techniques are needed, centered on visits to the farms, and calls for a dependence on memory, with the additional sources of error that this entails.

Finally, the general scarcity of qualified manpower suggests that farm economists will have to implement all phases of the approach, rather than specialize in investigation, planning, or extension. Their contribution will be in liaison with local organizers of the advisory services, planning the content for extension programs.

## IMPROVED TECHNOLOGY IN TRADITIONAL AGRICULTURE

The opportunities for the development of traditional agriculture are certainly less dramatic visually than postindependence politicians had hoped and probably are more mundane than prominent agriculturalists have believed. J. Hutchinson, speaking of the need for progress in agriculture as the basis for development of the Ugandan economy, said, ". . . this is beyond the scope of a subsistence agriculture with a couple of cash crops grafted on to it. What this country is going to need is modern farming and we can't provide it until we have worked it out experimentally."[7] Quite the contrary: Nurturing the successful graft of cash crops is perhaps the only route to development—introducing new ones and improving productivity of both the established ones and the subsistence crops, allowing a reorientation of resources to marketed production.

Potentially the opportunities for development are wider than this. M. P. Collinson has distinguished five types: structural change, system reorganization, new enterprises, intensification, and increases in scale.[8]

### Structural Change

Structural change has been an evolutionary feature of agriculture dominated by the unique relationship of the industry to land as a factor in production. It has been associated mainly with land ownership reform arising from community pressures due to increasing population density. Historically such change has been centrally directed, since usually only the authority of government has been able to cope with the cross-pressures in these situations. Kenya has seen two examples of pressures and reform over the last fifteen years: registration and consolidation in the central highlands and the resettlement parts of the former white highlands, both arising from high densities of population.

Structural reform as a means of introducing technical change has had some notable failures. Even politically inspired reforms have been called into question when they are not in response to local pressures—for example, many "transformation" schemes, though these still form the core of agricultural development efforts in several African countries. (See Chapter 5.)

## System Reorganization

Optimizing the combination of enterprises by relating the available farm resources to market opportunities is a feature in the management of sophisticated production systems in advanced agriculture. It is less appropriate to the diverse objectives of the peasant farmer in traditional agriculture. However, the introduction of changes of any type into a balanced system implies a degree of reorganization which systems techniques are useful in measuring. The use for systems analysis shifts from optimizing the combination of market enterprises to evaluating the impact of changes on the existing resource/product balance. It has no independent contribution to make until there is a complex of market opportunities facing the farmer.

## New Enterprises

The addition of cash crop enterprises to existing systems has been the core of the development of traditional agriculture to date. Most of the opportunities were fostered by the metropolitan powers seeking controlled sources of raw material. Overseas interest established a marketing, transport, and research infrastructure which has proved a valuable heritage to African governments. Independence brought its own surge of new opportunities as crops tacitly confined to plantations and estates were opened up to smallholders. Tanzania offers the examples of coffee, flue-cured tobacco, pyrethrum, and (lately) tea.

New crops are readily acceptable to smallholders, particularly in farming systems with "surplus capacity," where food production does not absorb all family labor. Difficulties of assimilation are increased when the resource requirements of foods and the new crops clash.

Future possibilities for new crops are difficult to assess. Clearly, the spate of opportunities released by independence will not be repeated; and although many of the present possibilities can be expanded further, many also face declining markets, with falling prices. With further new enterprises the developing countries will no longer be small suppliers moving into an established trading crop. The crops themselves will be new and the extent of their market will be limited by the penetration they can achieve. Their rate of expansion is likely to be slow and uncertain.

## Intensification

Intensification is used here to describe measures designed to increase the productivity of the land already used by the peasant farmer. It is separate from an extension of scale, which implies use of an increased amount of land.

The usual management practices associated with intensification are better seed, correct time of planting, proper plant populations, and more thorough cultivation operations. These are usually complemented by the use of purchased inputs, particularly fertilizers and insecticides. Package programs which include a number of changes selected for their complementarity have become an established feature of agricultural development. Measures for intensification formed the core of extension effort through the colonial period and, bearing in mind increasing population densities, higher productivity per acre remains a long-term objective throughout agriculture. Even in areas without an absolute scarcity of land, increasing population upsets the arable/fallow balance, and soil fertility deteriorates. Because the use of purchased inputs of manure and fertilizer is an aid to fertility maintenance, and intensification may offer greater productivity without an increased use of land, which aggravates the fertility problem, it is likely to be the most important means for the further development of traditional agriculture. With the spread of new seed varieties in some crops as an exception, intensifying practices must be interpolated with the existing system by inducing the farmer to adopt them. The difficulties in this process form a good deal of the content of this book, and some of the reasons for past failures are presented in the later sections of this chapter.

## Extension of Scale

Increasing the scale of the system is a substitution of capital for labor on labor-intensive operations and implies the use of a larger area of land. We have noted the anomaly that under increasing population density, although an extension of scale is a short-term means to higher income, it accelerates the fertility loss in traditional systems characteristically dependent on an arable/fallow sequence to maintain yields per acre.

There has been little machine technology developed for traditional African agriculture. The limited and uncertain nature of the market gives no incentives for the heavy expense usually associated with machinery research and development. Efforts have been limited to attempts to penetrate the market with implements and mechanized processes developed elsewhere. The surge of interest in intermediate technology has appropriately stressed the need to get away from advanced ideas of machinery. It has, however, faltered by touting machines rather than diagnosing problems and devising mechanical techniques as a solution. There has been confusion between the role of a machine in advanced and traditional agriculture; in the one it is a cost saver; in the other, a system expander. However, because machines have the capacity to reduce effort, they are attractive to the farmer and often afford a short-term incentive to support other, more complex changes requiring internal reorganization of the farming system.

Intensification and mechanization will both play a part in the future development of traditional agriculture. Because of the long-term fertility maintenance problem, the key role falls to intensification: increasing the productivity of the existing resource endowment of the peasant farmer. For a successful improvement approach to development, all improved technology must be oriented to problems created by needs and resource constraints of the existing systems of farming.

The sources of improved agricultural technology as a whole are fewer in traditional agriculture, and dependence on central government is increased. The inventiveness of individuals is stifled, and the interests of manufacturing industry are limited by the uncertain extent of the market.

Industry contributed 39 percent of the research effort in British agriculture in 1965-66, the major part from chemical, feed, and machinery manufacturers anxious to expand or merely maintain their share of the market.[9] Machinery, and to a lesser extent chemicals, have found increasing applications as capital substitution has penetrated further into agriculture, stimulated by the high returns to labor in other parts of the economy. But the limited market for inputs in traditional agriculture has restricted the contribution of manufacturers to supplying samples for test on experimental stations and for government-sponsored demonstrations.

Enterprising individuals within agriculture in advanced economies also play a significant role in the development of technology. The highly specific, multivariate nature of this kind of technology makes formal research expensive. The high educational and motivational characteristics of some individuals generate a solution which is often adopted by the industry concerned and generalized for a market of other farmers growing the same combination of crops. Such inventiveness is rare in traditional agriculture. The specialized opportunities, the educational and the motivational basis are all missing. The community, often organized to preserve individual security, requires conformity for successful operation.

The brunt of the research effort falls squarely on the central government. The overseas demand for raw materials has encouraged research on export crops as part of the infrastructural development necessary to build up the supply. Such crops have been the focus of the limited financial and manpower resources available. Very recently there has been a growing consciousness of the need for work on food crops, particularly the major grain staples. The overhead of machinery research has been too high to allow anything but a limited contribution, even from government. Research efforts in East Africa seem to have been more intensive than in West or Central Africa. In the colonial era the level of effort seems to have been dictated by the density of expatriate settlement. Kenya dominated East African research programs; but again the Tanzanian example is not a typical, though the position is somewhat fluid, with centers closing and new ones being established and with staff altering from time to time.

In Tanzania six multidisciplinary research centers are coordinated by the Ministry of Agriculture. Each center has a range of specialists, usually one or more agronomists and plant breeders, a soil chemist, a pathologist, and an entomologist. Some centers have a pasture research specialist, an engineer, or an economist, with a complement of junior staff for laying out and supervising the experiments. These six centers cover a country of some 360,000 square miles. Zonal centers reach into the different ecologies within their areas through substations. The experiments are designed and analysed by the specialists and are administered by executive cadre staff in charge. Much of the work is done on the stations, though more and more emphasis is being placed on programs of external field trials. A. M. Scaife has described one such program, and it serves as an example of the coverage being achieved in this type of area.[10] Scaife worked in 1963-66 on maize, the major food grain staple and the only crop, in addition to cotton, with a significant experimental program in the 110,000 square miles under the authority of the Western Research Center, located near Mwanza on Lake Victoria, an area with 800,000 farms and a population of over 4 million people at the 1967 census. From his three-year program involving 132 fertilizer experiments Scaife distinguishes sixteen response areas which are aggregated into seven areas for recommendations at three levels of nitrogen and phosphate. Each of his recommendations covers an average of over 110,000 farms, an example of the blanket research characteristic of programs in developing economies.

## EFFECTIVENESS OF GOVERNMENT-SPONSORED AGRICULTURAL RESEARCH

The blanket nature of agricultural research is a result of the scarcity of funds and skilled manpower of developing economies. "Blanket technology" is a term used somewhat skeptically by research workers rooted in the individualization of advanced agriculture and the refined experimentation of British research programs. But from several points of view there seems little justification for any more sophisticated approach, for in terms of crop acreage rather than numbers of farms, a different picture emerges. Although the average zone Scaife draws includes over 110,000 farmers, it covers perhaps 300,000 acres of intercropped maize, representing an output of about 60,000 tons of maize, wholly for subsistence. This would represent the output of some 200 medium-size arable farms in the United Kingdom. In these terms, the research effort is relatively intensive. Also, the recommendations for particular farms in advanced agriculture depend mainly on the cropping history of the farm in question. With the arable/fallow sequences of traditional agriculture, cropping history will be uniform over large areas.

Criticism of the blanket nature of the research effort is superficial, and indeed this sort of coverage is particularly appropriate where cropping opportunities can be demarcated on a geographical basis. In other respects, however, efforts have been poorly adapted to local needs and conditions. Lines of research were dictated by the individual export crops required by the metropolitan areas. Emphases sprang from a demand orientation. While the grafting of cash crops onto peasant farming was absorbing surplus labor capacity, the process moved smoothly. Once surplus capacity was exhausted, further increases were dependent on revisions in management routines and on reallocation of farmers' resources. Under these conditions a demand-oriented approach has been inadequate, needing to be balanced by a consideration of supply conditions to highlight barriers to the expansion of production. No such reorientation has occurred, and this failure has contributed more than anything else to skepticism with improvement as a strategy for development.

The rundown conditions and poor morale of many extension services bear witness to the lack of sympathy between farmers and advisers, a situation created partly by inappropriate program content.

Many economists, often pleading for a voice in the design of research, have given examples of recommendations inconsistent with the needs and capacities of the smallholder. E. Clayton has summarized many of the relevant points in discussing opportunity costs and decision making in peasant agriculture.[11] He shows how criteria which agronomists take as axiomatic often involve the sacrifice by the smallholder of alternative objectives. He points out in particular the optimizing of yields through the correct time of planting, and the optimizing of the quality of produce. Time of planting is a particularly common feature of recommendations from research programs, but in a labor-limited peasant system it inevitably means restriction of output. Clearly, all farmers can plant up to the recommended date, while those planting additional acreage afterward will increase total production despite a drop in yield levels. Again, in a system with reliable food supply as a major priority, spreading the planting reduces the variability of results. The laws of probability dictate a much wider variation for rainfall totals over short periods of time.

Tanzania affords an example here. For Sukumaland the coefficient of variation of December rainfall is 53 percent (over a nineteen-year period). This is the recommended planting time for most crops. If we can assume that yields are closely correlated with rainfall in this area—K. J. Brown gives a highly significant $R^2 = .745$ for cotton yields and planting rainfall[12]—output could be down to 47 percent of average levels once in every six seasons. Since the coefficient of variation for the total December, January, and February rainfall is 37 percent, the staggered planter can expect 63 percent of his average level in all but very poor years. The peasant farmer's strategy is to increase his average plantings to give a higher expected output that will cover his requirements in the worst years. With the rigidity imposed on acreage planted by a specified time of planting and the lower reliability to be insured against, such a strategy may be unfeasible from recommendations based on criteria of maximum yields per acre.

Several factors made research routines inflexible and perpetuated an orientation in research programs unsuited to the economic and motivational barriers against increased supply from traditional farming systems. Two can be traced to expatriate domination of research activities; two others, to the traditions of experimental method in agriculture.

The expatriate personnel dominating the research services of developing economies had advanced agriculture as their whole background—indeed, local specialists are themselves indoctrinated in the traditional mold by a higher education built around overseas syllabi. Even in Britain itself agricultural researchers have been criticized for their isolation from the farmer. But within Britain, a sophisticated advisory staff and sophisticated farmers were quite able to interpret results for themselves. Workers in developing agriculture retained their insularity, and a plethora of features of expatriate life and local conditions widened the gap, preventing any effective dialogue between researchers and farmers.

1. As an educated elite, researchers have little in common with the peasant farmer. For expatriates this isolation is emphasized by their enclave existence as a minority group and by language barriers.

2. Senior extension staff, the products of a similar environment, accord with the criteria of the researchers and are similarly isolated from both the farmers and their junior staff.

3. Contact extension workers are unqualified to diagnose problems at the farm level. Nor does the quasi-military routine o. advisory service organization encourage a feedback from the field to the laboratory. On the contrary, the flow is noticeably one way, with solutions passed down and assumed to be good for the farmer.

4. As D. G. R. Belshaw and M. Hall point out, traditional farmers do not articulate their needs as well as those in advanced agriculture.[13] Not that they cannot, but because there has been no encouragement for real rapport in the teacher/pupil attitude fostered by advisers.

Expatriate research personnel also brought their experimental criterion from the land-scarce environment of British agriculture, concentrating in their work on maximizing the physical yields per acre. K. Dexter has criticized the narrowness of this criterion even for advanced agriculture; and Davidson and Martin have noted that farmers are much more concerned with returns to labor and capital, and that high productivity from these factors often means a reduced yield per acre.[14] In traditional agriculture, with family labor clearly limiting the overall size and composition of the system, yield per acre is an inappropriate base for evaluating potential changes in practice. The return to labor use is all important. The return to capital, because of its scarcity and special significance for development, is also important. Finally, with the priority placed on the security of food supply from year to year, the reliability of the results from changes is an important qualification of average expected returns. Increased yields per acre may or may not be consistent with these three criteria. Research has concentrated, appropriately enough, on adaptive work—but adaptive only in a limited sense: to the natural resources, the climate, and soil of these areas. It has ignored the economic and motivational peculiarities of traditional agriculture.

While there are problems which are purely of the natural environment—the incorporation of disease control into new varieties is an example—most are dependent on the particular economic circumstances. Even apparently straightforward botanical changes may have a significant impact on farmers' resource allocations. For example, M. Kiray and J. Hinderink have described the effect of replacing Yeri cotton varieties in Turkey with Akala and Deltapine.[15] The new varieties mature within a period of three weeks, thereby concentrating the labor requirement for harvest and creating a dependence on hired casual labor which the old varieties avoided by having a longer maturity period.

Two aspects of traditional experimental methods, also built up in advanced economies, raised barriers to reorientation of efforts to local needs. The direction of all production through the market determined the selection of research lines. Experimentation has concentrated on testing alternative practices and different levels of practice in very intensive and detailed programs. The lack of a market as a denominator between the subsistence and cash crop sectors of traditional farming systems means that there is no guide as to profitable lines of work. The market evaluation of subsistence production becomes a complex matter of the opportunity costs of family labor. There has been no ready method of identifying crops in the subsistence sector which, with improved practices, would improve the productivity of the whole system. Agricultural research has remained entrenched in the testing of practices and levels of usage for the export crops, readily identified as important by familiar market forces.

F. Yates illustrated the principle that the intensity of experimental effort should be governed by the marginal principle: that the cost of the last increment of experimental effort should equal the net gain in total output when the results were implemented.[16] He used the figures shown in the first four columns of Table 13 and emphasized that these marginal returns may be very much smaller than total returns. The fourth and fifth columns, giving the capital/output ratios of the alternatives, are pertinent for developing economies where research resources are scarce.

If this sort of relationship really does hold between marginal and total revenue, then, in the resource and manpower starved context of the research efforts of developing economies, effort should be spread

TABLE 13

The Marginal Principle in Research Evaluation
(pounds)

| Research Costs | | | Capital Output Ratios | |
|---|---|---|---|---|
| Over-head | Program Costs | Expected Net Gain | Net Gain/ Total Cost | Net Gain/ Program Cost |
| 5,000 | 4,000 | 87,000 | 9.7 | 21.7 |
| 5,000 | 2,000 | 85,000 | 12.2 | 42.5 |
| 5,000 | 1,000 | 78,000 | 13.0 | 78.0 |

Source: F. Yates, "Principles Governing the Amount of Experimentation Needed in Development Work," Nature, CLXX (1952).

among several lines of research, rather than optimizing results in any particular line. The comparison is particularly striking when based on program costs alone, since these would be the flexible element with a change of emphasis to low-intensity, multiline investigation.

At the same time, the very methodology of experimentation is based on returns to land as the vital factor in the production process. Research workers have been confined by this narrowness and have been too preoccupied with the administrative complexities of work in developing agriculture to question the utility of their stock-in-trade techniques.

The insular character of British agricultural researchers and the enclave nature of their presence in developing economies have interacted with these aspects of focus and technique to create an inflexibility of approach which has hindered reorientation of research effort to the problems of traditional farmers. The result has often been inappropriate recommendations as the content of extension programs, with a subsequent inevitable deterioration of relationships between farmers and the authorities.

Although the blanket nature of research efforts is inevitable with the resources available, and indeed is appropriate to conditions in the economy—both the structure of traditional agriculture and the structure of the extension services—any orientation to farm problems has been by happy coincidence. Farm management investigation can provide a service in problem identification by describing and analyzing economic relationships in the existing system in order to aid reorientation of the research effort. At the same time the conflict within research circles between the desire to refine experimental procedures further and the wish to make them more relevant to farmer conditions should be resolved in favor of increasing the usefulness of the results. Agricultural experimentation is likely to form the major, and often the only, source of means for the next steps in the development of traditional agriculture. Better ecological and economic definition of homogeneous zones to be covered by less sophisticated programs probing a wider range of opportunities within each zone could greatly increase the contribution of the resources available.

There are important interactions between farm economics and agricultural research. (The final chapter of the book summarizes the type of contribution which can be made by farm management economics to research orientation.) The most important consequence for the application of farm economics is the inevitability of the blanket nature of the technological opportunities which gives solutions relevant for large numbers of farmers. It reinforces the evidence from an examination of the structure of the agricultural sector and the institutional capacity of the economy, particularly of the extension services, that replicability should be an attribute of advisory content. This in turn reinforces the case for the use of the representative farm in planning extension for defined types of farming areas; the possibilities as well as the problems are likely to be homogeneous.

A secondary consequence of agricultural research as the major source of technological change suggests a logical siting for farm economists within the research team covering a contiguous group of ecological zones. His function is very much that of a link man: between farmers and researchers, on the one hand, and researchers and extension staff, on the other.

## NOTES

1. F. E. Emery and O. A. Oeser, Information, Decision and Action: A Study of the Psychological Determinates of Changes in Farming Technique (Carleton, Victoria: Melbourne University Press, 1958).

2. H. Mynt, The Economics of Developing Countries (London, 1964).

3. UNESCO, World Education Statistics (Paris: 1970).

4. M. Upton, Agriculture in South-Western Nigeria, "Development Studies," 3 (University of Reading, 1967); E. Bowden and J. Moris, "Social Characteristics of Progressive Baganda Farmers," East African Journal of Rural Development, II, 1 (1969).

5. D. H. Shepherd, "Advisors' Opinions on Advisory Methods," NAAS Quarterly Review, LXIX (1965).

6. H. D. Ludwig, Ukara: A Special Case of Land Use in the Tropics, "Africa Studies," XXII (Munich: IFO, 1967). (In German.)

7. J. Hutchinson, "The Objectives of Research in Tropical Agriculture," Empire Cotton Growing Review, XXXV (1958).

8. M. P. Collinson, "A Survey of Innovations in Traditional Agriculture in Tanzania," East African Agricultural Economists Conference paper (1970).

9. J. Ashton and R. F. Lord, eds., Research, Education and Extension in Agriculture (Iowa City: Iowa State University Press, 1969).

10. A. M. Scaife, "Maize Fertiliser Experiments in Western Tanzania," Journal of Agricultural Science, LXX (1968).

11. E. Clayton, "A Note on Research Methodology in Peasant Agriculture," Farm Economist, VIII, 6 (1957).

12. K. J. Brown, "Rainfall, Tie-Ridging and Crop Yields," Empire Cotton Growing Review (January, 1963).

13. G. D. R. Belshaw and M. Hall, "The Analysis and Use of Agricultural Experimental Data," East African Agricultural Economists Conference paper (1964).

14. K. Dexter, "Report on Discussion Group D," Proceedings of the Conference of the International Association of Agricultural Economists (1967).

15. M. Kiray and J. Hinderink, "Interdependencies Between Agroeconomic Development and Social Change," Journal of Development and Social Change, IV, 4 (1968).

16. F. Yates, "Principles Governing the Amount of Experimentation Needed in Development Work," Nature, CLXX (1952).

CHAPTER

# 5

# THE ECONOMY: II. GOVERNMENT POLICY FOR AGRICULTURAL DEVELOPMENT

Clearly, any approach adopted for farm economics must be applied within the context of government policy. Many of the exogenous conditions for the planning phase in particular will be dictated by policy objectives, price or output control, foreign exchange needs, and import substitutions. But farm economics has a dual role, and although this study concentrates on its contribution to extension strategy, there is a circularity. The potentials indicated by the analyses at the micro level will themselves be a factor in policy decisions, especially in choosing between alternative agricultural development strategies of improvement or transformation. This chapter supports the assumption of the study that improvement is the appropriate general strategy for agricultural development and identifies a particular role for transformation.

The essential difference between the two is that improvement aims to speed up the evolution of the farming system <u>within</u> the existing structure of agriculture. Transformation sees structural change—the amalgamation or consolidation of holdings, imposition of formal landholding rights, or modification of the settlement pattern—as a prerequisite to development potential. Both strategies have seen failures. Chapter 4 attributed many failures in improvement to inappropriate content. The failures in transformation are often attributed to poor organization; and faith in transformation, as a general strategy for development, tends to be sustained by the knowledge that organization can be improved. However, the real shortcomings of transformation seem more fundamental.

Transformation has been promoted as a general strategy for agricultural development in two African countries in particular, Tanzania and Nigeria. It was a politically inspired switch from the improvement efforts of the colonial government, which rarely achieved conspicuous success. Disillusion with these plodding, detailed, and diffuse programs stimulated politicians to look for alternatives. They found them in a scheme approach with strong visual appeal and tended to corelate modern methods, using machinery particularly, with improved results. Table 14 illustrates the emphasis placed on transformation in Nigeria and Tanzania during the 1960's.

Although the comparison omits recurrent expenditure, which is heavier under the improvement approach, the size of the scheme investments, in relation to those by the established agricultural services, clearly demonstrates the new emphasis. Such a significant shift against the existing establishment shows the level of political commitment and an impatience for results fostered by the nibbling, creeping blanket of ordinary extension.

Support was not only political however. M. E. Kreinin has noted in Nigeria:

> Support for the schemes came from experts who saw little hope for progress in the present village framework; a land tenure system that contains no legal boundaries and no security of tenure, deprives the farmer of an asset against which to raise loans and of an incentive to investments in long term improvements.[1]

TABLE 14

Capital Expenditure on Transformation and Improvement in Tanzania and Nigeria (£1,000)

| | | Nigeria | |
|---|---|---|---|
| | Tanzania | West | East |
| Transformation | 8,506 | 9,740 | 16,640 |
| Improvement | 1,734 | 3,340 | 6,620 |

Sources: J. C. Wells, "Nigerian Government Spending on Agricultural Development, 1962-67," Nigerian Journal of Economic and Social Studies, IX, 3 (1967). Government of Tanzania, First Five-Year Plan (Dar es Salaam, 1964).

Similarly, in Tanzania, the 1960 IBRD report was probably the origin of the subsequent emphasis on transformation. Although its report was hedged with the need for pilot schemes, even the scale of these readily absorbed the whole of government agricultural investment for several years.

The experts seem to have been seeking structural reform as a means to economies of scale, vaguely associated with large fields and machinery. Nevertheless, without the political enthusiasm they could not have penetrated the agricultural establishment in the way they did in Tanzania. The new policy amounted almost to a vote of no confidence in the existing department, and drained it of executive and managerial manpower. Even though formal "villagization" was dropped as policy in Tanzania as early as 1965, the "irreversibility of commitment" noted by Robert Chambers was certainly partly responsible for a lag in re-coordinating agricultural development within one ministry until 1969.[2]

Superficially, two levels of failure can be identified. First in organization, usually the scapegoat, so that escape clauses were added, stressing the potential of properly managed schemes. The supervision of schemes of this sort demands a management function quite distinct from the usual role of the extension staff, even those of graduate level, who were given charge of the settlements. Given a reorientation in training, such staff could cope with the job, supported by effective stores and marketing operations. The transplanting of agricultural extension workers, with overstretched supervisors and service groups, was too great a change; and poor organization has certainly been an important contribution to failure.

However, other planning failures soon emerged from the day-to-day problems of logistics and administration. It is a further example of the hazards of planning down from above. Political emphasis was laid on raising the living standards of the people. Concrete-based houses, water supplies, and school and medical facilities were the social goals; consequently, the overhead investments in the original schemes were extremely high. By the end of 1965 total projected cost in the Tanzania village settlement schemes was over £1,000 per family settled.[3] Cropping patterns were devised to give the returns necessary to cover these high levels of investment. In one example, with cotton and groundnuts as the core of the planned system, repayment levels required each family to grow ten acres of cash crops. The tractor force was geared to cope with this acreage. In fact the family labor force could normally only harvest 30 percent of the production envisaged from this acreage. Once this limitation made itself felt, the over-investment was not only in social facilities but also in machinery and equipment bought to cover the original targets—providing an even greater cost burden.

The schemes represented an attempt to opt out of the prevailing economic and technological constraints. The politicians pitched their social goals beyond the productive capacity of the available resource unit. The experts generalized about economies of scale—which certainly are a justification for structural changes, but pinpointing opportunities requires careful and sophisticated diagnosis in the light of the ruling economic relationships among all the factors in the production process. The structural content in transformation was often as inappropriate as the husbandry content of the improvement approach, and both were derived from inadequate research and planning.

Of course, the whole problem of rural development is how to reach welfare goals and how to foster improved agricultural practices effectively. But these structural changes ignored economic constraints, at both the national and the farm level, and cut across the limitations of known technology. Clearly, varying the means for solution does not remove the problems of fertility maintenance, seasonal uncertainty, the high cost of purchased food supplies, or the limiting effect of family labor capacity.

It is of course possible, by the manipulation of market opportunities and infrastructural investment, to create special circumstances under which special resource relationships become highly productive. Indeed, for a government committed to a policy they believe is basic to their economic philosophy, the line between public investment in infrastructure and scheme investment to be recovered is highly susceptible to political pressure. It is greatly to the credit of the Tanzanian government that they withdrew sixty-two of the projected sixty-nine settlements only three years after initiating their transformation policy. These "special arrangements" are clearly impossible to replicate for a general strategy of agricultural development. Indeed, they are so capital-intensive that even a handful of schemes becomes the whole of the development effort in agriculture, absorbing all available funds and manpower.

To be effective, a transformation approach, just as an improvement approach, must be based on knowledge of the needs of the farmers that are being met by the existing system. Further, it must be shown that the approach creates the opportunity to satisfy these needs more efficiently, in order to give an incentive for participation by the farmers. Guided in this way, transformation is certainly feasible. Again, however, except under very particular circumstances, it is unlikely to be popular and therefore unlikely to be successful. Chambers has coined the phrase "social nakedness," which nicely describes the feeling of farmers asked to withdraw from their traditional community organizations, which are intimately connected with their agricultural activity, and submit to government authority, which is normally regarded with trepidation and uncertainty.[4] Experience has shown that the type of settler attracted to this sort of scheme is more concerned to avail himself of the free food and pocket money which often characterize the first year's activities.

"Social nakedness" gives the clue to the circumstances in which transformation can contribute to development. In situations where the existing community structure is disintegrating, usually due to the imbalance created by increasing population densities, structural change may be the only way forward. Inevitably there are cross pressures in such situations which can be dealt with only by government authority. Action is often a necessity, for inactivity can result in the loss of social and—subsequently—political stability. Examples of this type of pressure, and the structural reforms to alleviate it, are the consolidation of the central highlands in Kenya, where the uprising in the 1950's created the opportunity for government action, and since independence, the resettlement of the white highlands, also in Kenya. Chambers has noted other types of situations which demand restructuring of communities: refugee problems, displacement by dam construction, and urban unemployment.[5] All these require essentially new initiatives and investment in settling displaced persons. The difficult cases are where government needs to preempt the threat of local instability by a policy of structural change. The appropriate timing for intervention is when the economic incentives for the change outweigh loyalties to the existing community structure, bringing a tolerance of the uncertainty and "social nakedness" consequent on reform.

A second justification for a structural element in agricultural development is the exploitation of exceptional opportunities. Irrigation development means either that land and water need supplementing by imported labor and capital, or that land and labor need supplementing by imported capital and water; the Gezira and Mwea in Kenya are prominent examples. Alternatively, a new market opportunity related to particular soils requires labor and capital for its exploitation; flue-cured tobacco schemes are a particularly suitable example in Tanzania and Malawi.

These two types of situations create a role for structural investment in agricultural development. The need to cover the often considerable overhead requires production as close to the margins as possible. The evidence suggests rapidly diminishing returns to increments of capital investment on the individual farm unit, certainly within the technology available for the Sukuma land area of Tanzania. Table 15 shows the results from accelerated developments of a trial farm unit from 1962 to 1965.

The final increments of capital, approaching the margin under existing technology, would certainly be more profitably invested as initial increments on other farms. The emphasis should be placed on the coverage of a large number of farms with low extension costs, a strategy we have been at pains to prove feasible within the conditions of developing agriculture. But population pressures of the level of the central Kenya highlands are not characteristic of the rural economies of Africa; they are localized features, and in the same way exceptional opportunities to balance resources are scarce. In the mass of the rural population, existing community organization presents a secure framework for productive activity; outside pressure to change the framework causes resentment and the very instability, both social and political, which structural reform at the right time can alleviate. Without the pressures of land scarcity, there are rarely the incentives for consolidation and tenurial reform, so that transformation as a general strategy is forced to look to mechanization and central management for economies of scale, which can rarely be created. Three vital aspects prevent transformation from being an effective general strategy for agricultural development under the conditions we have described as characteristic of most rural African economies.

Inevitably, significant overhead costs are incurred in the physical planning and layout of schemes and in the capital investment required for machinery. The dramatic change in the pattern and methods of productive activity demand a manager, and the need to cover high overhead costs requires intensive management. Together with the tact required in stabilizing this type of unit, the management task will limit the range of suitable personnel available and the size of the individual scheme. Large-scale farms the world over command high levels of agricultural management. Large-scale expatriate mixed arable farms in Kenya would rarely have more than 400 acres under crops.

Plantations or estates would normally consider 500 acres of crop enough for a graduate supervisor. Certainly there is no basis for believing that the managers available to governments for supervising transformation projects could cope with more than 1,000 acres, with the assistance of subordinate staff. The consequences of this requirement for qualified manpower are fundamental. Tanzania in

TABLE 15

Decreasing Returns to Increments of Credit on a Trial Farm Unit in Tanzania, 1962-65

| Criteria | Average Local Farmer | 1962-63 | | 1963-64 | | 1964-65 | |
|---|---|---|---|---|---|---|---|
| | | Plan | Actual | Plan | Actual | Plan | Actual |
| Level of Farm Costs (E. Af. shillings) | 170 | 226 | 239 | 848 | 817 | 1,549 | 1,346 |
| Percent Return on Working Capital | 793 | 803 | 766 | 350 | 326 | 192 | 162 |
| Percent Return on Extra Working Capital | — | 725 | 680 | 161 | 122 | 17 | 96 |

Note: In 1962, extra return is evaluated against the average local farmer.

Source: M. P. Collinson, "Experience with a Trial Management Farm in Tanzania, 1962-65," East African Journal of Rural Development, II, 2 (1969).

1964 proposed sixty-nine schemes ranging from 2,500 to 3,750 acres. In its 1960 report the IBRD listed the agricultural officer strength of the Ministry of Agriculture as fifty-eight.[6] It had certainly fallen below this by 1967. The sixty-nine schemes would have required the whole graduate strength of the extension services as managers, and in addition would have absorbed the whole field officer (diplomat) cadre as supervisors. Management is a very different skill from extension, and the requirements of transformation programs cut right across the manpower output gearing of the education system—which as we have seen, takes a long period to adjust. Because the scale of individual schemes is confined by the need to draw management from qualified staff only, there can be no extensive gearing. Schemes will absorb only a limited number of the contact worker cadre at present in extension services, and only within the confines of the management capacity of the available qualified staff.

The second vital aspect is a blocking of the diffusion process. Improvement, working within the same context as the traditional farmer, is faced with the same constraints and conditions. Each innovator serves as a catalyst, advertising his successes throughout the community. Benefits to investment in advisory services are multiplied by the rate at which the changes spread to other farmers. Transformation alters the structural context, with a view to exploiting the advantages obtained by this change. With a new range of problems facing the decision maker, the solutions available may be—and indeed, if transformation was justified, should be—different from those facing the peasant farmer. The new possibilities are irrelevant to the peasant farmer still saddled with the structural disadvantages as a context for his farming. Diffusion of the practices employed on schemes is inhibited, and there is little multiplication of the direct benefits of the investment.

This aspect is widely evidenced by what research workers have called the "over the station wall" phenomenon. Local farmers living around a research station, often working as casual labor on the experiments, steadfastly adhere to traditional practice. The machinery and fertilizers and abundance of labor used in achieving the results are completely foreign to their own circumstances, and they can rarely see the results as relevant. Adegboye, Basu, and Olatunbosun have reported the lack of impact of western Nigerian settlements on surrounding farmers.[7] Even further, the prospects for diffusion from schemes carefully tailored to local farming conditions are limited to the immediate area by the focal nature of the schemes themselves.

These aspects interact: the managerial unit in transformation requires intensive capital investment and prevents the extensive coverage usually achieved by gearing down to a large number of farmers through less qualified manpower. The lack of coverage itself precludes the multiplication of benefits associated with diffusion and further reduces the productivity of the capital employed. In terms of returns to qualified manpower, perhaps the limiting factor in African agricultural development, transformation can be demonstrated as inferior unless there is heavy pressure on land with identifiable economies for structural change or unless exceptional opportunities exist.

The case of Sukumaland illustrates the gap. There is particularly strong weighting against transformation, since any economies of scale are rapidly limited by the hand harvesting of cotton, the only cash crop in the area with an established market infrastructure. With the cultivation operations limiting the existing system, mechanization allows a 30-40 percent increase in output until the secondary labor peak, due to cotton picking, prevents further expansion. The example shows both improvement and transformation operation within the existing market opportunities and available technology. The characteristics and assumptions of the alternatives are outlined first.

The adoption model and planning results described in Part III of this book are used to calculate the cost and return flows to an improvement strategy. A project life of twenty-five years covers the initial adoption model and a diffusion history. In the adoption model, extension contact with innovators revises their farming system over an eight-year supervision period. The diffusion assumptions are a two-year awareness lag, then sixteen years to full coverage, with proportions of farmers adopting each year based roughly on the normal distribution. It is assumed that once a farmer becomes an adopter, like the initial innovators he will revise his farming system over an eight-year period. Thus the final 1.5 percent laggards, adopting the initial innovations in year 18, complete their own revision in year 25.

The "unit" for comparison is based on the agricultural graduate, acting as local organizer for the improvement strategy and as manager for transformation. The gearing between ranks in the service for the improvement strategy is taken from the example area (discussed in Chapter 20), which raises some points on extension organization. The gearing and the associated costs for both improvement and transformation are set out in Table 16.

Total supervision costs are 320,000 shillings per annum on the improvement unit and 134,000 shillings on the transformation unit. The main difference is in salaries for the larger number of supervisors and contact workers in improvement, together with much higher travel

TABLE 16

Government Supervision Costs for Improvement and Transformation

| | Improvement | | | Transformation | | |
|---|---|---|---|---|---|---|
| Staff level | Graduate | Supervisor | Assistant | Graduate | Supervisor | Assistant |
| Number | 1 | 4 | 20 | 1 | 2 | 6 |
| Personal Salary (E.A. sh. 1,000) | 35 | 50 | 125 | 35 | 25 | 37 |
| Office, Travel (E. A. sh. 1,000) | 25 | 75 | 10 | 25 | 7 | 4 |

Source: Compiled by the author.

expenses for supervisors. Administrative costs above the levels costed out are assumed to be common to both approaches. Each contact worker in improvement advises fifteen, giving a unit of 300 farmers. We have seen transformation limited by arable acreage to about 1,000 acres, or 125 farmers.

For transformation, modest overhead costs have been assumed at 6,000 shillings per family, well below the 21,000 shillings experienced in the early settlements in Tanzania. This includes physical planning and layout, light bush clearing, central facilities for water, and concrete-house floors. Mechanization consists of three tractors and equipment. With the short cultivation season, each tractor is limited to 350 acres disked and ridged, with farmers cultivating their own special plots for sweet potatoes and rice. The scheme develops to full potential in three years, and after two years consolidation is managed by a growers' committee with no direct government supervision. The government staff moves on to a new project and two projects have been included, running consecutively, against the eight-year supervision period for the improvement unit. Project life is twenty-five years, the repayment period for structural overheads on the scheme. Table 17 sets out the cost and return flows for the improvement unit and Table 18 for the transformation unit; both tables are carried up to year 18.

Three sets of flows have been discounted. An interest rate of 8 percent, the usual rate for short-term lending in Tanzania, has been applied to the costs and benefits over the full twenty-five-year project life. Table 19 compares the net present values of flows from transformation and improvement units, and from improvement with diffusion through the community.

The comparison demonstrates how the falling productivity of capital on the individual unit and the poor coverage due to the management limitations penalize the transformation approach. The level of benefits from diffusion illustrates its importance to the rate of development. In addition to superior net present value, improvement has a shorter payback period and a lower capital debt.

The final criticism of transformation must be on political grounds. African socialism has made an issue of egalitarianism. It urges vigilance against the emergence of class in rural society. Accepting the assumption that diffusion is inhibited by transformation, its failure to penetrate the rural population is disastrous in this context. In an improvement strategy, diffusion will spread benefit throughout the community. Even after ten years benefits will have been felt by over 75 percent of the rural population. Transformation,

## TABLE 17

### Cost and Revenue Flows for an Improvement Strategy

| | Year | 1 | 2 | 3 | 4 | 5 | 6 | 7 | 8 | 9 | 10 | 11 | 12 | 13 | 14 | 15 | 16 | 17 | 18 |
|---|---|---|---|---|---|---|---|---|---|---|---|---|---|---|---|---|---|---|---|
| **Per Farm** | | | | | | | | | | | | | | | | | | | |
| **Additional Gross** | | | | | | | | | | | | | | | | | | | |
| On Farm Returns | (shs.) | 22 | 118 | 372 | 521 | 766 | 1,148 | 1,564 | 1,871 | 1,871 | 1,871 | | | | | | | | |
| On Farm Costs | (shs.) | 0 | 35 | 105 | 150 | 195 | 355 | 505 | 600 | 600 | 600 | | | | | | | | |
| **Per Unit** | | | | | | | | | | | | | | | | | | | |
| Farm Returns | (sh. 1,000) | 7 | 35 | 110 | 156 | 230 | 342 | 470 | 558 | 558 | 558 | | | | | | | | |
| Farm Costs | (sh. 1,000) | 0 | 11 | 32 | 45 | 58 | 107 | 152 | 180 | 180 | 180 | | | | | | | | |
| Extension Costs | (sh. 1,000) | 320 | 320 | 320 | 320 | 320 | 320 | 320 | 320 | | | | | | | | | | |
| Diffusion Curve | (% adopting) | 1.5 | 1.5 | 2.5 | 3 | 5 | 7 | 9 | 11 | 14 | 12 | 9 | 7 | 6 | 5 | 3 | 2 | 1.5 | 1.5 |
| Number Farmers | | Innovators | | 500 | 600 | 1,000 | 1,400 | 1,800 | 2,200 | 2,800 | 2,400 | 1,800 | 1,400 | 1,200 | 1,000 | 600 | 400 | 300 | 300 |
| Diffusion Benefits | (sh. 1,000) | | | 11 | 72 | 279 | 632 | 1,273 | 2,335 | 3,958 | 6,232 | 9,117 | 12,464 | 16,144 | 20,030 | 23,850 | 27,238 | 29,899 | 31,966 |
| Diffusion Costs | (sh. 1,000) | | | | 18 | 74 | 173 | 342 | 655 | 1,137 | 1,830 | 2,721 | 3,770 | 4,934 | 6,212 | 7,488 | 8,610 | 9,487 | 10,171 |
| **Unit Flows** | | | | | | | | | | | | | | | | | | | |
| In | | 7 | 35 | 110 | 156 | 230 | 342 | 470 | 558 | 558 | 558 | | | | | | | | |
| Out | | 320 | 331 | 352 | 365 | 378 | 427 | 472 | 500 | 180 | 180 | | | | | | | | |
| Net | (sh. 1,000) | -313 | -296 | -242 | -209 | -148 | -85 | -2 | +58 | 378 | 378 | | | | | | | | |
| Cumulative | | -313 | -609 | -851 | -1,060 | -1,208 | -1,293 | -1,295 | -1,237 | -859 | -481 | -103 | +275 | | | | | | |
| **Unit and Diffusion** | | | | | | | | | | | | | | | | | | | |
| In | | 7 | 35 | 121 | 228 | 509 | 974 | 1,743 | 2,893 | 4,516 | 6,790 | | | | | | | | |
| Out | | 320 | 331 | 352 | 383 | 452 | 600 | 814 | 1,155 | 1,317 | 2,010 | | | | | | | | |
| Net | (sh. 1,000) | -313 | -296 | -231 | -155 | +57 | 374 | 929 | 1,738 | 3,199 | 4,780 | | | | | | | | |
| Cumulative | | -313 | -609 | -840 | -995 | -938 | -564 | +365 | 2,103 | 5,302 | 10,082 | | | | | | | | |

Source: Compiled by the author.

## TABLE 18

Cost and Revenue Flows for a Transformation Strategy

| Year | | 1 | 2 | 3 | 4 | 5 | 6 | 7 | 8 | 9 | 10 | 11 | 12 | 13 | 14 | 15 | 16 | 17 | 18 |
|---|---|---|---|---|---|---|---|---|---|---|---|---|---|---|---|---|---|---|---|
| Per Farm | | | | | | | | | | | | | | | | | | | |
| Additional Gross | | | | | | | | | | | | | | | | | | | |
| On Farm Returns | (shs.) | 420 | 1,000 | 1,800 | 1,800 | 1,800 | 2,220 | 2,800 | 3,600 | 3,600 | 3,600 | | | | | | | | |
| On Farm Costs | (shs.) | 420 | 503 | 760 | 760 | 760 | 1,180 | 1,263 | 1,520 | 1,520 | 1,520 | | | | | | | | |
| Per Unit (125 farms) | | | | | | | | | | | | | | | | | | | |
| Farm Returns | (sh. 1,000) | 53 | 125 | 225 | 225 | 225 | 275 | 350 | 450 | 450 | 450 | | | | | | | | |
| Farm Costs | (sh. 1,000) | 53 | 63 | 95 | 95 | 95 | 147 | 159 | 190 | 190 | 190 | | | | | | | | |
| Extension Costs | (sh. 1,000) | 134 | 134 | 134 | 134 | 134 | 134 | 134 | 134 | 134 | 134 | | | | | | | | |
| Structural, Machinery Costs | (sh. 1,000) | 834 | — | — | — | — | 834 | — | — | — | — | 84 | — | — | — | — | 84 | — | — |
| Unit Flows | | | | | | | | | | | | | | | | | | | |
| In | | 53 | 125 | 225 | 225 | 225 | 275 | 350 | 450 | 450 | 450 | 450 | 450 | 450 | 450 | 450 | 450 | 450 | |
| Out | | 1,021 | 197 | 229 | 229 | 229 | 1,115 | 293 | 324 | 324 | 324 | 274 | 190 | 190 | 190 | 190 | 274 | 190 | |
| Net | (sh. 1,000) | -968 | -72 | -4 | -4 | -4 | -840 | -57 | 136 | 136 | 136 | 176 | 260 | 260 | 260 | 260 | 176 | 260 | |
| Cumulative | | -968 | -1,040 | -1,044 | -1,048 | -1,052 | -1,892 | -1,835 | -1,699 | -1,563 | -1,427 | -1,251 | -991 | -731 | -471 | -211 | -35 | +225 | |

Source: Compiled by the author.

TABLE 19

Comparison of Cost and Revenue Flows from Improvement and Transformation Strategies

| Strategy | Source | Net Present Value (sh. 1,000) | Highest Capital Debt (sh. 1,000) | Payback Period (yrs.) |
|---|---|---|---|---|
| Transformation | Unit | 292 | 1,892 | 17 |
| Improvement | Unit | 888 | 1,295 | 12 |
| | Unit and Diffusion | 71,494 | 995 | 7 |

Source: Compiled by the author.

on the other hand, will reach only 6 percent of the same population over twenty-five years—indeed, with population growth the number untouched by development at the end of the period will be greater than at the beginning.

A strategy based on transformation can create the type of elite which the socialist philosophy of many independent African states explicitly seeks to avoid. There is a good deal of literature on the dangers of dualism.[8] Less formally, the African socialists have been concerned to guard against sectarian development, and particularly to avoid strategies promoting the individual, fearing the exploitation of the rural majority as a consequence. It is a real fear; M. Kiray and J. Hinderink have recorded the emergence of a landless class due to the introduction of improved farm practices which are productive enough to allow wage employment.[9] They note that wage laborers, in an extreme case 81 percent of the community, are worse off than unimproved farmers who still have their own farms. But this is not the inevitable consequence of an improvement approach to agriculture development. On the contrary: with the intensity of investment required and the limited funds available, it is more inevitably a consequence of transformation. The improvement strategy envisaged is individual only in a very limited sense, in the contact between adviser and farmer. Given an area-based planning sequence, aiming at replicable content appropriate to the majority of farmers in a community, the government has full control of the level of benefit being offered to the individual. Complementing these controls, it has the authority, given popular support, to control the power derived from individual wealth—and, indeed, the accumulation of wealth itself. Popular support is more likely to be sustained where the benefits of public spending penetrate the rural areas.

Patently, there will be transformation schemes, and we have identified a place for them. Because they are so resource-intensive and involve preferential treatment of minorities, their usefulness is limited to circumstances in which the pressures arising from local situations demand structural changes. Transformation is too selective to offer an equitable general strategy for agricultural development. But, unless improvement can be made more effective, circumstances creating pressures will become more and more general as population density runs further and further ahead of agricultural change. Ultimately, transformation may become the only answer, with capital and manpower requirements so much greater than an effective improvement strategy at the present man/land ratios of most of traditional African agriculture.

## NOTES

1. M. E. Kreinin, "The Introduction of Israel's Land Settlement Plan to Nigeria," Journal of Farm Economics, XLV, 3 (1963).

2. Robert Chambers, Settlement Schemes in Tropical Africa (New York: Praeger, 1969).

3. N. Newiger, "The Village Settlement Schemes in Tanzania," in H. Ruthenburg, ed., Smallholder Agriculture and Development in Tanzania, "Africa Studies," XXIV (Munich: IFO, 1968).

4. Chambers, op. cit.

5. Ibid.

6. IBRD, The Economic Development of Tanganyika (Washington, D.C.: 1960).

7. A. C. Basu, R. O. Adegboye, and D. Olatunbosun, "The Impact of Western Nigerian Farm Settlements on Surrounding Farmers," Nigerian Journal of Economic and Social Studies, XI, 2 (1969).

8. H. Mynt, The Economics of Developing Countries (London: 1964).

9. M. Kiray and J. Hinderink, "Interdependencies Between Agroeconomic Development and Social Change," Journal of Development and Social Change, IV, 4 (1968).

CHAPTER

# 6

# THE APPROACH FOR FARM ECONOMICS

The role of the discipline of farm economics in diagnosing weaknesses in existing farm systems and evaluating possible improvements to increase resource productivity is universal. It is a role which takes on enhanced importance in traditional agriculture, where the gulf between farmers and researchers is particularly wide, and is responsible for the inappropriate orientation of a good deal of past and present research and extension effort.

Traditionally, largely through historical circumstance, the ecological environment has been the sole interest of researchers and advisers, although it is only a part of the production environment of the farmer. By providing description and analysis of the economic and social environment, farm economics can fill two complementary gaps in what should be a cycle between farmer and researcher, via the advisory services. It will identify the problems of the farming system, allowing more pertinent research programs, and evaluate the impact of possible improvements on existing resource allocations as a basis for the selection of appropriate content for extension programs.

The problems of improving productivity in traditional farming are unlike those of advanced agriculture, where planning techniques concentrate on combining enterprises to the best advantage of the particular farmer's resource position. To farm economists in advanced agriculture, recombination is the major source of improved productivity. Workers have shown allocative efficiency to be high within traditional technology. Similarly, within the limited market opportunities available and the constant demand for family food, there is limited scope for adjustments. Planning in traditional agriculture centers on maintaining the balance in satisfying the complex set of farmer objectives while introducing more efficient technology. Changes in methods bring changes in resource allocation which, while they may promote marketed output, may also sacrifice the satisfaction of other priorities. Such changes may be unacceptable to the farmer. So while the planning process may be simplified by the limited production possibilities, it is complicated in a different way by the diversity of criteria to be considered. Production for the market can be maximized subject to ensuring the continuing satisfaction of the farmer's survival priorities, embodied in subsistence production and social organization.

We have described the approach as a balance of investigation, planning, and extension. The key feature of this balance in conditions of rurally dominated African economies is that farm economics has no direct involvement in extension. Farm units are so small and the low risk ceilings of peasant farmers so limit the degree of acceptable change that the use of manpower qualified to apply farm planning techniques is not cost-effective. Further, individual farm planning is necessary only in advanced agriculture, where each farm has a unique structure of fixed assets which, to a large extent, dictates its production pattern. Traditional agriculture, on the other hand, is characterized by groups of farm units which are homogeneous in important attributes influencing the pattern of production. Given a limited stock of improved technology, the way to higher productivity is the same for large numbers of farmers; and both planning and investigation can validly be approached on a group basis. Individual farm planning is both unjustifiable and unnecessary.

Without direct involvement in extension farm economics in traditional agriculture will center on investigation and planning to provide content for area-based extension programs, content which meets both farmers' and government objectives.

Its role as a link between farmer and research suggests the experimental stations as the proper location for farm economists. The ecological framework for technical research programs forms a base on which economic and social factors can be superimposed for a full description of the production environment facing the peasant farmer. The scarcity of qualified manpower suggests that one or two economists for each center will be the most that can be expected. Bearing in mind the imbalance in many programs during the 1960's, a point to be stressed is that each economist will have to cover the whole approach—investigation, planning, and liaison with the extension and research services—with the division of responsibilities between economists on an area basis.

## EXTENSION

The improvements selected as a unit of investigation and planning will be communicated to the farmer through the hierarchy of the existing extension services. Although all types of potential improvement can be compared, new market opportunities are likely to be limited and slow-growing, depending mainly on the infrastructural development, which is geared to economic growth itself. In the medium term the further development of traditional agriculture is likely to be dependent on more efficient production of crops already established. With the fertility problem arising from increasing population densities, and until a good deal more is known about fertility maintenance in tropical soils, intensifying changes will be of first importance, leaving enough fallow to allow traditional rotational practice. However, because the fertility problem is not obvious to the individual producer, and because extension of scale is more easily managed, intensification needs to be an attractive alternative. Labor productivity of the whole system and reconciliation with nonmarket priorities are the key criteria in evaluating improvements.

The importance of diffusion to penetration of the rural community emphasizes the need for changes relevant to the majority of farmers and raises questions as to the use of scarce manpower and funds on satisfying specialized local markets, or on changes in the structure of traditional agriculture. This replicability is also important to keep the extension task within the capacity of the manpower available for work as contact officers. It is made possible by the homogeneity of traditional farming within ecological and tribal areas, giving problems and solutions common to large numbers of farms.

Farm economics can contribute to the field organization of extension in two ways. First, investigation describes the existing system, allowing the extension staff to understand farmer strategies and priorities. These provide the basis for communication between advisers and farmers, in the farmer's own terms. Second, the impact of a selected change on existing management routines can be analyzed, providing further advisory content. The adviser has an understanding of the reorganizational difficulties the farmer is likely to meet and can discuss them with the farmer before they arise. This alleviates much of the uncertainty felt by the farmer about both the demands the change will make on him and the know-how of the adviser.

At the same time, the description of the managerial complexity of changes to be promoted by the extension services allows for analysis of the intensity of supervision required. This will dictate the coverage of contact workers under specific field conditions, and thus the staff- and the farmer-gearing possible or required for a particular program.

## PLANNING

The planning phase is the fulcrum of the approach, particularly since there is no direct involvement in extension. The extension problem dictates the planning problem, which in turn dictates the pattern of investigation. As we have seen, the extension problem is to communicate to farmers changes which meet government objectives for increased production and are, at the same time, consistent with the farmers' own objectives. The planning problem is to select changes which reconcile the two sets of objectives. Increased market production is reflected in high cash income, but criteria derived from farmers' objectives must qualify this market criterion of profitability. Security of food supply is central to farmers' nonmarket priorities and provides three additional criteria: the adequacy of food supply throughout the year, the reliability of supply over the years, and preference for particular patterns of supply. Closely related to an assured food supply is the risk associated with major changes in method. The rate of change is limited by farmers' current income expectations and willingness to incur the costs involved. Uncertainty of the results of changes reduces their acceptability and is as easily generated by changes which appear large to farmers, even though they are designed to meet their felt needs more effectively. Thus the initial impression of the change proposed, regardless of ultimate appropriateness, is important.

The planning process interpolates available new technology into the existing system and measures the repercussions. The lack of any common base for the valuation of market and nonmarket satisfactions precludes wholly objective solutions. The impact of a change needs subjective evaluation in terms of the nonmarket priorities that investigation has shown to be important in the system. Profitability must be weighted by food supply and risk criteria, by the impact of the change on traditional custom contributing to the satisfaction of survival motivations, and by the impact on long-term fertility in order to give a full evaluation of alternatives as a basis for the selection of extension content.

The complexity of the planning sequence is associated with two factors:

1. The number of activities in the traditional system which can be identified as producing distinct products, which will include the same product at different times of the year.

2. The size of the stock of technology to be evaluated and the sophistication of the experimental design used to produce it.

The central analysis in the planning sequence will be aggregation of all distinct activities in the system to show the relationships between factors, especially land and labor in production, and the balance between resource and output flows. The analysis can justifiably be simplified by using representative farm techniques within types of farming areas which have homogeneous markets and methods of production. By this stratification large sources of interfarm variation are isolated. Solutions can be identified for large numbers of farmers facing the same opportunities, and this makes the use of sophisticated techniques and highly qualified manpower viable.

The narrow criteria of agricultural researchers and the experimental designs used demand a good deal of groundwork to relate the stock of potential improvements to the planning criteria. Most research programs seek to minimize the effect of microclimatic variations by single site experiments, and most programs include labor as an unspecified variable. Both features isolate the results from factors of vital importance to the peasant farmer, and it requires, a good deal of manipulation to adapt results for planning. The more sophisticated the design, the more complicated the manipulation can be.

## INVESTIGATION

The investigational phase is also dominated by two factors:

1. The type and level of complexity in planning is reflected in the range and detail of data required from investigation of the traditional farm system.

2. Illiteracy creates a dependence on either objective- or memory-based collection techniques, both of which are expensive.

Farm economic surveys are usually limited to the economic characteristics of the farm business. Because of the nonmarket priorities of farmers and the need for additional criteria in planning, the scope of investigation must be widened. The food economy of the farm and family, the reciprocal obligations and customs between household and community, and household and individual members, and the husbandry techniques used to forestall erosion and maintain fertility supplement the economic content of surveys usual in advanced agriculture. There can be no shortcutting the need for investigation of the full range of farmer priorities. The field is very much interdisciplinary, and techniques from anthropology, household budget investigations, and sociology may be useful.

Shortcuts do become possible and indeed, given the high costs of data collection and the limited investigational resources, are almost inevitable in the degree of detail required in describing the farming system. Three dimensions of accuracy must be reconciled in investigation: the level of detail required in the description of the system, sampling errors, and errors arising from memory bias.

Reconciliation resolves into the question of visit frequency and coverage of an adequate sample. Increased visit frequency allows greater detail and reduces memory biases but reduces the coverage of farmers by each enumerator; an adequate sample needs more enumerators, with a consequent escalation in survey costs. This is the central issue of investigational design and is discussed at length in Part II.

Investigation is simplified by the possibility of identifying types of farming areas which serve as a first-stage sample and thus removing natural conditions, market opportunities, and methods of production as sources of interfarm variations, which plague the grouping of farms in advanced agriculture. The homogeneity within these areas allows the description of nonquantifiable general attributes by local, informal interview, a presurvey stage which also lays the framework for detailed survey design.

In the field the sequence follows from investigation through planning to extension. Parts II and III follow this order, detailing the useful methodology for each phase of the farm economic approach in African peasant agriculture.

# PART II

# THE INVESTIGATION PHASE

CHAPTER

# 7

# THE PLANNING AND INVESTIGATION TASKS

An approach for the application of farm economics in guiding farmers' resource use in traditional African agriculture has been synthesized from the discussion of conditions in this type of economy. Neither the approach presented nor the conclusions which follow on methodology have been field-tested as an integrated hypothesis but have crystallized over several years of farm economics research. The illustration of techniques tends to be piecemeal, taken from various survey and planning examples. The full significance of the social, agronomic, and food supply aspects became apparent only gradually, and examples are also drawn from studies by many researchers in various disciplines.

The planning task, as defined by the synthesis, is to build a model of the resource relationships in the existing system and show its productivity in both market and nonmarket output; it is also to relate the agronomic, social, and motivational characteristics of the system to the model and to analyze the conditions these impose on factor and product relationships. The model is used to interpolate possible changes in farming practice to see how they affect productivity, the system, and the conditions which bound it.

There are two central problems to the planning, and hence to the investigational, task: to isolate groups of farms which can be covered together, in order to make the exercise viable, and to decide the detail in which simulation of the system is required for planning and is achievable in terms of the planning and investigational resources usually available. We deal here with the possibility of farm classification, and in Chapter 8 with the technique for deriving a representative farm unit. The latter half of this chapter outlines the second problem which resolves into the cost/accuracy compromise. The remainder of the investigational section details the compromise in relation to eight data categories.

## FARM CLASSIFICATION

The obviously prohibitive cost of dealing with every unit of the farm population in agricultural economic investigation and planning has stimulated the search for a reliable basis for generalization. Historically, the case study has been a favorite tool for detailed examination of on-farm production relationships. The development of statistical techniques has allowed the use of sample surveys, providing a more formal basis for cost-saving investigative techniques. The representative farm is in fact the case study derived from a sample survey in an effort to ensure that results can be reliably generalized.

All work on farm classification represents an effort to provide a better basis for the generalization. In 1928 F. F. Elliot drew attention to the need for "type groups," and there has been extensive research on farm classification.[1] Problems have centered on the criteria to be used in farm grouping, and complete definition of the appropriate criteria has been frustrated by the wide variety in sources of complexity in the agriculture of predominately industrial economies. W. Wilcox writing in 1938, denied the possibility of useful cross classification; and R. Hurley the chief of the Agricultural Division of the American Bureau of Census, writing in 1965, seemed to repudiate the idea of the average farm.[2]

Since work is carried out within obvious natural differentials of climate and soil, a great deal of confusion has arisen from the general-purpose nature of usual data collections. A focusing of prominent agricultural economists on the problem in the United States crystallized many of the key issues involved.[3] An article by J. F. Thompson in the resulting bulletin makes the point that when the use for the typical farm is known in advance, it may be easy to list the aspects in which it should be typical.[4] Other writers, such as H. O. Carter, have since stressed the need to tie the representative farm to specified empirical problems.[5] Indeed, recent progress in the use of the representative farm has come from two specific applications. First is the use of a typical resource endowment pattern for linear programming, as a short cut to whole-farm planning on individual farms wanting comprehensive extension advice. Barnard's 1963 article is a prominent example in this field.[6] Second is deriving aggregate supply functions for commodities by the use of benchmark farms; here the work of R. H. Day has been important.[7]

Barnard was one of the first to lay down criteria as a basis for homogeneous subgroups within which typical models could be drawn up for use with linear programming techniques to optimize enterprise combinations:

1. The quantity of resources, noting specifically the importance of homogeneity in the capacity of fixed asset structures.

2. The quality of resources.

3. Contracts and quota and rotational limitations.

4. The inputs required for enterprises and expected level of outputs.

5. The cost of inputs and the expected price of outputs.

Barnard makes an important comment about variables which are used for aggregation which may have other differences dependent upon them and uses as an instance the size of dairy herd and the type of dairy unit. Earlier writers were preoccupied with size as a basis for classification; this linking of technology and size is now seen as a key factor for the valid grouping of farms.

American workers in supply-response research have made important contributions on the necessary conditions for grouping. R. H. Day laid down the full restrictive assumptions for grouping sample farms to avoid aggregation biases:

1. Proportional variation in constraint sectors (which include fixed, quasi-fixed, behavioral, and policy bounds).

2. Proportional variations in net return expectations (which may include proportional variation of output and input matrices or proportionality of price expectations).

Day acknowledged these as sufficient conditions for exact aggregation and suggested that further work be directed to simple approximations for adequate aggregation. Subsequently L. M. Day noted that over a given price range, some resources will not be effective constraints on a solution.[8] Other workers have made comparisons to show that homogeneity in the limiting resource, at least over some range of price possibilities, is the only necessary condition to minimize aggregation bias, when compared with other traditional criteria for classification. Sheehy and McAlexander demonstrate that conventional groupings of farms with different limiting factors will lead to an overestimation of supply possibilities, factor deficits on below-average farms being offset by surpluses from farms with different constraints.[9] They show that grouping on the basis of homogeneity in the limiting factor gives a lower level of aggregation bias than do increasingly complex conventional classifications; the final one they give as an example includes farm type (dairy or nondairy), cropland, herd size, and labor force. Importantly, they note that variation in variables unrelated to the objective—in their case the determination of the response of milk supply to price changes—is much higher in the subgroups of the homogeneous restriction model than in the conventionally grouped classes. This stresses the point that classification without a specified objective is too complex.

Only when the variables of interest are predefined is useful grouping possible. Frick and Andrews demonstrated this principle by using fifty-one farms as a complete universe, programming each individual farm and aggregating to provide a base line for the measurement of error in alternative grouping criteria.[10] The aggregate supply response derived from conventional classifications gave biases ranging from 17.7 percent for a model built from the means of the variables of the fifty-one farms population to 15.3 percent when the universe was disaggregated into six size groups. The bias in the classification based on homogeneity in limiting resources was 6.6 percent. Frick and Andrews also stated four important problems arising from disaggregation by homogeneity of the limiting resource:

1. It ignores size, which, if correlated with the level of technology, requires further disaggregation.

2. The method requires detailed data on input/output coefficients in order to develop the limiting resource constraints. Such information is not readily available from census returns.

3. It tends to restrict the analysis to the resource base and organization at the time of the survey, particularly critical in supply response work, which by its nature is concerned with changing market circumstances.

4. When handling more than one product in supply response work, developing the order in which resources become limiting for each product, and for products jointly, would result in a large number of benchmark farms.

In noting problems of technologically based differences in scale, Frick and Andrews confirm the importance of Barnard's criteria—the capacity of asset structure—and Day's emphasis on proportionality in constraint vectors, particularly in fixed assets. Size differences are important only when caused by changes in resource relationships associated with different technology. Criticisms of the static nature of the analyses are specifically related to supply response work. Innovation and market movements alter factor/factor or factor/product relationships before policy can be implemented. Our application aims to identify the innovations appropriate to farm systems; and with agricultural research and government extension virtually the sole source of innovation, technical change is contained by the planning objective. Market movements can disrupt the analysis, and the linking of micro-planning with international or national action related to commodity markets is an important aspect taken up in the planning section. Possible innovations will require reappraisal as and when movement in price can be anticipated and the planning sequence will be recursive. Little change in the relationships in local farming would be expected in the short term.

Most of the obstacles to using representative farm techniques derive from the problems of selecting criteria for grouping the farm population. These problems are created by the proliferation of market opportunities and technical possibilities in advanced agriculture which distort the pattern that would result from natural advantages of climate and soil. Our earlier examination of the structure of traditional agriculture showed it to be dictated by natural factors and by historical circumstances of tribal affiliation, and to be characterized by large numbers of farmers faced with the same needs and opportunities and having the same limited techniques to reconcile the two. Even Day's full and sufficient conditions for valid use of representative farms are met under these circumstances. Within natural and tribal areas both factor/factor and factor/product ratios are constant, subject to managerial and motivational differences which can be validly averaged. The amount of available labor is the basic determinant of the scale of activity, which is uncomplicated by differences in technology. Frick and Andrews have pointed out that the identification of limiting factors in advanced agriculture, as a basis for grouping, is itself a complex exercise often requiring programming procedures.[11] Given the degree of homogeneity described in traditional agriculture, limiting factors will be the same over large numbers of farmers. Information from secondary sources, the percentage of total area cultivated, the density of population, the incidence and timing of hired casual labor, and traditional reciprocity of labor within the community aid identification of the limiting resource.

Within these areas identified as homogeneous in market opportunities and technology there will still be sources of variation.

Individual farmers may prefer different foods to fulfill essentially the same role. Patches of soil may offer particular opportunities to individuals. Farming will not be identical even on neighboring farms. In this respect agricultural experimentation suffers in the same way as economic planning, and local variations in the level of benefits from improvements must be expected. As with blanket research in agriculture so with blanket economic planning: the general level of benefits must be high enough to override the effects of local sources of variation. Where major variations begin to assert themselves because of major climatic or soil differences, uniformity is lost and the boundary of a type of farming area is defined.

Although major variations in soil type will normally form the boundary for a type of farming area, altering the crop opportunities and the possibilities for improvement may or may not lead local variations, as in areas with patchy soil characteristics, to demand subgrouping. If the different soils create different opportunities and farms are located selectively, subsamples will be needed, each with its own investigation and planning sequence. If farms are located across several subtypes, the advantages of each type will be reflected in the cropping pattern. In this case the zone is effectively homogeneous. The general rule is that farmers' opportunities must be the same.

We have previously stressed similar opportunities and homogeneity in asset structure as separate characteristics of traditional agriculture, but it is particularly important there should be no interaction between opportunities and asset structure. Where investment in an enterprise requires several times its expected net output, an inflexibility in resource reallocation is created, certainly over the medium term. A sequence of such investment decisions is responsible for the unique planning needs of the individual farm in advanced agriculture. Where investment in assets remains limited, or existing assets are general farm assets and not specific to particular enterprises, they are not limiting farmers' opportunities. Under these conditions, improvements in different enterprises are equally relevant to all farmers and can be comparatively evaluated as easily as different improvements within the same enterprise.

From the discussion of the structure of traditional agriculture and the brief review of problems in farm classification, three criteria have been chosen for grouping farms into type of farming areas:

1. A pattern of climate and soil over which production opportunities and improvement possibilities are the same.

2. A common tribal background, giving homogeneity in motivational patterns, social tradition, and agricultural practices.

3. Limited variation in the man/land ratio.

As a result of tribal isolation, different tribal areas within the same natural conditions may have developed in different ways and may need distinctive improvement paths. Similarly, historical circumstances within the tribe may have led to differences in methods or population concentrations within the tribal area. The three criteria are discussed more fully with Sukumaland as an example.

## Climate and Soil Offering the Same Production Opportunities and Improvement Possibilities

Pragmatically, the best indicators of significant variations in climate and soil are the cropping patterns of the existing agriculture, even though these may often override considerable natural differences. Direct use of climate and soil criteria is inevitably inconclusive, for these are meaningful only in terms of the effect they produce on the ground. Climate and soil data can be tied to existing cropping in order to give an indication of likely changes in productivity of crops grown throughout the area.

Information on cropping patterns, and on climatic and edaphic factors, can be obtained from secondary sources, including censuses or surveys and meteorological records. This can be supplemented by local discussion with agricultural staff. Research staff in the general area will be working within the context of ecological zones and will be able to demarcate the areas in terms of climate and soil for which their experiments and the currently available improved practices are relevant.

Sukumaland proper is a very large area covering six districts of Tanzania with an area of 20,000 square miles and a population of 1.8 million and over 300,000 farms. The tribe has spread beyond this area as increasing population densities have stimulated migration. The Sukuma have penetrated, and in many cases now dominate, areas formerly under the control of other tribes. We limit our consideration here to the six districts of Sukumaland proper.

Figure 2 shows the six districts of Sukumaland within the area bounded by 2°-4° S and 31°-35° E and, to the north, by Lake Victoria. The numbered soil types are as follows:

FIGURE 2

Sukumaland: Physical Characteristics

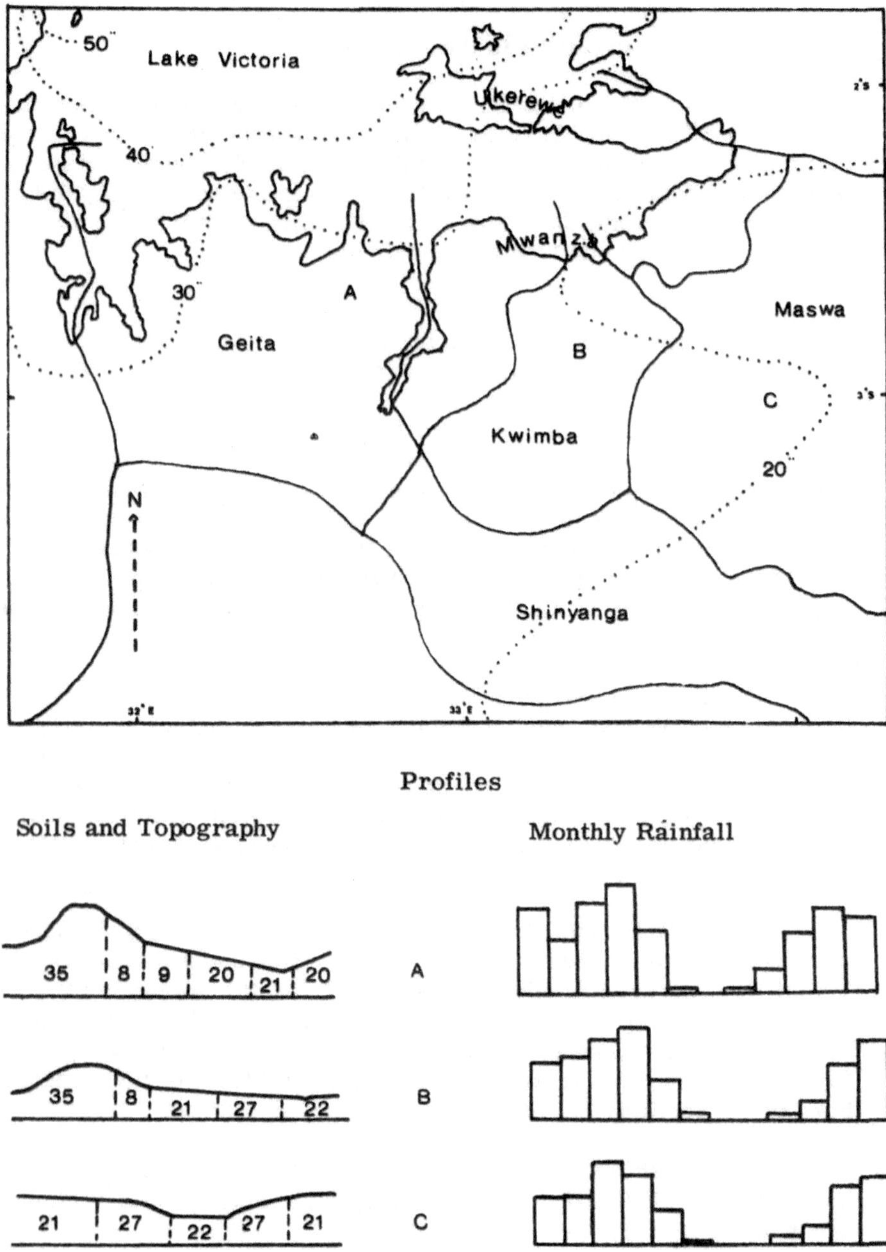

No. 8 Red-yellow/red, gritty/sandy clay loams.

No. 7 Yellow/red loamy sands.

No. 20 Strong brown to pale yellow loamy sands with laterite horizon.

No. 21 Light gray to white mottled loamy sands with laterite horizon.

No. 22 Black to dark gray clays with impeded drainage.

No. 27 Dark gray to grayish brown compacted loamy sands.

No. 35 Shallow, stony soils with rocky outcrops.

Development from A through B to C involves successive stages in pene-plantation, with the steep topography characterized by granitic inselbergs and light, easily cultivated soils, gradually giving way to undulating country with heavier soils.

Table 20 shows the proportion of farms growing the major crops at the points A, B, and C indicated on the map; the population figures refer to the district as a whole and the cropping pattern to farm survey samples.

The pattern of cropping is similar over the three districts. The type of legumes grown varies according to preference, with groundnut, cowpeas, and bambarra nuts dominant; they are all used to flavor dishes based on grain or root staples. Increasing population density in Geita and Kwimba areas has forced farmers into slightly more land-intensive systems. Rice is relatively important as a source of grain, especially in Kwimba, and cassava is more widely used as a fallow crop and food reserve. In Maswa, with a lower density of population, there is little cassava or rice, though there is much more general planting of sorghum. Sorghum in the southern and eastern part of the area, and bullrush millet in the northern, were the traditional grain staples. Maize has replaced sorghum to a large extent, though not as completely as millet in the north. As is shown in Figure 2, rainfall in the south and east is lower, and the season shorter, than in the north and west. Also, as is shown by the 10 percent probability isohyets on the map, rainfall there is less reliable. Sorghum is more drought-resistant and has retained its place in the system in the southeast as an insurance against crop failure, particularly where cassava is not grown as a reserve staple. Cassava has not entered the system,

## TABLE 20

Percentage of Farmers Growing the Main Crops
in Three Areas of Sukumaland

| Area | Total Population (thousands) | Population Density (per sq. mi.) | Cotton | Maize | Legumes | Sweet Potatoes | Cassava | Sorghum | Rice |
|---|---|---|---|---|---|---|---|---|---|
| A Geita | 371 | 106 | 93 | 97 | 85 | 90 | 92 | 12 | 42 |
| B Kwimba | 306 | 130 | 92 | 96 | 91 | 91 | 63 | 13 | 70 |
| C Maswa | 431 | 52 | 94 | 100 | 89 | 85 | 14 | 66 | 10 |

Note: These survey samples, drawn from frames of registered members of the cooperatives, are biased. They exaggerate the number of farmers growing cotton. Larsen, using a different frame, gives an average of 85 percent growing cotton in four Sukuma districts.

Sources: M. P. Collinson, "Usmao Area," Farm Economic Survey No. 2 (Dar es Salaam: Tanzania Dept. of Agriculture, 1962); "Maswa Area," survey No. 3 (1963); "Lwenge Area," Survey No. 4 (1964) (mimeographed); and Tanzania Central Statistical Bureau, 1967 Population Census (Dar es Salaam, 1968).

partly because of the abundance of land and partly because of the heavier soils in the south which tend to inhibit root development.

These are the limited differences in cropping pattern over a very large number of Sukuma farmers who have basically similar opportunities. The only crop with an effective market infrastructure in the whole of this area is cotton. Mwanza, the largest town, is supplied with milk and some fruit and vegetables from its immediate hinterland; but its penetration as a market is very restricted. There is a marketing outlet for rice and maize through the same channels as cotton, though it operates solely with food surpluses because neither crop can compete with cotton for the use of critical labor in the period of November to January.

This general picture of uniform crop opportunities is verified by the agriculture research programs in the area. Cotton production is limited to the north, east, west, and northwest by higher ground, higher rainfall, and low radiation. To the east and southeast there is a lack of rainfall, and to the southwest there is a large swamp of the Malagrasi River headwaters. This leaves a funnel of expansion of cotton production south into Unamwezi tribal areas, although the rate of expansion is presently inhibited by sucking pests which thrive because of drier conditions and suitable hosts in the natural vegetation.

Technological possibilities are extremely limited, and with one or two exceptions they are general over the whole area. Insecticide regimes on cotton differ; in the west and south the general regimes against bollworm is supplemented to control the sucking pests and stainers. Fertilizer recommendations for both cotton and maize refer specifically to the light sandy soils in the north, requiring nitrogen supplemented by phosphate for cotton. With the addition of dust for maize stalk borer, these are the only recommendations for purchased inputs in the area.

## Tribal Background

Tribal background differentiates characteristics secondary to natural conditions which might have created peculiarities in the system of agriculture. Usually, because the tribe has evolved as a community, its members have common objectives. Within these will be a diversity of preferences which will be randomized and peripheral to the basic relationships of the farming system. Nevertheless, tribal history may have created subdivisions which have interacted differently with the natural conditions, or may have been expanded into areas of different natural conditions. Similarly, an area of varying natural resources may be covered by a single tribe, or various tribes may inhabit an area with the same natural conditions. These distinctions are fairly easy to draw in practice. Where tribal expansion has been into areas of different natural conditions, new technology may have evolved, requiring a subarea for further investigation. Equally, tribal expansion, although motivations remain the same, may break down the rigid social tradition of the tribe, especially where different clans are moving into the same new area. This also may open the way for a modification of the traditional farming system.

The approach to identifying differences of this sort is purely pragmatic, and a history of tribal boundaries, particularly expansion areas, determines the main zones. There are three main factors indicating a variation in technology: the tools and equipment used for cultivation, weeding, harvesting, or processing of the major crops; the method of seedbed preparation for planting annuals; and the operational sequence and frequency of opening virgin bush or cutting back heavy regrowth. All these aspects can be readily enumerated from secondary sources. Less significant, but nevertheless important in the selection and evaluation of acceptable innovations, are changes in social custom. For example, the inflexibility of sex differentiation by crop or operation may have been broken in new areas, where the slightly different natural conditions demanded a new work pattern and participation by all. With a checklist of those social factors which are likely to influence resource allocation, differences here can also be described from secondary sources.

Sukumaland is a good example of the expansion of a tribe into new areas. With the conflict between grazing for livestock and increasing requirements for arable land, especially after the introduction of cotton, the traditional heartlands rapidly became overcrowded and migration west into Geita, south into Nzega, and east into Maswa occurred on a large scale between 1948 and 1967. Table 21 illustrates the movement.

Capital equipment and agricultural practice are homogeneous over most of Sukumaland. The single exception is the use of ox plows in the south and east. It is interesting to speculate on causes: the flatter topography, the shorter season and a need for more timely cultivation, heavier soils, and most emigrants being cattle owners looking for grazing—all may have contributed. The important point for the identification of type of farming areas is the flat cultivation and broadcasting which have followed from ox plowing, moving away

TABLE 21

Population Growth of Districts in and Around Sukumaland, 1948-67

| | Population (thousands) | | | Annual Growth Rate (percent) | | |
|---|---|---|---|---|---|---|
| District | 1948 | 1957 | 1967 | 1948-57 | 1957-67 | 1948-67 |
| Mwanza | 179.4 | 177.5 | 236.6 | -0.1 | 2.9 | 1.5 |
| Kwimba | 238.4 | 242.1 | 305.6 | 0.2 | 2.4 | 1.3 |
| Ukerewe | 86.4 | 94.4 | 109.2 | 1.0 | 1.5 | 1.3 |
| Shinyanga | 214.1 | 255.6 | 321.1 | 2.0 | 2.3 | 2.2 |
| Geita | 139.5 | 270.2 | 371.4 | 7.6 | 3.3 | 5.4 |
| Nzega | 187.9 | 205.3 | 298.1 | 1.0 | 3.8 | 2.5 |
| Maswa | 245.4 | 292.3 | 43.6 | 2.0 | 4.0 | 3.0 |

Notes: Mwanza excludes Mwanza Town.
Ukerewe is populated by a distinct tribe with close affiliations to the Sukuma.

Source: Tanzania Central Statistical Bureau, 1967 Population Census (Dar es Salaam, 1968).

from the traditional Sukuma technology based on five-foot ridges and row planting. M. P. Collinson reported that exactly half of Maswa farmers owned ox plows (98 percent using flat cultivation), the remainder hiring equipment.[12] D. Rotenhan reported that 92 percent of Shinyanga farmers sampled owned ox plows.[13]

Both researchers report that northern areas are dominated by the hand hoe. Three zones can be identified: hand hoe dominant, ox cultivation dominant, and mixed. Other things being equal, coverage of the hand hoe and ox zones would allow two planning sequences suitable for use in the mixed zone. Given other differences, the mixed zone would require more intensive sampling to give adequate data for both hand hoe and ox systems.

## Man/Land Ratio

Although areas may be following the same agricultural system, the urgency of problems and appropriateness of alternative solutions may differ where population density is high. Once a critical density is reached, the existing technology cannot maintain fertility and a downward interaction begins. Investigation may cover areas of varying population density farmed under the same opportunities and methods, but the solutions considered will be limited by the need for intensification in areas where a downward fertility spiral is established.

Many sources of secondary data can provide information on the long-term resource balance. Most countries carry out a population census from which present and future densities can be estimated. Under tribal authority, land tenure patterns, renting, and then the sale of land usually develop in a sequence. Differences in tenure practice within a general area should be noted from discussions with local officials. In the course of initial data analysis, the amount of slack in the system, in terms of preferred but land-extensive or labor-intensive foods, can be assessed; and, given the rates of population increase, a rough time scale for fertility decline can be worked out to show the urgency of the problem. Other things being equal, however, differences in man/land ratio do not influence investigational procedures which seek to define the position of the system on the downward fertility spiral. Where population pressure has already changed the system—either the dominant crops or the methods used—it will have been reflected in the first two criteria.

Sukumaland demonstrates a whole range of population densities within what is substantially the same system of agriculture, but the speed with which land is being absorbed emphasizes the need for intensification. Pockets such as Ukerewe and Ukara islands, with population densities over 400 per square mile, have already demonstrated how intensive the traditional system can become, although these areas enjoy better rainfall conditions than are found over most of the mainland. It is the clash between the social tradition of cattle keeping and the spread of cotton production which is precipitating the crisis in other areas.

With the gradual breakdown of tribal insularity under national government, "border" problems in the identification of types of farming area will increase. Compound systems in fringe areas will be a mixture of features from the systems they surround. The key to investigation will often be an analysis of the two discrete systems from which the compound one is built. The planning phase will focus on local experimental findings on the potential of the alternative crops in the areas concerned.

Identifying types of farming areas represents the drawing up of a frame of first-stage sampling units. Resources will normally be limited, requiring the selection of priority areas for further investigation by presurvey and survey stages. Ranking of priorities will depend on a rough appraisal of potentials based on assessment of the technological opportunities available, their potential return to investment in extension effort, and compatability with farmer needs. Usually such a ranking will be closely guided by the focus of past crop research for the area, and again the experimental stations will afford the best outline of the possibilities. Final ranking of course begs the major purpose of the whole exercise, but some ordering is necessary. Apparent economic potential as a criterion will be weighted by government policy objectives. There may be political reasons for concentrating efforts in certain areas or for a regional balance to meet welfare objectives or relieve local pressures.

The identification of homogeneous farming areas and their ranking by priority form a very important first step for investigation and planning. They represent the only framework for central coordination of the efforts of manpower qualified in this field and located in the rural areas. Lack of a framework of this sort, however crudely defined, was responsible for a great deal of the ad hoc nature of data collection in East Africa over the 1960's, and for a failure to see where and how widely any collected data could be validly applied.

## THE ACCURACY/COST COMPROMISE

Within a particular type of farming area problems of measurement predominate. These might be called the problems of farm economics investigation proper. While a part of the aim of investigation is to confirm the homogeneity of the population regarding those attributes used to delimit the area and to describe how they influence resource allocation, the major interest is to quantify the levels of resource use, resource relationships, and productivity among farmers following a closely similar system. More formal tools are needed: statistical techniques in sampling and data collection techniques for recording the information. Statistical sampling techniques for surveys are well documented;[14] our particular interest is in relating alternative data collection techniques to the other conditions influencing survey design: the sampling requirements and the complexity of the farming system.

The starting point is the level of accuracy required in simulating the existing farming system. In his 1949 book, F. Yates stated a principle for deciding the level of precision in sampling survey work, which holds equally well for our application: "The level of accuracy should be such that the sum of the cost of the survey [investigation, in our case] and expected losses due to errors in the results should be minimized."[15] Subsequent contributions, usually at the theoretical level, on formal methods for calculation of this sum have been limited to single-variate applications.[16] Calculation of the loss element is fraught with data problems which are drastically compounded for multivariate applications.

The principle holds well in our application. Planning should seek to minimize the sum of investigational costs and the loss in output due to a failure to identify the most acceptable innovations, which causes lower rates of adoption and reduces the potential rate of return to public investment in extension effort. However, the rudimentary techniques and limited experience in establishing the level of benefit arising from extension—with problems of field measurement of yield increments, high interseasonal and interfarm variation, and the ex ante estimation of adoption rates, let alone demonstrating differential levels of benefit from alternative extension programs—illustrate the data problems already stressed.

We have already noted the importance of balance in the coverage of different types of farming areas and the intensity of investigational effort within each area. At some stage in investigational programs the returns to increasing coverage will fall as systems with lower potential are studied, and greater returns will be realized from intensified investigation of the higher-potential systems already covered. However, the initial emphasis has been laid on coverage to realize the immediate benefits of a program with acceptable extension content. For our example of Tanzania, assuming the 178 zones identified by L. Berry to be distinctive types of farming areas and on the most optimistic assumptions of staffing and survey output per year, coverage of the more important crop areas would take seven or eight years.[17] The more detailed cost/accuracy compromise, within the defined areas, will bound the possible alternatives in the balance between coverage and intensity of investigation.

In practice, the level of detail in simulating the existing system must usually be a compromise because of limited manpower and funds for investigation. A problem arises in the size of sample to be used: small samples increase the sampling error and, when relying on verbal responses, the frequency of visits to be made to the farm;

reducing the frequency of visits increases the dependence on memory. Almost inevitably full simulation of detail is impossible. Many factors form bases for the differentiation of discrete productive activities: related minor food types, the same food type that matures at a different time to give a food flow, minor specialized soil types or location, different planting times, variations in cultivation methods, and differences in crop history. Discrete activities need identification before the survey; and a decision is required on whether separate coverage, to give a more useful planning model, is justified in terms of the increase in cost. The basis for the decision must be the importance of each activity in its contribution to the satisfaction of farmer objectives and in the importance of its demand for resources. It is either grouped with a similar activity or investigated independently.

The clash in survey design usually takes the following form. The greater the detail sought on the particular farm, the more frequently are visits required. For example, recording inputs and outputs on a plot basis poses difficult investigational problems and requires very frequent visits. Within a given manpower unit, the more frequently visits are required, the smaller the size of sample which can be covered and the larger the sample error. There is a spectrum of possible visit frequencies ranging from daily visits to a single visit at the end of the season. For our purposes a distinction will be drawn between frequent-visit collection techniques and limited-visit techniques, when the farm is visited from one to three times in the season. Reducing the visit frequency sometimes, but not inevitably, increases memory bias. The art of survey design is a balancing of the three sources of inaccuracy: simulation of system activities, sampling error, and memory bias. Beyond the point of balance, the quest for more detailed simulation is self-defeating because of increasing inaccuracy arising from the other two sources.

However the balancing act called for is not straightforward for two reasons:

1. There is no universal level of acceptable sampling error.

2. Memory bias is not consistently related with visit frequency, but varies with the nature of the attribute or event under recall.

Both these qualifications are examined further.

## Sampling Error

Precision in collected data is usually measured by the percent standard error showing the expected dispersion of the sample mean in a distribution of means of many similar samples around the population mean. The precision required will vary for every farming system, depending on the level of improvement offered by possible innovations. With large potential increases in productivity, data can be less precise; where increases are small, the data will have to be more precise. Sensitivity analyses on the results of planning will highlight components, with an important effect on the solution. The differences required in such components to alter the solution will provide a guide as to the level of sampling error to be sought in data collection. Realizing the necessary accuracy will be a sequential process, but a permanent on going collection program will rapidly crystallize the compromise between sampling precision and possible losses. A guide to initial investigation in areas of significant potential from improved husbandry is a standard error of between 7.5 and 10.0 percent of the mean.

In an interesting example of the cost/accuracy compromise for soil surveying, Robertson and Stoner have demonstrated the importance for costs of the decision on required precision.[18] Their argument for sufficiency in precision has familiar parallels in the history of farm economic investigation, where preoccupation with the minutiae of collection has led to neglect of the required accuracy of data. They discuss the trend to rigid and detailed specifications for soil survey and land classification, decrying the tendency to appraise a project report on the way it meets these specifications rather than on the needs of the problem at hand, holding that this prevents cost-effective surveying. They give an example of reclaiming saline swampland and postulate that the authorities need only a statement that 60 percent of the land is saline to appreciate the extent of the problems. They then query whether the validity of a plan would be seriously affected if the true percentage were between 50 and 70 percent, and whether it would be operationally of greater value to say that it lay between 55 and 65 percent or even 58 and 62 percent. Are the extra cost and time incurred worth the precision gained? Under certain cost assumptions and a specification that the project will proceed with 70 percent of observations in a certain class, Robertson and Stoner show graphically the cost of reducing standard errors. Extracted from their graph, Table 22 shows the rapid escalation of costs for increasing precision.

## Memory Bias

The cost/accuracy problem has often manifested itself on a narrower plane as a compromise between sampling error and observational error, usually arising from memory dependence. The

TABLE 22

Soil Sampling: An Example of Escalating Costs to Increase Precision

| Sample Size | Cost Total | Cost Unit | Percent S.E. Precision Level | Increase in Percent of S.E. Precision | Increase in Total Cost |
|---|---|---|---|---|---|
| 50 | £2,000 | 40 | 6.5 | – | – |
| 100 | £2,000 | 20 | 4.7 | 1.8 | – |
| 200 | £4,000 | 20 | 3.3 | 1.4 | £ 2,000 |
| 500 | £6,000 | 12 | 2.1 | 1.2 | £ 2,000 |
| 1,000 | £10,000 | 10 | 1.5 | 0.6 | £ 4,000 |
| 5,000 | £50,000 | 10 | 1.3 | 0.2 | £40,000 |

Source: V. C. Robertson and R. L. Stoner, Shell Symposium on New Possibilities and Techniques for Land Use Surveys with Special Reference to Developing Countries (1970). (Mimeographed.)

importance of this clash is emphasized in traditional agriculture when the population being investigated is illiterate. The dependence on memory as a source of information is augmented as the need for the control of response errors increases. Without farm records or completed postal questionnaires, survey design revolves around the frequency of visits required over the production period and the implications for memory bias of a reduced frequency in order to cut down costs. Memory performance centers on the time elapsed since the event, called the reference period, and the characteristics of the event in question.

S. S. Zarkovich notes the increasing difficulty of reliable recall as the length of the reference period increases but emphasizes that this is not the only important factor.[19] Tables 23 and 24 show that data are improved by questioning based on a longer period.

If the measurement twice weekly in Table 23 is taken as a standard of accuracy, then the monthly interview comes closer on all items than the weekly interview, and closer in three out of four items than the weekly measurement. Objective measurement per se is not

TABLE 23

Consumption of Specific Items per Person per Week:
Two Periods of Reference and Two Techniques
(ounces)

| | Interviewing | | Measurement | |
|---|---|---|---|---|
| | Week | Month | Once a Week | Twice a Week |
| Rice | 16.21 | 15.14 | 15.61 | 14.87 |
| Pulses | 1.56 | 1.19 | 1.42 | 1.19 |
| Sugar | .23 | .16 | .16 | .14 |
| Salt | 1.07 | .76 | .86 | .82 |

Sources: P. C. Mahalanobis and S. B. Sen, "Some Aspects of the Indian National Sample Survey," Bulletin of the International Statistical Institute, XXXIV, 2 (1954).

a guarantee of accuracy but is importantly related to the reference period and data characteristics. The phenomenon is further illustrated in Table 24, where shop classification and recording is adopted as a standard of accuracy; of three periods of reference the year, being the longest, gives the best estimates.

Zarkovich accounts for the phenomena in these examples by the characteristics of the data. Shorter reference periods for consumption data will create higher variances. What one family eats one day, they may miss altogether another, and the day as a reference period reflects this variation. But with a year as a reference period, this type of variance is averaged and lost, giving a comparatively unbiased basis for estimation.

Zarkovich makes three other points important in evaluating the accuracy of alternative visit frequencies in relation to a particular attribute.

## End Effect

Both recollection and chronological ordering are important when part of a sequence of events must be isolated into a given reference period. The placing of events at the periphery of the reference period

TABLE 24

Weekly Purchases of Specific Foodstuffs
by Different Reference Periods
(rupees per family)

| | Period of Reference | | | Daily Est. on Shop Basis |
|---|---|---|---|---|
| | Day | Week | Year | |
| Pulse | 0.35 | 0.31 | 0.23 | 0.21 |
| M. Oil | 0.77 | 0.62 | 0.48 | 0.42 |
| C. Oil | 0.14 | 0.19 | — | 0.14 |
| Salt | 0.07 | 0.11 | 0.08 | 0.07 |
| Gur | 0.07 | 0.10 | 0.07 | 0.02 |
| Pan Supari | 0.07 | 0.12 | 0.08 | 0.03 |
| Leaf Tea | 0.07 | 0.05 | 0.02 | 0.01 |
| Tobacco | 0.35 | 0.20 | 0.13 | — |
| Bidi | 0.35 | 0.37 | 0.29 | 0.21 |

Source: A. Ghosh, "Accuracy of Family Budget Data with Reference to the Period of Recall," Bulletin of the Calcutta Statistical Association, V (1953).

is particularly difficult and produces some transfer into and out of the period. This transfer, being associated with the start or finish of a reference period, is called end effect. While distinct, important events may provide little difficulty, recurring routine events may easily be misplaced. The degree of transfer is reduced where reference periods can be related to natural cycles, such as the agriculture season in our application. It is thought to be particularly bad for short artificial reference periods, but the seriousness of the transfer will vary from item to item.

Zarkovich says:

> There is some indication that with long reference periods and characteristics connected with frequently recurring events (which cannot be remembered separately) the

> respondents make an effort to establish some kind of average per unit of time and multiply up by the number of units of time in the period of reference. . . . In other words, no attempt is made to recall each event separately.

With short reference periods memory is relied on, and it appears that respondents transfer more events in than out; thus survey characteristics are positively biased. Zarkovich draws the parallel between this and the well-known phenomenon of edge effects in sample yield cuts.

### Open and Closed Reference Periods

An open period is one with both ends arbitrarily located in the past and is thus particularly susceptible to edge effect. A closed period has its cutoff points clearly distinguished in the memory, and may be effectively related to natural or organizational cycles. Periods are never closed completely but it is believed that the better they are closed, the more accurate the resultant data.

### Conditioning

Zarkovich stresses that the influences of conditioning can be either positive or negative. Initial doubts about whether a survey is in their interests may give a high rate of refusal from respondents or a deliberately poor quality of response. With increasing experience their fears may be removed, thus improving the quality. Respondents are conditioned to a level of effort in recall by the initial insistence or persuasion of enumerators. More often conditioning may create biases, the respondents gradually losing interest in the survey and cooperating less and less. It is possible that this conditioning effect is present in all repetitive surveys.

These facets of memory performance will be considered in discussing alternative collection techniques for the right types of data important in farm economic investigations.

In summary, there are five aspects to be related to the attributes under investigation in designing the sampling and data collection program.

### Importance to the Farming System

This is central to a decision on whether the design must meet precision targets for a particular attribute.

## Objective Measurement or Memory Dependence

Where objective measurement is possible (as with chain and compass methods in area measurement), the relative precision of alternative, memory-dependent techniques, as well as other objective methods, must be evaluated for relative cost against accuracy requirements in planning. Clearly, survey design cannot be structured around an objective technique for a minor attribute. Memory techniques, perhaps giving a degree of observational error, may be useful where other attributes dictate a survey design which rules out objective measurement.

## Data Variance

The inherent level of variation in the population decides the sample size needed for a required level of precision. There has been a failure in farm economic survey work to appreciate the concept of sampling precision and a tendency toward "safety in numbers," partly because of the failure to crystallize data uses in advance of collection. Clearly, variation in a truly homogeneous type of farming area will be as great over a group of adjacent farms as throughout the whole area. Necessary sample size is not dependent on the size of the population but on its variability, a conclusion repeatedly ignored by expressions of the need for a 1 percent or 5 percent sample of the population common to many investigations. Zarkovich states that a sample of fifty units will usually give a "reasonable" picture of the variation in a homogeneous population with attributes which are normally distributed.

The two other important classification characteristics describe the nature of the event occurring on an individual farm, giving a guide to sources of error when collection techniques are memory-dependent.

## Frequency of Occurrence

Frequent events are unlikely to be remembered individually unless they are important to the respondent. Infrequent events will usually be remembered individually.

## Regularity of Occurrence

Events which occur regularly create a pattern of experience. Response can be based on average levels multiplied by the frequency of occurrence within a specified reference period. Irregular events present problems when they are frequent and thus unlikely to be remembered individually, since no pattern is formed. An example often quoted is in eating meals. The number of lunches eaten is easily remembered because it is a frequent and regular meal. But the foods eaten for lunch may present more difficulties once the period of recall increases beyond the current day; many items are eaten frequently but irregularly.

It is attributes which are frequent but irregular which pose the major problems to memory-dependent collection techniques. Labor inputs, central to the model of the existing system, have these characteristics; and conclusions as to the feasible techniques for collecting this vital data will be important to the range of feasible survey designs.

The need to reconcile the detail in simulation, sampling error, and memory bias in survey design requires an initial appraisal of the structure of the system under investigation and of the likely level of memory performance on attributes which are important in the planning model. Investigational procedure within an identified type of farming area has two stages: presurvey and the survey itself. The survey itself will be concerned with the measurement of required attributes that vary between farms, and its major preoccupation will be quantification. The presurvey stage is equally important to the whole investigational phase in creating the framework on which the survey can be designed for best effect. The presurvey stage is used to identify productive activities in the system which should be represented in the planning model because of their role in satisfying farmer objectives. Such activities, and the attributes linked to them, need investigation as discrete subpopulations within the general population of farms in the area. Identification of these distinct productive activities allows the measurement of land, labor, and output data to give effective representation in the planning model. Presurvey has three other roles:

1. To describe the characteristics of attributes known to be important in limiting collection techniques, in order to guide survey design.

2. To outline general aspects of each data category known to be important to the format of the survey questionnaire.

3. To describe those attributes which are general to the area but do not require measurement and therefore have no survey content.

The presurvey and survey content of the eight data categories is identified as each category is discussed. The first three categories are agronomy, the food economy, and social customs influencing resource allocation (headed "general attributes"). These categories, being common to the area as a whole, are enumerated mainly during the presurvey stage on a general level, though a few aspects are confirmed or measured during the survey. The remaining five categories are classed as individual attributes, reflecting their susceptibility to interfarm sources of variation and the need for sampling. The presurvey content in these categories is limited to probing the factors which influence survey design and questionnaire construction. Most of the content requires measurement in the survey proper.

Before the data categories are examined, Chapter 8 discusses the alternative techniques for deriving representative units within groups of farms rated as homogeneous. The conclusions have further implications for flexibility in collection techniques and sampling schemes.

## NOTES

1. F. F. Elliot, "The Representative Firm Idea Applied to Research and Extension in Agricultural Economics," Journal of Farm Economics, X (1928).

2. R. Hurley, "Problems Relating to the Criteria for Farm Classification," Journal of Farm Economics, XLVII, 5 (1965).

3. U.S. Southern Cooperative, Farm Management Bulletin, 56 (1958).

4. J. F. Thompson, "Defining Typical Resource Situations," U.S. Southern Cooperative, Farm Management Bulletin, 56 (1958).

5. H. O. Carter, "Representative Farms as Guides for Decision Making," Journal of Farm Economics, XLV, 5 (1963).

6. C. S. Barnard, "Farm Models, Management Objectives and the Bounded Planning Environment," Journal of Agricultural Economics, XV (1963).

7. R. H. Day, "On Aggregating Linear Programming Models of Production," Journal of Farm Economics, XLV, 4 (1963); "More on the Aggregation Problem," Ibid., LI, 3 (1969).

8. L. M. Day, "The Use of Representative Farms in Studies of Interregional Competition and Production Response," Journal of Farm Economics, XLV, 5 (1963).

9. S. J. Sheehy and R. H. McAlexander, "The Selection of Benchmark Farms," Journal of Farm Economics, XLVII (1965).

10. G. E. Frick and R. A. Andrews, "Aggregation Bias and Four Methods of Summing Farm Supply Functions," Journal of Farm Economics, XLVII, 3 (1965).

11. Ibid.

12. M. P. Collinson, "Maswa Area," Farm Economic Survey no. 3 (Dar es Salaam: Tanzania Dept. of Agriculture, 1963). (Mimeographed.)

13. D. von Rotenhan, Land Use and Animal Husbandry in Sukumaland, "Africa Studies," XI (Munich: IFO, 1966). (In German.)

14. F. Yates, Sampling for Census and Survey (London: 1949); S. S. Zarkovich, The Quality of Sample Statistics (Rome: FAO, 1964).

15. F. Yates, op. cit.

16. Grundy, Heely and Rees, "The Economic Choice of the Amount of Experimentation," Journal of the Royal Statistical Society, ser. B, XVIII (1956).

17. L. Berry, Economic Zones of Tanzania (Dar es Salaam: Bureau of Resource and Land Use Planning, 1968).

18. V. C. Robertson and R. L. Stoner, Shell Symposium on New Possibilities and Techniques for Land Use Surveys with Specific Reference to Developing Countries.

19. Zarkovich, op. cit.

CHAPTER

# 8

# BUILDING THE REPRESENTATIVE FARM MODEL

Within areas identified as homogeneous in farming system, representative farm units are used as a vehicle for the evaluation of proposed improvements. There are alternative techniques for deriving "typical" farm models, and the choice of technique has important consequences for the flexibility of method in investigation and the accuracy of planning. The broad alternatives are the selection of a particular farm to represent a group and the synthesis of sample averages into a typical farm unit.

Within types of farming areas we have noted two sorts of attributes: those common to most of the population, influenced by sources of variation which the selection of areas has sought to isolate, and those affected by local sources of variation-microclimatic and local soil differences and motivational and managerial differences. These sources influence measures of resource endowment, use, and productivity, which are central to the derivation of a representative farm unit within an area homogeneous in general natural conditions, market opportunities, and methods of production.

Representative farms have received little explicit attention in the literature, probably because of the continuing argument over appropriate criteria for farm classification. The alternative possibilities of selection and synthesis both deserve some discussion; we will look first at selection, centering on the question of the criteria to be used in the choice of a farm to represent the group.

## SELECTION OF A REPRESENTATIVE FARM

El Adeemy and J. D. MacArthur have shown a procedure for the selection of typical farms and for testing their representativeness using percentage deviations from the mode on the four criteria used in selection.[1] The farm with the lowest average percentage deviation is offered as most representative and compared with the sample over a range of eleven attributes. A similar procedure is adopted here in considering selection as a possible method of deriving a model for planning. The examples will use a sample of forty-two farms from area A in Figure 2, for which particularly full data is available.

The selection of criteria for choosing the representative farm will depend on the planning problems being faced. In seeking to evaluate new technical practices in terms of their impact on the system, the emphasis is placed on both the results achieved—the increase in productivity—and the resource reallocations implied. Both the causal variables of the resource base and the "effect" variables of output and performance are important as benchmarks in evaluating the technical changes available. Adeemy and MacArthur used deviations from the mode in order to relate the results to the farming population. By comparing the inflated modal farm with the sample totals, the deviations across the eleven criteria used were inevitably high and representativeness was distorted, with the sample totals reflecting the positive skew of the distributions (most deviations were negative).

This method of checking representativeness is more logically based on the percentage deviation from the mean. Attributes reflecting size in our sample have similarly positively-skewed distributions. The differences in size are caused by (a) a population of two-generation families where grown children have not yet broken away from their parents and established their own farms and (b) a proportion of more commercial farmers hiring seasonal casual labor. Size rather than content of change, is important in considering the level, for the level must be consistent with the debt ceiling of the majority in order to give the fullest possible penetration of the rural population by the diffusion process. As we shall see later, this is taken as an independent condition of the model which is established and can be varied to change the scale of the system at will. However, resource and product relationships based on the arithmetic mean retain their validity over the changes in scale because they are not technologically based. Thus scale criteria will not distort relationships, even though based on the mean rather than the mode.

Five fields are important to analysis and planning using the representative farm technique in traditional agriculture, and they center on attributes likely to vary within the area. In addition the selected farm should have the tribal affiliations and asset structure common, but not necessarily universal, in the area: both are general attributes which are readily identified.

## Cropping Pattern

Any wide variation in cropping patterns will have been removed when the type of farming area was identified. Also, the influence of cropping patterns on other important and locally variable attributes is so strong that high deviations in these would almost inevitably result. Nevertheless, extremes should be explicitly avoided. In the example area most innovations available are for cotton, and it is important that the selected farm have a typical proportion of its resources in cotton production. The criterion adopted to reflect cropping patterns is the percentage of the cultivated area in cotton.

## Labor Supply And Use

Regression analysis on the sample of forty-two farms pinpoints the labor variables which significantly influence both size of farm and farmer performance. Farm income was related to five independent variables which accounted for 79 percent of the variation. Gross return per acre was used to account for differences in land quality and management ability. The four labor variables were available family labor[xx], hired casual labor[x], seasonal use of labor[xx], and the rate of work in cultivation[x]; all of these are significant, availability and peak period usage being highly significant. In addition to the quantity of labor available, important interfarm sources of variation are identified as (a) the willingness to use available labor, particularly at peak periods of the season, and (b) the rate of work achieved by the family on key operations. Both of these stem from the managerial and motivational differences between farm families.

Four criteria were adopted to cover this field:

1. Total labor available.

2. Total labor used.

3. Percentage of available labor used in the critical months of November and December.

---

[x]Significant (in the statistical sense) at the 5% level of confidence.

[xx]Significant at the 1% level of confidence.

4. Rate of work achieved by labor in the cultivation operations, the most labor-intensive operations of the critical period.

## Labor Profile

A further aspect of labor as the resource limiting the system is sufficiently important to warrant its own group of criteria—the timing of labor use. The central analysis of the planning phase will be how labor allocations are altered as a result of innovation possibilities. It is important that the allocation of the model unit be typical. Where seasonality is not so pronounced, this group would be of less consequence. Some details of the labor profile of the example farmers are set out in Table 25.

Only the critical months in the system are useful selection criteria, and an initial analysis of the labor profile is needed to pinpoint these months. The monthly frequency distributions of raw data vary in character. There is distortion in the coefficiency of variation of the raw data caused by scale factors giving a positive skew to the distributions and by bimodality, particularly in May, September, and October distributions. In May it results from microclimatic factors causing early maturity and thus high labor requirements, in particular for cotton farmers. Other microclimatic factors create timing differences in September and October; in September the cotton harvest finishes on some farms, while in October some farmers begin preparations for the next season that are facilitated by early local rain. Others wait until November.

The third horizontal row shows a more valuable set of data based on the percent of total labor used in each month, which removes the scale effects. This is supplemented in the fifth row by the frequency with which a month is limiting on individual farms. The importance of a month is indicated by three criteria:

1. The percent of total labor used in that month.

2. A low coefficient of variation, indicating that the majority of farmers agree on the importance of that month.

3. The frequency with which, over the sample of farmers it is the month of greatest effort.

Clearly, the three indicators, which are not mutually exclusive, must be examined together. Based on knowledge of the system,

## TABLE 25

Details of the Labor Profile of Forty-Two Traditional Farmers

| Factor | Nov. | Dec. | Jan. | Feb. | Mar. | Apr. | May | June | July | Aug. | Sept. | Oct. |
|---|---|---|---|---|---|---|---|---|---|---|---|---|
| Average Man-Days Used | 52 | 56 | 41 | 24 | 23 | 18 | 23 | 49 | 46 | 40 | 17 | 9 |
| Coefficient Variation | 46 | 50 | 44 | 58 | 52 | 72 | 91 | 64 | 59 | 64 | 92 | 94 |
| Percent Total Used Each Month | 13.3 | 13.9 | 10.3 | 6.4 | 5.9 | 4.6 | 5.5 | 12.3 | 11.5 | 10.1 | 4.2 | 2.2 |
| Coefficient Variation | 31 | 29 | 21 | 52 | 36 | 59 | 72 | 38 | 32 | 55 | 97 | 71 |
| Frequency Month Is Limiting | 6 | 15 | 1 | 1 | 1 | 0 | 1 | 5 | 3 | 5 | 1 | 0 |

Source: Compiled by the author.

November, December, June, and July are adopted as the critical months; out of these four November and December are dominant, since the cultivation operations covered in this period are fundamental to the system. The main component of labor use in June and July is cotton picking, which is liable to large interseasonal variations in yields; data on cotton yield per acre for a sample of farmers from the same population showed a yield of 461 pounds per acre in 1963-64, compared with an average of 690 pounds for this sample in 1965-66. While the relation of neither yield to the interseasonal average in this area is known, and while there are overheads to the picking operation which make unlikely a pro rata increase in labor input requirements for increased yield, it is clear that the labor profile over the harvest period in this year could be exaggerated by up to 50 percent. This is a good example of the need for careful choice of selection criteria, for June and July are still important, if we are to avoid the selection of farms with very high yields which may feature these as limiting months. The criteria for this group are the percentage of total labor use over the season used in the two periods November and December, June and July. Calculation of these two criteria will be by the algebraic summing of the deviations within the pairs because of the substitutability of labor between the months. To accumulate the deviations, ignoring the signs, would be a poor simulation of the circumstances of the system.

## Scale

Although pure size differences are unimportant as long as factor proportionality is maintained, the size of the acceptable innovation step will be governed by current income levels, interacting with individual risk preferences. Although the criteria relating to labor as the limiting factor will check exaggeration of scale, acreage cropped is included as a further criterion directly measuring the results of the interaction of labor supply and managerial efficiency.

## Output

Although output is a dependent variable, it is a measure of current resource productivity. As such, it acts as a benchmark against which innovations can be measured and typicality is thus important. In addition, output per acre acts as a proxy variable for land quality and helps to measure a resource characteristic which is usually and necessarily neglected, because direct measurement is complex and expensive.

Three criteria are chosen for the example: total cotton production, total food grain production, and gross return per acre cropped.

In the five fields outlined, eleven criteria have been detailed. The means of these were calculated and each farm related to them to give its deviation from the mean. This deviation was expressed as a percentage in order to allow aggregation of deviations of different attributes on a common base. The deviations on the different attributes were averaged within their respective groups, and the group means were added and then divided by five to give an average score for each farm. Farm 2 was selected as the most typical farm on all criteria, with an average deviation of 13.2 percent. The percentage deviation from the mean data for each criterion for Farm 2 and for Farms 13, 37, and 11 (each most closely representative when selected on single groups of criteria—13 on output, 37 on labor supply and use and 11 on the labor profile) are presented in Table 26. The selection of farms on criteria scores in particular groups will allow discussion of the possibility of proxy criteria, of which those centered on labor as the limiting factor are of particular interest.

The results for Farm 2 show the inevitable pattern. In order to get reasonable typicality over a range of criteria, precision on any individual criterion is sacrificed. The result underlines the importance of choosing the selective criteria with the planning objective in mind. The very high deviations in other subgroups for farms selected by criteria of a single group pinpoint the difficulty of trying to shortcut the procedure. This is particularly true for the two labor groups. Farm 11 would be the farm with the lowest score over the two labor groups, yet over the five groups deemed important for our application it averages a score of over 48. Typicality of the limiting factor is not an adequate basis for selection of a representative farm for our application. Closer examination of Farm 11 emphasizes the delicacy of the task of choosing the selection criteria. Food crop production is the main distortion in Farm 11, caused by very high food-crop yields exaggerated by a high proportion of the total area under food crops. June and July labor use was selected to weight out extraordinary yield levels but relates particularly to cotton harvesting. Food crops are harvested in May; and for Farm 11, May labor use shows a deviation of over 350 percent from the mean. This would produce a gross distortion of the whole labor profile of the productivity of labor employed on food crops if it were used as a model for improvement evaluation.

Typicality is badly distorted by use of wrong or insufficient criteria and by the wrong weighting of criteria used. The five groups were

TABLE 26

Results of Typical Farm Selection Based on Various Criteria
(% deviation from the mean)

| Farm | 2 | | 13 | | 37 | | 11 | |
|---|---|---|---|---|---|---|---|---|
| Groups Selected for | All Criteria | | Output | | Labor Supply | | Labor Prof | |
| Crop Pattern | | | | | | | | |
| Percent Acreage in Cotton | 19 | 19.0 | 2 | 2.0 | 8 | 8.0 | 11 | 11 |
| Labor Supply and Use | | | | | | | | |
| Total Available Labor | 5 | | 34 | | 30 | | 18 | |
| Total Labor Use | 30 | | 23 | | 2 | | 7 | |
| Percent Used at Peak | 24 | | 79 | | 12 | | 16 | |
| Cultivation Rate of Work | 6 | 66 | 5 | 141 | 6 | 50 | 13 | 54 |
| | | 16.5 | | 35.2 | | 12.5 | | 13 |
| Labor Profile | | | | | | | | |
| Nov./Dec. Use as Percent of Total Use | 1 | | 1 | | 5 | | 6 | |
| June/July use as Percent of Total Use | 38 | 39 | | 69 | 62 | 67 | 8 | 14 |
| | | 19.5 | | 34.5 | | 33.5 | | 7 |
| Scale | | | | | | | | |
| Acreage Cropped | 3 | 3.0 | 38 | 38.0 | 76 | 76.0 | 39 | 39 |
| Output | | | | | | | | |
| Cotton Production | 20 | | 6 | | 23 | | 29 | |
| Food Grain Production | 1 | | 15 | | 5 | | 409 | |
| Gross Return per Acre | 4 | 25 | 7 | 28 | 6 | 34 | 78 | 516 |
| | | 8.3 | | 9.3 | | 11.3 | | 172 |
| Sum of Group Means | | 66.3 | | 119.0 | | 141.3 | | 242 |
| Average of Group Means | | 13.2 | | 23.8 | | 28.3 | | 48 |

Source: Compiled by the author.

given the same weighting in this exercise, but it is not clear whether this can be justified. Even within groups the treatment of individual criteria will distort selection. For example, summing the deviations on labor profile criteria, ignoring the signs, would have placed Farm 11 thirty-fourth instead of first in the population of forty-two farms for this group. The high level of error and the uncertain residual variances, from exogeneous climatic factors in particular, together with the delicacy and insight required in the choice of selection criteria, argue against selection as a method of deriving a model of the existing system for use in planning.

There are three other criticisms of selection as a basis for planning procedure:

1. It represents a considerable extra amount of work in calculating deviations. While this is unimportant for a computer, in the field calculations may have to be done manually.

2. There is no logic in finding means for the sample attributes, then trying to mirror these in one farm, claiming that this is more representative than the means themselves. Each step in selection is a departure from the average already established for the sample.

3. Most important for our immediate purpose, selection of a typical unit across a wide range of criteria demands that all data be collected for every sample farm. The failure to demonstrate that particular, easily collected criteria can be used to proxy the range of attributes which planning demands should be typical imposes limitations on the choice of sampling techniques. This inflexibility is particularly damaging when investigation is expensive in terms of available manpower and funds.

Selection, as a method of deriving a model of the existing system, is not considered further.

## ALTERNATIVE TECHNIQUES FOR SYNTHESIZING REPRESENTATIVE FARM MODELS

We have noted the crop/land/labor relationships over the crop calendar as the focal analysis in planning. It is this profile—and the relation of production, rotational practice, consumption patterns, and social customs to it—which will reveal the improved productivity and resource reallocation required for alternative improvement possibilities.

The use of averages or other measures of central tendency in synthesizing a model unit from survey data brings with it the problem of aggregation bias. It is the same phenomenon noted in supply response work; in this case interfarm differences in timing create different peak requirements on particular farms, which are damaged when averaged—and peaks on one farm are offset by relatively slack periods on another, so that the whole labor profile is flattened. The point is illustrated in Table 25, where, in a summary profile averaged out from individual farm summary profiles, December, with 13.9 percent, is the month with the highest proportion of labor use. However, December is only the peak month for 35 percent of the farmers; and 90 percent of sample farmers used more than 13.9 percent in a single month. The average proportion used in the critical month was 18 percent, and an average 13.9 percent represents a 23 percent distortion of peak requirements in the system. Aggregation causes underestimation of two aspects important in our planning application:

1. The amount of labor willingly made available by the typical farmer when the season demands it.

2. The usual peak requirements of the system and, consequently, an overestimate of the current productivity of seasonal labor used.

Figure 3 illustrates the point further. The data used are from trial management farms in Tanzania, where every effort was made to achieve recommended times of planting, with none of the inherent flexibility of the traditional systems seeking to adjust to seasonal contingencies. The data are also on an enterprise basis and use inter-seasonal variations to demonstrate the aggregation effect. Bias of this magnitude in representation of the limiting factor is clearly as distorting as selection itself.

Averaging also raises the problem of incidence in the population. In the sample of forty-two farmers used earlier, only 60 percent weeded their cotton a third time, and an even lower percentage grew rice. Straightforward averaging absorbs this type of difference, and a considered decision is needed on how to cope with this type of variation. The basis for decision is the concern to plan for the majority of the farmers of the area. If incidence covers the majority on attributes such as rice growing, it should be included at full value. On attributes subject to exogenous sources of variation (for example, a third weeding of the cotton crop) which is affected by microclimatic variation), it may be more valid to take an average over the full sample, assuming that a third weeding is required over only a proportion of the crop. Both management and climatic sources of variation

FIGURE 3

Crop Labor Profiles on Trial Farm Units
Showing how Varied Timing Alters Incidence of Peaks

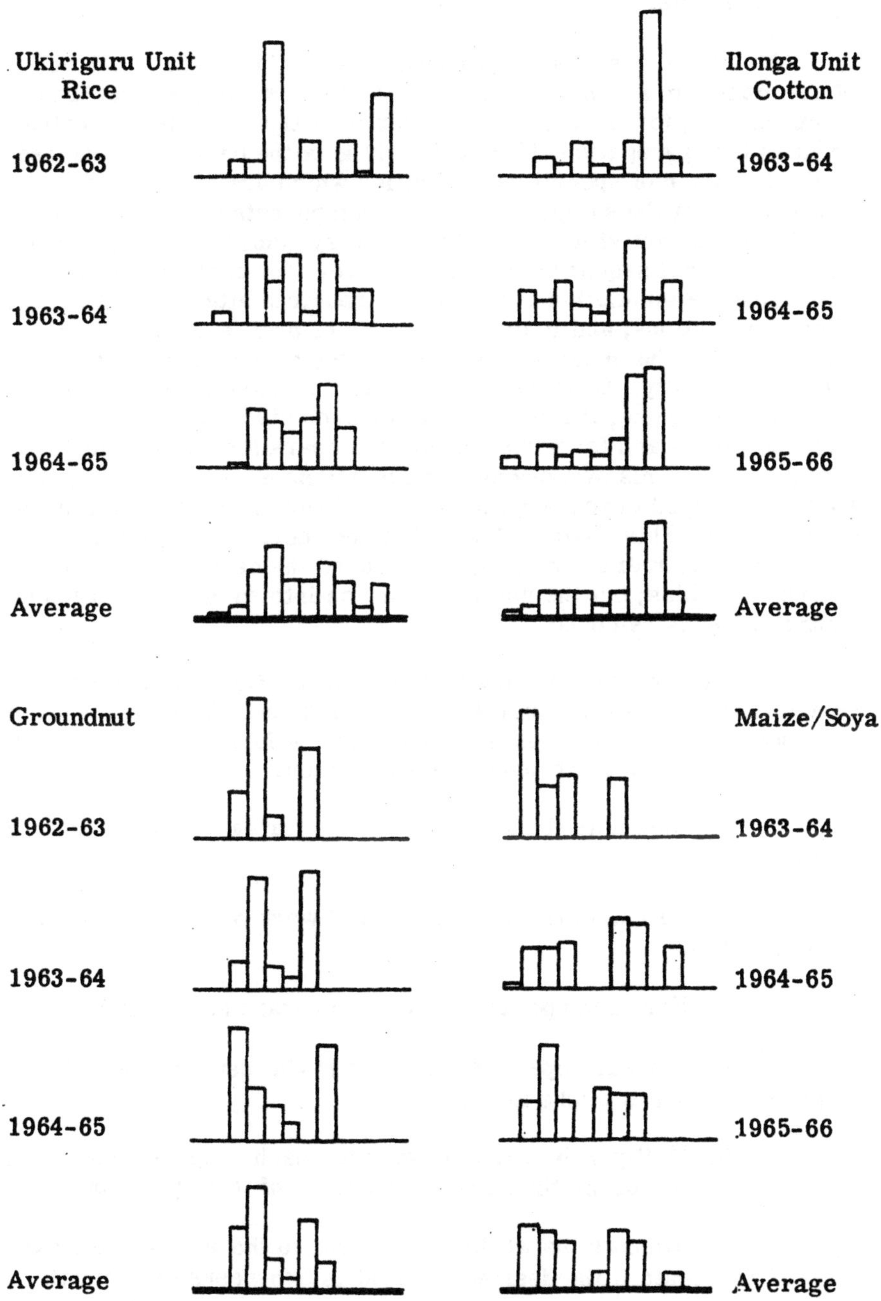

are involved, and solution requires a confirmed judgment of their relative importance. The averaged profile easily slides over this area of analysis.

An alternative to straightforward averaging is the construction of the model from components, each of which is sampled, perhaps independently, for the population under investigation. Such construction reduces the aggregation effect and highlights the decisions required on the inclusion of specialized activities which are recorded only on a proportion of the sample units. The components of a detailed labor profile are the activities identified in the system, the acreage, calendar, operational sequence, and rates of work for each operation. Improving the simulation of the profile requires attention to the timing of crop operations; and in the description of profile construction which follows, the mode is used only for the timing variable to give greater typicality. Construction reduces the aggregation bias by enumerating timing, the center date for each discrete operation, as a component. Each farmer spreads each operation round a center date peculiar to his own seasonal circumstances. On averaging, this spread is ranged over a similar spread of center dates, different for each farmer. This flattens the profile and increases aggregation bias. Clearly, averaging the spread around the center date still allows some bias; but the main source, the interfarm distribution of centre dates, is removed.

The data to illustrate the technique of profile construction comes from a trial farm unit.[2] The detailed records for two seasons are used to represent records for independent farms from cross-section data. There are six steps in construction:

1. Calculate the average acreage for each crop or activity identified.

2. Calculate the average rate of work per acre for each crop operation.

3. For each operation take the average spread in weeks.

4. For each operation take the modal center date and range the spread in weeks around it.

5. Multiply the rates of work for each crop by its acreage and divide the result by the spread, in weeks, of that operation.

6. Allocate the weekly labor used to the weeks in the calender and aggregate if the desired interval is two weeks or a month.

Table 27 shows a data array for some of the main crop operations for the construction of the profile for the trial farm. The acreages, rates of work, spread in weeks, and center date are the averaged values for the two seasons.

There is no dramatic improvement in representation in this model; observations for the two seasons are insufficient to give a range of timings. This is particularly so because management on the trial farm was carefully controlled around fixed planting times recommended by the agronomists. There was not the flexibility to adjust plantings to seasonal contingencies that is characteristic of the traditional system. All changes in timing were forced into the farming pattern by the uncontrollable influence of the weather and by the circumstances of the family. As individuals take different planting decisions, so the range of operation center dates increases; and over a viable sample there will be significant improvement in representation by selecting the center date and spread adopted by most farmers.

Construction in this way is of course possible from detailed daily labor records, with real benefits for investigation in the flexibility given to collection techniques, particularly in quantifying work rates. It opens up the possibility of measurement by other methods: by work study or by tapping the managerial know-how of the farmer. In many traditional systems, particularly where seasonality is a feature of the natural conditions, labor input data is the only information which is continuous over the season. The possibility of using collection techniques which focus on the rate of work as a component, rather than having to observe the flow of input over the whole period, allows surveys with fewer visits, giving greater coverage and lower costs per area covered. It further increases the scope for sampling schemes related to the variation in each component, as opposed to the scheme covering farms being adapted to the most variable attribute to be measured. Limited visit techniques, as we shall call them, are most useful in situations where data are urgently required; but they are feasible only in relatively simple farming systems. They sacrifice some of the advantages of detailed collection on each sample unit—for example, the type of ex post analysis of this chapter would be impossible. The usefulness of limited and detailed collection techniques will be compared in subsequent chapters in the light of the problems associated with particular data requirements.

The advantages of cheaper collection apart, construction is adopted as the method for deriving a synthetic model of the existing, traditional farming system within a type of farming area, and so for the evaluation of possible farming improvements. It is adopted as

TABLE 27

Sample Data Array for Main Crops and Operations in Profile Construction

| Crop Acreage | Seed | Ridge Plant | Top Dress | Weed | 2nd | 3rd | Pick | 2nd | 3rd | Process |
|---|---|---|---|---|---|---|---|---|---|---|
| Cotton 4.10 | | | | | | | | | | |
| Man-days per Acre | 4.3 | 9.8 | 1.4 | 2.7 | 3.2 | 1.5 | 10.0 | 6.5 | 7.4 | 9.6 |
| Spread (weeks) | 5.0 | 5.5 | 5.0 | 2.5 | 7.0 | 4.0 | 4.0 | 3.5 | 3.5 | 9.5 |
| Center Date (month/day) | 12/5 | 12/11 | 2/13 | 1/12 | 2/15 | 4/17 | 5/14 | 6/9 | 7/9 | 8/15 |
| Maize 1.08 | | | | | | | | | | |
| Man-Days per Acre | 7.9 | 10.8 | 1.6 | 4.0 | – | – | 3.8 | – | – | – |
| Spread (weeks) | 3.0 | 2.5 | 1.0 | 3.0 | – | – | 3.0 | – | – | – |
| Center Date | 1/13 | 1/13 | 2/20 | 2/28 | – | – | 6/4 | – | – | – |
| Groundnuts .90 | | | | | | | | | | |
| Man-Days per Acre | 6.9 | 19.5 | – | 6.4 | – | – | 15.0 | – | – | 20.2 |
| Spread (weeks) | 1.5 | 1.5 | – | 2.0 | – | – | 1.5 | – | – | 2.0 |
| Center Date | 12/30 | 1/3 | – | 3/12 | – | – | 4/20 | – | – | 8/9 |

Source: Compiled by the author.

TABLE 28

A Constructed Labor Profile

| Crop | Acreage | Nov. | Dec. | Jan. | Feb. | Mar. | Apr. | May | June | July | Aug. | Sept. | Total |
|---|---|---|---|---|---|---|---|---|---|---|---|---|---|
| Cotton | 4.10 | 16 | 44 | 18 | 15 | 5 | 6 | 42 | 39 | 31 | 16 | 13 | 245 |
| Maize | 1.08 | — | — | 20 | 5 | 1 | — | 1 | 3 | — | — | — | 30 |
| Groundnuts | .90 | — | 12 | 14 | — | 6 | 14 | — | — | 15 | 3 | — | 64 |
| Rice | .30 | — | 2 | 4 | 9 | 7 | 1 | 6 | 6 | — | — | — | 35 |
| Cassava | .20 | — | 1 | 3 | 6 | — | — | — | — | — | — | — | 10 |
| Sweet Potatoes | .22 | — | 2 | 4 | 4 | 2 | 1 | — | — | 2 | 1 | — | 16 |
| Bambarra | .25 | — | 3 | — | — | 1 | — | 5 | — | — | — | — | 9 |
| Cowpeas | .15 | — | 2 | — | — | — | 1 | — | — | — | — | — | 3 |
| Total | 7.20 | 16 | 66 | 63 | 39 | 22 | 23 | 54 | 48 | 48 | 20 | 13 | 412 |
| Percent of Total Used | — | 3.9 | 16.0 | 15.3 | 9.5 | 5.3 | 5.5 | 13.1 | 11.7 | 11.7 | 4.8 | 3.2 | 100 |
| 1962-63 | — | 7.8 | 10.4 | 16.3 | 9.8 | 4.2 | 5.5 | 12.6 | 18.6 | 10.5 | 4.2 | 0 | 100 |
| 1963-64 | — | 4.6 | 13.9 | 12.3 | 7.5 | 3.3 | 4.8 | 14.6 | 8.7 | 5.0 | 12.3 | 8.8 | 90.0 |
| Averaged | 5.8 | 5.8 | 12.6 | 13.9 | 8.6 | 3.7 | 5.1 | 14.0 | 12.5 | 7.0 | 9.3 | 5.6 | 97.5 |

Note: 1963-64 and the averaged profile had small percentages in October, not included here.

Source: Compiled by the author.

an alternative which reduces the aggregation bias inherent in averaging observed data across a sample of farms and offers greater flexibility in sampling and data collection techniques. The body of the investigational section which follows describes the eight types of data required for planning and examines the possible means of compromise between accuracy and cost for each category.

## NOTES

1. A. S. El Adeemy and J. D. MacArthur, "The Identification of Modal Type Farm Situation in North Wales," Farm Economist, XI (1970).

2. M. P. Collinson, "Experience with a Trial Management Farm in Tanzania, 1962-65," East African Journal of Rural Development, II, 2 (1969).

CHAPTER

# 9

# GENERAL ATTRIBUTES

Three data categories are included as general attributes. Capital might reasonably be added to the group, since the asset structure is common to the majority of farms in a homogeneous type of farming area. However, because it requires significant quantification by survey, it is grouped with labor and land in the group headed "resources." The three general categories, are enumerated mainly in the presurvey stage, though specific factors for quantification are included in the survey itself. Pre-survey investigation will be largely by interviews with local personnel and are best carried out by senior staff, either the economist or good field supervisors with an understanding of the needs of the planning sequence. This allows discussion loosely structured around the attributes on which information is sought. The survey itself will be done by junior staff with more formally structured questionnaires.

## AGRONOMIC DATA REQUIREMENTS

There are two objectives in this category, of which the first—the enumeration of husbandry practices—is purely descriptive. The second, establishing whether the system is being pressured by increasing population, is more complicated.

### Husbandry Practices

A description of the husbandry practices of the system is the key to understanding why the farmers do what they do. The long-term objective of maintaining the productive capacity of the land as the basic resource is common to all agriculture; we have discussed the short-term objectives of traditional farmers, with considerable emphasis on the security of food supply. An understanding of the methods by which both long- and short-term objectives are achieved is fundamental to an understanding of the system. Changes urged by the extension services may be unacceptable and positively detrimental where they undermine the role of existing methods. At the same time, changes which offer a solution to a fertility or reliability problem overtly recognized by the community and understood by the extension staff will be more readily absorbed.

Practice may vary with soil type, and the initial requirement is to distinguish soils used differently on the farm. Differences may be in cropping, when crops and crop associations on specific soils should be classified, or in methods; the sequence of operations over the season should be described for each crop or association, as should the method of work for the important operations. Other practices may vary with soil type, and any further description should make a clear distinction as to type.

Rotational practice will be fully discussed below as an indicator of the system's relation to population density. A range of farm practices supplements rotation in maintaining soil fertility and needs enumeration; these practices include soil conservation measures and the use of green crops, animal manure, or crop residues.

A second group of practices will ensure the reliability of food supplies, by conserving soil moisture, by the prevention of pests and diseases, or by hedging against microclimatic contingencies. Such practices will include the staggered planting of staple crops, inter-cropping of drought-resistant crops or crop varieties, the planting of reserve crops, replanting practice, and the use of a certain number of seeds per hole.

Finally, information should be sought on any phenological indicators used by local farmers in seasonal and daily management decisions on soil type, fertility status, and the timing of crop establishment.

All these data will be general throughout the area under investigation and can be collected by interview from individuals with local experience. Practices will be inherited; and farmers themselves may not articulate why they are used, particularly in the case of soil conservation and food insurance practices. Discussion should be with educated local people to whom the concepts of erosion and reliability can be outlined. On conservation and fertility maintenance, much may be achieved by a locally based agronomist who is familiar with rainfall and soil conditions and works from first principles, for the understanding of physical and chemical relationships in tropical soils remains very limited. Practical questions as to whether weed cover in the dry season or the multilevel intercrop canopy contributes to soil stability can be usefully answered, though the degree of protection would remain a matter of conjecture.

The only survey content provided by this essentially descriptive group of questions will be to confirm the incidence of the practices over the population. This will be picked up in the course of collecting labor input data, where crop, soil type, and operation will be key headings in the labor questionnaire. The description of practices and classification of crop/soil type peculiarities will make an important contribution to the design and layout of the survey questionnaire. Changes of soil and method will be one basis for the definition of subpopulations significant enough to justify independent representation in the planning model.

### Soil Fertility of the System

This subcategory seeks evidence that the system is under pressure from increasing population density. Various terms—the man/land ratio, critical density, falling fertility spiral—cover the same phenomenon; medical innovations raise the rate of population increase, and the technology of traditional agricultural communities cannot meet the change in pace in order to restore a balanced man/land relationship, with the result that farming systems come under pressure. Land is rarely absolutely scarce, but the rotational systems depend on a fallow period, which is reduced when more and more people require and for support. When fertility falls, labor use must be increased to maintain production or food priorities must be adjusted. Labor-intensive foods may be sacrificed to improve productivity, so that, initially, preference patterns are distorted and ultimately nutrition patterns may suffer until an increased death rate restores the man/land relationship.

E. Baum has provided an interesting example; and although his sample sizes are small, the data are internally consistent.[1] He uses R as a measure of the intensity of the arable/fallow sequence.

The semipermanent cultivators with a high R value have a lower yield per acre and per unit of labor use. Despite their very high clearing requirements, the shifting cultivators produce 220

TABLE 29

Some Characteristics of Shifting Cultivation and Semipermanent Farming, Kiberege Strip

| | Shifting Cultivation on Escarpment | Semipermanent Farming in Valley |
|---|---|---|
| No. of Holdings | 6 | 14 |
| R Value | .15 | .55 |
| Crop Acres (per Man-Equiv.) | 1.50 | 2.30 |
| Rice yield (lbs./acre) | 1,500 | 1,060 |
| Labor Input (Man-days/acre) | 110 | 100 |
| Labor Input as Percent of Total | | |
| Cultivation | 31 | 19 |
| Planting | 19 | 16 |
| Weeding | 20 | 32 |
| Harvesting | 30 | 37 |

Notes: Labor input data refer to four and eight holdings respectively.

$$R \text{ value} = \frac{\text{years of cultivation}}{\text{years of cultivation and fallow}}.$$

Source: H. Ruthenburg, ed., Smallholder Agriculture and Development in Tanzania, "Africa Studies," XXIV (Munich: IFO, 1968).

pounds of rice with sixteen man-days of labor; semipermanent farmers, with twenty-one man-days. Pressure of population in the valley bottoms has enforced sedentary agriculture, and the higher level of labor use has been a necessary response to falling fertility.

Identification of falling fertility is vital in deciding extension content, which must be consistent with the needs for long-term resource balance. Scale-increasing innovations will aggravate the rate of fertility loss but even so may be of short-term value in wooing the community to the advisory services as well as creating greater investment potential.

A fertility crisis will not be revealed by description and analysis of the relationships in a single production period nor, because of the interseasonal variation in yield levels, by anything except a very long-term study. Interseasonal and microclimatic variations in yield which mark any overall trend, and the slow rate of decline in fertility, similarly obscure the phenomenon from any one generation of farmers. It is important to assume that the problem is present and to investigate its severity. Information on the level of family labor effort will come from the main analysis of the labor profile. Planning sequences will show the possibility of increasing food supply by greater use of family labor and will allow projections of the time scale over which, assuming a rate of fertility loss, more permanent solutions are needed.

Population pressure will be manifest in three facets of the system which are more easily identified than the gradual decline in yield levels: rotational practice, methods of land acquisition, and a history of changes in the cropping pattern. Each is examined in turn.

Rotation

The evolution of rotational practice follows a pattern dictated by the scarcity of land. The pattern has been split into three phases—household shifting, arable/fallow sequences, and crop rotation proper—though a farming system is as likely to be in the process of moving between phases as to be in a single phase. This is particularly true where a system contains markedly differing soil types. For example, heavy valley-bottom soils may lose fertility slowly, allowing a long arable/short fallow sequence, while there has been evolution to crop rotation with manuring on the poorer soils in the system. Where systems cover distinctive soil types, investigation of rotational practice may need duplicating by types.

Before examining the content and possible collection problems associated with each phase, there are three general points it is useful to cover in the course of discussions on rotational practice.

1. It is useful to establish the mechanics of rotation under shifting cultivation: whether farmers change field by field or shift to a new fallow area, starting a new farm. This will highlight possible survey problems with farmers who have fields isolated from their homesteads.

2. Allied to this is evidence of the decision base in the mind of the farmer. The use of fertility indicators is covered in enumerating farm practice. More specifically here, is the move or crop change made in response to low production? Are there positive indicators of fertility loss—the dominant weeds, or loss of grasses which would be established over the dry season on fertile soils? Or is there a positive sequence after the pattern of classic Western rotations, where a field will carry a crop next year, or will stand fallow next year, in its turn?

3. The rotational cycle influences resource allocation over and above the single production period to be studied by the survey. Under shifting cultivation, the clearing of primary or regrown vegetation will occur on a proportion of the sampling units. Labor use of this nature becomes meaningful only in the context of the rotational cycle.

Household Shifting. The shifting of households is characteristic of primitive agriculture, when man was moving out of hunting and gathering society. Initially it involved deeper penetration into primary vegetation, then developed into circular migration as untouched areas were reduced. With household shifting a long fallow period is characteristic, the same ground being cultivated perhaps for a whole generation. Baum gives an example where a six-year short rotation, three in rice and three in grass fallow, is superimposed on a fifteen-year cycle covering two, and sometimes three, areas.[2] This gives a thirty-year fallow where three areas are cycled.

Local practice can be enumerated by interview. Of interest are the timing of shifts between areas, whether the shift is into areas of primary vegetation or regrowth, and whether the land is already held in right by the family or will be acquired by possession. These can usefully be confirmed over the sample of farmers in the survey. Data are all memory-dependent and, as with all other historically oriented questioning in the survey, it is useful to employ a calendar of prominent community events, constructed in the course of presurvey investigation, as a benchmark to local responses. In extreme cases, shifts might be outside the experience of younger operators because of their infrequency. However, they are a major feature of family life and would be recalled by operators who moved as a member of their fathers' households. Respondents' location of household shifts in time will be approximate but will not be obviously biased, and variation between farms is unlikely to be large. Baum gives data for six households with an average R value of .15. A coefficient of variation of about 30 percent from this small sample indicates the likely homogeneity in a well-defined community.

Within a random sample of farms, subsamples at varying stages in the rotational cycle can be identified by the time each household shifted last. A comparative analysis of labor use and output between subgroups will give estimates of the rate of fertility loss, the increases in family labor use, and the level of productivity at which shifts are stimulated. Such analysis does not justify increasing the sample size to improve precision within subgroups. If a major source of variation is being removed, a worthwhile comparison should be possible; but the number of subgroups is best kept to three or four.

Arable/Fallow Sequences. As is seen in Baum's example, an arable/fallow sequence may be superimposed on a system of household shifting, and the possibility of this "tiered" structure should be investigated before the pattern of a rotation is presumed to have been finalized. Once new land becomes scarce, families retain rights over land which they have cleared and used. Permanent settlement is established and the cultivated area is shifted around the area held in right. Nevertheless, the community as a whole is vested with ownership, and it has been a clear principle of customary tenure that the community may reassert its rights over fallow land when density of population demands reallocation. Arable/fallow sequences are thus a transitional phase between shifting cultivation proper and permanent, continuing cultivation of the same land. It is a phase which currently dominates the major part of traditional agriculture in Africa.

Elements of both previous and subsequent phases may appear in areas where, although the farming system is homogeneous, population density varies; hence the man/land ratio is stressed as a criterion in identifying types of farming areas, particularly for planning sequences. D. Rotenham, working in Sukumaland, has compared three areas of different population densities.[3] Interpolating into the population trend for 1963, the year of his work, the three sample areas in Shinyanga, Kwimba, and Ukerewe Island have population densities of 80, 118, and 415 per square mile, respectively. Rotenham commented on the main points emerging from the study of 309 cropping histories.

With relatively fertile soils in Shinyanga, long sequences of the same crop are grown until falling production or rampant weeds stimulate a change. There was no evidence of any cycle. In Kwimba, with poorer soils, there was the beginning of a more regular basis for change. Two-year fallows were frequently recorded, and a period of cultivation was often finished with cassava. On Ukerewe Island, the most densely populated area, a regular cassava/cotton rotation had emerged, with cotton the first crop planted after fallow. The pressure on the land has evolved a genuine crop rotation, and even an element of double-cropping. When the rice crop is harvested, the moist soil is made into mounds for sweet potatoes, often interplanted with maize and legumes.

Three general points form useful signposts for investigation of the degree of transition of the system through the arable/fallow sequence:

1. The introduction of a cash crop with pure stands provides a focus for questioning. Particularly if it has a high nutrient requirement, it may have become a pivot for any incipient crop sequence.

2. The emergence of a crop with a dual role as a famine reserve and a fallow, particularly where it has a low food preference, indicates a growing fertility problem. This also may be a pivot for an emergent crop sequence.

3. The emergence of crop sequences which strive to utilize available water and soil conditions more fully, within the single production period, is further indication of pressure in the later stages of the transition.

An outline of current rotational practice will be possible from interviews with individuals and agricultural officials in the area. It should stress the types of fallow found in the area and seek out any incipient crop sequences by focusing on fallow crops, hungry crops, or examples of double-cropping. The incidence of any sequences which emerge should be confirmed by the survey. It will be important to measure the ratio of cultivated to fallow land as a benchmark against which historical data, from secondary sources, can serve as a measure of the rate of increase of the population cultivated. This measurement and the enumeration of sequences present collection problems.

The cultivated area is measured in the course of enumerating crop acreages. The total area held under right presents a formidable measurement problem. Physical measurement of the total area under rights, especially where the area is fragmented and the cultivated area is a small proportion of the total, is arduous and greatly increases the response burden on the farmer. Care is necessary to ensure that the concept of fallow areas is put to the farmer in a meaningful way. Various researchers report that respondents consider only the cultivated area as their farms and that surrounding fallows, which are often available for communal grazing, are overlooked.[4] Because of the limited use for this measure, three proxy measures are adequate and avoid the problem of objective measurement of the fallow areas during the survey.

1. Given a counting of family size in the survey, a known population density allows the total available acreage per farm to be estimated. Where appropriate, further estimates of uncultivable land or soil types can modify this figure.

2. Aerial photographs can be used to identify areas which have been taken out of primary vegetation and put under cultivation. Taken at the appropriate time of the year they can also identify standing crops. The ratio can be worked out from this source.

3. Farmer estimates of the total area under his control may be reliable. As with all questions, care is required to couch them in terms which he comprehends. Multiples of his present area under crop have been used to estimate total farm area.

Under true arable/fallow, confirmation of the sequence in the survey is difficult because no formal pattern has developed. Shifts are normally based on declining crop performance and are made field by field.

Memory of field sequences appears good from the results given by Rotenhan and others. M. Upton found he was able to ask the farmer directly how long a piece of land had been under cultivation.[5] However, difficulties are created by the fluidity of plot boundaries: a field currently under a particular crop or mixture may be split into several plots of distinctive crops for the following season; and although present field boundaries may be easily determined by eye, past boundaries often bear little relation to them. Thus the base on which to build a history of cropping is shaky, and may well mask any simple patterns which exist. The use of specific points—where the enumerator is standing with the farmer, for example—helps to avoid this confusion but may be difficult for the farmer to understand. General questioning of the farmer in relation to a specified soil type taps his farming experience, rather than his memory, and describes the sort of sequence and time schedule he <u>would expect</u> to follow on that soil type. Similarly, the farmer is able to remember the approximate season a particular field he is cultivating was opened from fallow. Recording this date for all fields on all farms in the sample during acreage measurement gives sufficient information to show the crops planted on each soil type in the various years after opening. The sample will provide observations to highlight any sequence that is followed. There should not be too much effort to crystallize a regular pattern which is unlikely in a true arable/fallow sequence. The sequence may be a continuous process with a little bit of the farm being fallowed and a little opened each season. D. Pudsey demonstrated the "creeping" nature of rotation in seasonal crops on eleven holdings in Uganda.

The analysis of decreasing yields by subgroups in the sample is theoretically feasible. Given field-by-field shifting, however, yield data must be available on a plot basis for such grouping. This creates problems in relating production to particular plots and presents practical administrative difficulties requiring a distinctive survey design. Such a design could not be justified for the sake of this analysis alone, which is peripheral to the main planning sequence. (The difficulties of plot recording are discussed in the category on output data.)

Crop Rotations. Regular cropping sequences are likely to emerge only when the system has come under heavy pressure from rising population density. H. D. Ludwig's description of Ukara Island, with agriculture based on the same range of cropping opportunities as in Sukumaland, though with higher rainfall and no marked dry season, with a population density of about 580 per square mile, shows perhaps the ultimate potential in human carrying capacity for the area.[6] Almost inevitably, as with permanently worked land anywhere, manuring has substituted for fallow, absorbing large quantities of labor, and family members work an average of ten hours each day. The field crop rotational cycle is of three years, during which Wakara farmers take three harvests of millet and one each of bambarra nut and sorghum, and a green manure crop is grown and worked back into the land. A separate rotation characterizes the rice-growing area at the lakeside, similar to the sequence Rotenham reports emerging on the as yet less densely populated Ukerewe Island. Ludwig reports Ukara to be little altered since first documented by D. Thornton and N. V. Rounce in 1936.[7]

With crop rotation on permanent fields a feature of the farming system, sequences are easily recorded either by discussion with individuals locally or by tracing the history of farmers' fields during the survey. Care is required to follow differences in soil and topography distinguished in the system.

TABLE 30

Evidence of Continuous Rotation in a System with a Fallow/Arable Sequence (acres)

| | Bananas | Pure Coffee | Food Crops | Total Area | Area Rested | Area Cleared |
|---|---|---|---|---|---|---|
| First Rains, 1964 | 2.28 | .33 | .91 | 3.52 | .07 | .36 |
| Second Rains, 1964 | 2.25 | .32 | 1.24 | 3.81 | .15 | .08 |
| First Rains, 1965 | 2.28 | .34 | 1.16 | 3.78 | — | — |

Note: Pudsey notes that the total area is increased over and above the margin between clearing and resting in the first rains, 1965, by one purchased plot and one borrowed plot being returned.

Source: D. Pudsey, A Pilot Study of 12 Farms in Toro," (Kampala: Uganda Dept. of Agriculture, 1966). (Mimeographed.)

## Methods of Land Acquisition

Within the community the ways in which land is acquired change in a way which parallels the evolution of rotational practice. As land becomes scarcer, communities exercise more stringent control over the acquisition of new areas by the individual; and this is balanced by an increasing awareness of his rights on the part of the individual. At the same time, more formal tenurial practice emerges, usufruct giving way to inheritance and finally to legal registration as the basis of land rights.

The ways in which extra land may be acquired form a scale on which the system is located, ranging from the clearing of virgin bush and requesting the community authority to renting and purchasing. It is important to qualify each step on the scale. If the individual wishes to clear new land, will he have to move away from his existing farm, or clear a piece on the settlement boundary involving a fragmented holding? If a farmer must ask the communal authority, will he ask for permission to clear new land or will it involve reallocation from land already held in usufruct by another family? Once renting and purchase of land are established, the rates to be paid will be an important supplementary question.

As with rotations, practices may vary within the system by soil types which are more or less desirable. R. S. Beck and others distinguish Kihamba and Kishamba land among the Chagga on Kilimanjaro: Kihamba land is on the mountain, to which the family has permanent rights, while Kishamba land is down on the plains, where rights are usufructory.[8] The distinction between types is important for planning when the development of crops may be associated with specific land types.

The incidence of different methods of acquisition can be enumerated during the survey and will be important in systems where renting, land purchase, registration, or fencing is beginning to encroach on traditional acquisition and tenure practice. Response is from memory, but events are so fundamental to individual livelihood that recall is easy. In areas where registered land purchase is subject to duty, biased response is possible; but bias can be minimized by thorough preparation in the area and by cross-checking on transactions with individuals of the same area.

## Recent Changes in the Cropping Pattern

The investigation of existing rotational and land acquisition practice allows us to interpret how close the system is to permanent farming. Only by relating this evidence to the reasons for recent changes in the cropping pattern can there be an assessment of whether the evolution of practices to meet the pressure has fallen behind, involving sacrifices of preferred foods or an increase in areas of staples to compensate for falling yields. Not only will this reveal the urgency of the problem, it will also pinpoint crops which were favored and for which advanced husbandry may offer greater potential than those now grown as substitutes.

The best indication of changes in the balance of crops grown in the system will be obtained from past survey and census information; and with three world censuses of agriculture, benchmark material is available for many areas. Subject to satisfactory sampling techniques for both investigations, a comparison over time will pinpoint new crops being grown or the emergence of a new dominance. New cash crops can be verified by local market information and new subsistence crops by local discussions on changes in traditional dishes and staples. This type of information can be usefully supplemented by discussion on the introduction of reserve crops and the use of green or animal manure over the last five or ten years.

The second and equally important step is to identify the reason, particularly for changes in food production pattern. Where a cash crop has entered the system, the reallocation of resources required may have involved a sacrifice in some aspects of subsistence production. Other reasons, to be investigated by interview, are the low levels of yield realized because poor fertility, pests, and disease depredations and changes in taste preferences. Survey content is limited to establishing the incidence of new crops and practices. It may be useful to follow up any recent and rapid changes on individual farms, in order to get a picture of the diffusion of the change through the community.

All the data required in this agronomy category may be obtained from local interview or secondary sources. Survey is used for confirming the incidence of events described over the farm population; and while the data are memory-dependent, there are no significant recall problems because the events are infrequent and important to family livelihood. Most aspects are picked up in the course of enumerating either crop acreages or labor inputs. Several possible analyses allowing an estimate of the rate of fertility loss require subgroupings within the sample, and larger samples would allow more precise testing of the differences between groups. However, sufficient idea of the fertility position of the system to decide on the weight it should carry in limiting extension content can be gained from the descriptive context. There is no justification for increasing sample size to allow more accurate analysis of the rate of fertility loss, unless this is a particular objective of the survey. The same argument holds in measuring total farm area on the sample units. Measurement greatly increases the respondent burden, and in areas with fragmented holdings or usufructory rights it may be an untenable concept for the farmer. Approximations give adequate answers for the use of this information in our planning application.

## DATA ON FOOD SUPPLY PATTERNS

The food economy of traditional agricultural systems is the central manifestation of farmers' nonmarket priorities. The results of decisions taken will be highlighted by relating labor and crops under the homogeneous capital structure in the main analysis. The complementary side is the pattern of food flow over and between seasons sought by farmers from their resource allocation decisions.

It has been stressed that innovations which require resource reallocation and jeopardize the satisfaction of nonmarket priorities will be relatively unacceptable, and such reallocations must be counted a cost in the planning sequence. Equally, changes which improve the satisfaction of nonmarket priorities are likely to be more acceptable and, given the retail food outlets, this includes the possibility of increased cash income allowing the purchase of preferred foods as an alternative to subsistence production. Similarly, extension effectiveness will be enhanced by the ability to discuss changes in terms of farmer needs. For example, a late-flowering variety which prolongs the supply of a preferred food may have infinitely more appeal than higher yields during a period when the food is plentiful.

A good deal of information on the food supply pattern is gleaned from output data. The quantities of foods produced on the farm reflect their relative importance to the family. In our application, the recording of output data is also important in measuring resource productivity and is considered under its own category heading. The interest here concentrates on recording the pattern of availability both over the season and between seasons, together with the preferences of the community and the combinations in which foods are taken. The pattern of availability over the season highlights times of the year when staples may be scarce and opens the way to investigating both the insurance measures adopted by the farmers against seasonal contingencies and the way in which scarcity may affect labor capacity and thus the following season's production. Our interest in preferences is limited to general preferences within the community. Individual taste preferences may vary from family to family; and both individual and community preferences are confounded by the dual roles played by substitute crops, which may be useful as insurance crops or may have a labor demand complementary to that of preferred foods. One farmer may prefer cowpea as a relish and supplement it by groundnuts as a substitute because of an inverse relationship in their proneness to local diseases, cowpeas being susceptible in wet weather and groundnuts in dry. A second farmer, preferring groundnuts, may supplement them with cowpeas for the same reason. Our inquiry aims only at distinguishing general preferences in the community, in order to allow some weighting of foods which would be sacrificed by alternative changes.

The identification of type of farming areas stressed homogeneity in cropping pattern and social tradition, guaranteeing close similarity in food supply patterns throughout the area. Because of the fundamental importance of food supply to the farm family (A. T. Richards expresses the belief that African farmers discuss their environments in terms of their associated dietetic variations)[9] and the rhythm of the seasons, a food availability pattern is fairly easily built up secondhand by local interview. It is important that the interviewees be practicing farm householders, for the pattern will be distorted by persons isolated from the community by education or position and tending to rely on purchased foods.

The types of food grown can be classed in two groups. The staples, which form the bulk of any meal, will be starch-based, either grains or roots. The relishes—legumes, concurbits, vegetables, fruit, and fish and meat—are used to flavor the staples. Our interest is in home-grown foods; and three types—grain starch staples, root starch staples, and relishes—will be referred to in discussing the category.

The first phase in presurvey investigation is to elicit from the respondent menu data for preharvest and postharvest periods. Initially the emphasis should be placed on listing what foods, combined into what dishes, the people of the community prefer to eat in each of these periods. Distinction should be drawn between those eaten fresh from the field (green maize, and many legumes and leaves fall into this category) and those stored for future use or processing. With the preferred foods for these two periods grouped into the three type classes, discussion should range around possible substitutes for each major food in the event of a scarcity.

The second phase is the construction of a food calendar which demarcates the period in the year when each food is available. Again, it is of interest to distinguish types by the form in which they are used: fresh, dry or processed. Table 31 shows the periods of availability for the main foods of the Bemba tribe of Zambia, both home-grown and gathered.

In building up this type of calendar it is useful to follow a sequence beginning with periods of availability of foods indicated as preferred, following with the substitutes, and finishing with supplementary foods which may be gathered or purchased as well as grown as reserve crops to fill the periods of scarcity. The period of availability will vary with the state in which the food is consumed: when fresh, it will be governed by natural conditions; when dry or processed, by the resources required to produce it. The first two phases will have identified gaps in the food supply pattern, and probing these will elicit the types and sources of supplementary foods eaten when others are scarce. Where the basic preferred staples are highly seasonal

TABLE 31

Seasonal Changes in Bemba Food Supply

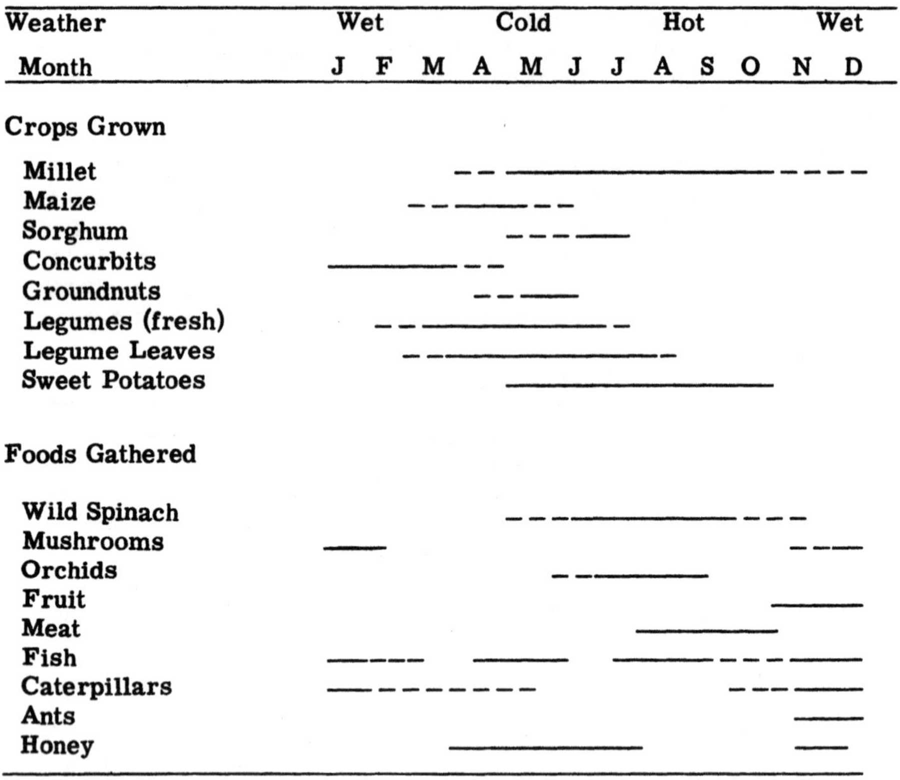


Note: Available__________ Scarce- - - - -

Source: A. T. Richards, Land, Labour and Diet in Northern Rhodesia (Oxford: Oxford University Press, 1939).

or present acute resource allocation or storage problems, supplementary foods, in the form of reserve crops, may be grown within the system. In some areas the basic staples will offer extended planting opportunities which may themselves be a solution to either resource allocation or storage problems.

The gaps in the calendar also provide the key to an estimate of the farmer's expectations of the reliability of the supply of the main foods in his system. A bad season for a particular crop will either cause failure or shorten the period it is available to the family. Not only is an idea of the frequency of this important in estimating the level of reliability required from crop innovations, but shortages may have repercussions throughout the system. Less food, or expediency foods of lower nutritive value, may reduce labor capacity. Where supplementary foods must be gathered, or reserves dug from the ground, labor demand will be increased. The contingency use of labor to satisfy immediate food needs may hinder the development of next season's cropping pattern. Where hungry gaps extend into seasonal peaks the repercussions of a bad season may take several years to work out of the system.

Investigation of the reliability of supply is more complex. There are two steps in the presurvey sequence: first, to record what types of contingency action will be taken in the event of low output of important staple crops, and second, to trace their incidence over the last few seasons. This second step can be repeated at the farm level during the survey.

Two types of decisions will be influenced by the results of a particular season. We outlined earlier how they may modify the production decisions of the farmer for the following period. The decisions on domestic use of food will normally be taken by the women in consultation with the farmer, particularly when contingency measures involving labor use are required. It will be apparent well before the current harvest if food scarcity is likely. Measures to stretch supplies may begin more than a year in advance of the next major harvest: these may include fewer meals each day, a reduced amount, the balancing of preferred with expedient substitutes, and pressing the family into foraging wild sources well before the stores are finished. Measures may differ for each main staple because the availability and sources of substitutes will be different. Each measure will be associated with a degree of failure in the crop, reflected in the reduced period of its availability.

Once the contingency measures associated with each crop have been enumerated, they are ranked according to which will be used first and therefore most frequently. From this ranking the respondent is asked to remember the year in which the community as a whole used the contingency measure resorted to only under most severe conditions. From this extreme, the frequencies of the use of less severe contingency measures can be established by ranging over the intervening seasons. This can be repeated for a series of important staples.

The rankings and frequencies derived from discussion in interview may be investigated more thoroughly by survey. Questionnaires are drawn up on the basis of the severity rankings found. An example would be green crops used as a relish with the main starch staple. The preferred and usual greens are cowpeas, but in seasons with a below-average cowpea crop, cassava leaves are used in September as a substitute to stretch stocks until the new harvest in November. In a poor year cowpea stocks might be exhausted in July, with no fresh cassava leaves available until September, so that wild spinach would have to be gathered as a supplementary relish in July and August. This sequence can be set up as a tabular survey questionnaire and approached by asking for the worst recent year for the crop concerned and enumerating the measures adopted on the farm for that year. Table 32 shows the format.

This type of questionnaire can be repeated for several main staples. Answers over the sample would be influenced by management and microclimatic variations. The typical pattern will be that found on most farms. The assumed ranking of the measures, derived in the presurvey stage, can be confirmed by ranking the severity of the seasons covered independently, elsewhere in the questionnaire.

The memory performance required to recall this type of information will vary with the seriousness of the contingencies. It is the serious contingencies with long-lasting consequences for the system which changes should seek to meet. At the same time,

TABLE 32

Questionnaire to Show the Incidence of Contingency Food Supply Practices in Recent Seasons

| Which Years Did You . . . | 1970 | 1969 | 1968 | 1967 | 1966 |
|---|---|---|---|---|---|
| Gather Wild Spinach in August | | x | | | |
| Eat Fresh Cassava Leaves in September | | x | x | | x |
| Eat Cowpeas Through Until the New Harvest | x | | | x | |
| Sell Cowpeas for Cash After Harvest | | | | x | |

Source: Compiled by the author.

innovations should avoid upsetting the existing mechanisms for meeting such contingencies.

The category gives limited survey content in confirming the menu data and enumerating the incidence of contingency measures to eke out failures in the supply of staple foods. All the data are memory-dependent and relates to past production periods. Timeliness in visiting during the current period is unimportant and, since there are no objective measurement techniques used, the category imposes no limitations on survey design. The construction of the detailed food calendar may bring to light productive activities with a particular objective which may justify definition of a subpopulation in addition to those identified in the description of crops and crop associations in the agronomy category. Special attention to long flows of staples in the light of staggered planting practice and to flows of foods consumed in a fresh state, may aid this. Thus presurvey enumeration in this category will allow construction of an effective questionnaire and may also influence the sampling scheme.

## LIMITATIONS OF SOCIAL TRADITIONS ON RESOURCE ALLOCATION

Social custom within traditional communities may exert an influence on the availability and use of resources and the distribution of crop proceeds. In cattle-herding communities, tradition may often inhibit the mobilizing of what might be a significant source of productive capital. Innovations which breach these traditions may be acceptable to both farmers and the community. This category identifies some of the more general customs and evaluates the type of limitation they impose, to ensure they can be incorporated as constraints in the planning model.

Some social customs safeguarding survival may be defunct when the rule of law prevails at the national level, as in the modern state; nevertheless, these should be accepted as planning constraints where they continue to pervade the community. Evaluating the current usefulness of social customs in fulfilling farmer motivations is a task beyond our investigation, though we can contribute by highlighting those areas where traditional social organization inhibits agricultural improvement. A second important objective for our application is in defining the foci for decisions on production activities within the household. This allows extension efforts to be directed to those individuals capable of making the changes, should they be convinced, and identifies family members appropriate as respondents for the survey.

Social traditions, especially in communities where they serve as a body of law, are two-way relationships. The obligation of the individual is paralleled by the community responsibility to him. In traditional tribal society this type of relationship often has two levels: between household and community, and between the head of a household and its members. These two levels are discussed in the context of control and use of land and labor, and the mobilization of wealth, with particular reference to livestock.

At the community level control is usually exerted by village elders, and often final authority is vested in the local representative of the tribal chief, supported by a council. With recent political developments, this structure has often been formalized and sometimes modified, with power diverted to the central government. The lower levels in the hierarchy have typically been divorced from the chief as a power source and married to the national administrative machine. Apart from an increase in political content, the machinery for administration at these lower levels has often changed little, the cost of reform being beyond the national pocket. It remains the responsibility of the local headman, perhaps retitled as a divisional executive officer, to deal with land allocation and tenurial disputes, and to arbitrate within the local community. Household structure is extremely varied; authority for decision making will usually rest with the senior male, but the degree of delegation of decision making will differ with the size and composition of the extended family group. Questions for community and household levels are covered under the headings of land, labor, and livestock.

## Land

### Community and Household

The agronomy category will have provided a description of land acquisition practice and the degree of control excert by the community. Some elaboration is needed here to establish the possibility of the family unit increasing the area at its disposal where innovations allow an increase in scale. Discussion with individuals in the area and examination of local records of land transactions should give an impression of the flexibility in land supply.

### Household and Member

Within the household it is important to determine how far production decisions are delegated by the senior member. Two patterns are common: Either the head of the household controls all lands and decides the pattern of cropping, both for the season and for interseasonal shifts, or each individual controls his own plots and contributes to the communal food supply. A frequent intermediate pattern is of communal plots on which all members work under the authority of the head of the household, and individuals own plots where they grow what they choose.

The pattern prevailing is readily ascertained by interviewing local individuals. The degree of delegation in decision making is important in organizing the field survey, though there is no survey content. Where decision making is widely delegated to individual family members, it may increase the number of respondents the enumerator must deal with on the single farm unit, thus reducing his coverage and increasing survey costs.

## Labor

### Community and Household

The effects of social custom on labor supply, as the limiting resource in peasant agriculture, are particularly important. By and large, however, it is unusual to find custom a serious drain on labor at peak periods in the season. The community is only too aware of the urgency of bottlenecks to jeopardize achievement of the main objective of community and individual alike—an assured supply of foods—by diverting efforts to more general work. Reciprocal obligations may involve community members in helping individual households afflicted with illness at these periods, but most of the community requirements will be timed to complement the demand for agricultural labor. Some customs, such as the observation of funeral rights, will fall at random, while others may be regular throughout the year. Newer obligations, within the self-help philosophy of many African governments, may be superimposed on traditional custom but organized through the established channels for community effort. Local interviews can outline the likely importance of community commitments in reducing farm labor supply. Confirmation in the survey can be restricted to circumstances where commitments appear important enough to reduce farm activity.

When confirmation is required from the survey, direct questioning will be adequate; communal labor efforts are important events and should be corroborated by a number of adjacent farmers. A framework of last season's efforts can be constructed by interview with local officials, prior to the survey, to give a basis for questioning.

Household and Member

The most important customary influence on labor use and availability on the farm is the specialization of function by sex or by age group. Particular crops or operations may be the specific responsibility of a certain sex/age group in the family. With this situation the general constraints of labor availability may be complicated by functional rigidities inhibiting full participation by the family labor force. The pattern may be related to the decision pattern for land allocation where this is to individuals, although it may also be found in systems where labor is wholly communally organized.

Indications of crop or operational specialization by sex or other group within the family can be obtained by presurvey interviews. Where there is strong functional rigidity, the outline obtained by interview is useful in highlighting the crops or operational sequences to be followed up in the data analysis. No special survey enumeration is needed, since secondary analysis of labor input data classifies the participation of all the main sex/age groups in each activity. This confirms the presurvey information and gives a basis on which to quantify the distinctions for planning purposes.

## Livestock

Among cattle-herding people, livestock have a social role which justifies the inclusion of cattle with land and labor in this category. This social content is not common to all cattle-keeping communities; cattle kept solely as a source of manure will be enumerated in the agronomy category as a means of fertility maintenance, and the commercial aspects of livestock ownership will be covered in the special livestock category. However, among people with a herding tradition, a diversity of functions for their animals inhibits a purely commercial approach to the use of the capital they represent. This section outlines some of the main inhibiting factors and the process of investigation in determining these.

The social ramifications of cattle keeping may be tremendously complex, and investigation should be focused on those facets which are important to the success of available technical or marketing improvements. Effort should be made to establish what social roles would have to be altered if the type of improvement proposed were to be accepted. Consideration here is limited to two aspects. First, at the community level, there is the traditional practice of using community land for cattle keeping. This is particularly important because communal grazing presents a barrier to improvements which often involve sanitary measures (including isolation from other stock), breeding improvement (requiring separate herding of male and female), and grazing management. Second, at the household level, the use of livestock as a family wealth fund spanning generations creates obligations backward and forward in time, inhibiting the exploitation of the herd as a commercial enterprise or as a source of investment capital.

## Community and Household

Two aspects of livestock management—grazing rights and, arising out of this, herding practice—follow from traditional community authority over land allocation and require description to create a context for the development of livestock as an enterprise. Grazing rights will be communal or individual or a combination of the two. Communal rights create problems inhibiting change and imply access to all grass for all animals, with dangers of cross-contamination and hence a reduction in the incentive for improved animals. Communal rights may or may not extend to land which is occupied but standing fallow and even to crop residues, once the harvest is taken. The threat of animals wandering over the arable land has implications beyond the livestock enterprise itself, inhibiting permanent improvements for soil or water conservation. Individual grazing rights offer fewer obstacles to improvement. They imply restricted access, which may or may not be ensured by fencing. On the other hand, they create problems of access to water and in land allocation to farmers with stock and those without. Grazing land being the poorest, market forces reach it latest; and individual rights to grazing areas are likely to arrive late in the evolution of tenurial practice.

Within a communal pattern of grazing rights there will be a tradition of grazing management. This may be enforced by seasonal availability of grass, as when cattle are moved to valley bottom grass only in the dry season, or by the increase in stock densities as the cultivated area is extended to feed a larger human population. The complexity and formality of grazing rights will be a further indicator of the pressure on the system from increasing population, and may be useful in combination with the other indicators described in the agronomy category. Practices such as seasonal migration of cattle to wet season pastures in marginal agricultural areas will demonstrate the clash between the value placed on them traditionally and the new forces pressing the system. Similarly, within the context of communal rights will be a tradition of herding practice. Animals from several farms may be run together and herding shared by the families. All these practices offer barriers to the process of improvement.

Household and Members

Among cattle-herding peoples livestock are the family inheritance, a store of wealth which is tapped for traditional purposes, bride price being the best-known. On inheriting animals, the new owner may also inherit contracted obligations; for example, certain animals in the herd may be the property of his brother who is as yet unmarried but relying on the stock to pay for his wife. At the same time he will feel immediate obligations to pledge animals to his own sons for their future security. The pattern of rights over animals making up the family herd may be complicated by two other facets. Cattle as well as crops are susceptible to climatic vagaries and disease, and herds are periodically decimated, with consequences for the balance of rights in the remaining stock. At the same time, natural disasters affecting animals will not necessarily duplicate the pattern of disasters to crops. Thus the herd serves as savings to be tapped in exchange for food when domestic production is inadequate.

Both the community influence on livestock management and the rights in cattle within the household are best explored by interview with individuals from the area concerned. Practice will be uniform through the area, though where sedentary agriculture is established, not all the community will be stock owners. Some survey content on livestock will be required to cover aspects described in the livestock category, and pointers from the interviews may aid in survey organization. An example would be where grazing migration occurred within the season and so would cause miscounts of cattle unless specifically covered by direct questioning. Survey content should be minimized; in areas where cattle are the basis for taxation, questioning is likely to lead to considerable bias in response and loss of goodwill. Survey questioning of farmers should be limited to those aspects affecting the introduction of available and apparently viable innovations important to the long-term development of the area. Where, for example, the pattern of rights is complex and fluid because of periodic reductions in herd size, respondent burden may be very heavy indeed. The whole issue may be very delicate even within the household itself.

The general nature of the attributes outlined in these three categories allows investigation on a general level. Major sources of variation are isolated by the criteria used in defining the type of farming area, and formal sampling is unnecessary. Interviewees need careful selection. Although the individual farmer will operate by the practices and regimes being described, he will be unable to articulate the why and wherefore of many of them, since they represent an inheritance queried by members only when pressures for change accumulate within the community. Even prominent local individuals are unlikely to have the objective insight of an outsider, and stress is laid on structuring the discussions to reveal the manifestations of motivations and inherited practice which can be observed in the system, rather than discussing the concepts as such with the respondent. The outlines given set out why particular aspects are important to the planning of extension strategy and give examples of forms that these aspects might take. Different forms will characterize particular communities, but the category heads retain their relevance to the planning sequence.

There is little special survey content required for these three categories, since they are limited to confirming the generality of the practices described. Most of the measurement will be picked up in the course of collecting data in other categories; the quantities of food produced and the contribution of different groups within the family to labor use are examples. Through the description of several aspects could be more detailed through frequent-visit collection techniques, it is considered that adequate information on the contents of all three general categories can be obtained over a wide range of techniques. Additional detail does not warrant a limitation on the choice of survey design for the more vital data of land, labor, and output categories. Where these data dictate a frequent-visit collection technique, it may be worthwhile pursuing information in these general categories more closely.

## NOTES

1. E. Baum, "Land Use in the Kilombero," in H. Ruthenburg, ed., Small-holder Agriculture and Development in Tanzania, "Africa Studies," XXIV (Munich: IFO, 1968).

2. Ibid.

3. D. von Rotenhan, Land Use and Animal Husbandry in Sukamaland, "Africa Studies," XI (Munich: IFO, 1966). (In German.)

4. D. Thornton and N. V. Rounce, "Ukara Island and the Agricultural Practices of the Wakara," Tanganyika Notes and Records, I (1936).

5. M. Upton, Agriculture in South-Western Nigeria, "Development Studies," 3 (University of Reading, 1967).

6. H. D. Ludwig, Ukava: A Special Case of Land Use in the Tropics, "Africa Studies," XXII (Munich: IFO, 1967). (In German.)

7. Thornton and Rounce, *op. cit.*

8. R. S. Beck, "Coffee Farming on Kilimanjaro" (Dar es Salaam: Tanzania Dept. of Agriculture, 1963). (Mimeographed.)

9. A. T. Richards, *Land, Labour and Diet in Northern Rhodesia* (Oxford: Oxford University Press, 1939).

CHAPTER

# 10

# THE VARIABLE ATTRIBUTES: LAND

Those attributes influenced by interfarm sources of variation within an identified type of farming area form the main survey content. Measurement of their distributions within the population is the basis for representativeness in the planning model. In particular the relationships between land and labor, the two basic resources of peasant agriculture, are fundamental to simulation of the farming system. As quantitative and variable attributes, the available sampling and measurement techniques for land and labor parameters will play an important part in survey design.

Land is a simple category in the sense that only one group of data is of interest, the acreages of crops grown in the system. It is simple also in the sense that it can be covered by fully objective measurement techniques under a wide range of survey conditions. However, the more complex conditions create organizational problems that restrict survey design to regular and frequent visits over the productive period. In enumerating the sample units a good deal of other survey content will be related to crop areas. The plots form a framework within which the very much more difficult problems of labor and output measurement can be approached.

Although the area is designated as homogeneous on the basis of a common cropping pattern, several sources of variation affect the acreages of each crop grown. Differences of motivation, managerial ability, micro climate, and resource availability will vary the acreages of particular crops between farms. The sample taken must be adequate to show the extent of this variation in the farm population of the area. Motivational differences, preference differences, and particularly interseasonal and microclimatic differences may also vary the constituents of the cropping pattern from farm to farm. This is particularly important when the system covers crops susceptible to rainfall conditions. Rice, for example, may be both labor-intensive and an important source of food grain, but in some areas it may be grown only when rainfall is adequate to flood the fields. It may be necessary to increase the overall sample size to give an adequate sample on crop activities which are important but occur only on a proportion of the holdings. The number of observations of such a crop will also limit the number of observations the sample will give on labor use. Where crops are particularly susceptible to seasonal conditions, it may be useful to establish the history of that crop on each farm for the last two or three seasons, thus giving a better perspective of its importance in the system. Presurvey investigation is particularly useful in indicating the proportion of farmers likely to be following a particular crop activity, and the importance of the crop to those who do, as a basis for a decision on sampling detail and the size of the sample.

## DETERMINATION OF SAMPLE SIZE STANDARDS

Table 33 sets out some levels of precision obtained on crop acreages with limited numbers of observations and the proportions of total sampling units realizing observations on different crops.

If a 10 percent standard error is sought, it can be obtained for the major crops in this area with sixty-seventy observations. Once the distribution of an attribute in the population has been measured, albeit approximately, the mean and standard deviation can be used to estimate the number of observations required to give a desired level of precision. The likely point at issue in designing the survey is which acreages are important enough to justify this level of precision. This can be decided only by the significance of the contribution made to both the resource allocation and the consumption patterns of the community. The sorghum acreage in Maswa and rice acreage in Geita in Table 33 make an interesting comparison.

For sorghum, with twenty-six observations, the standard error was 18 percent of the mean, too high for our requirements. The standard deviation and mean of the observed distribution show that eighty-three observations would be required to meet the precision conditions. Our data show that 42 percent of the sampling units will give an observation on sorghum acreage. To realize a 10 percent standard error on this attribute will require a sample of about 200 units, or three times the size needed for the same precision on the

TABLE 33

Reliability of Some Crop Acreage Statistics, 1962-64

| Crop | 1962 No. Obs. | 1962 Mean | 1962 % Std. Error | 1963 No. Obs. | 1963 Mean | 1963 % Std. Error | 1964 No. Obs. | 1964 Mean | 1964 % Std. Error |
|---|---|---|---|---|---|---|---|---|---|
| Cotton | 82 | 3.23 | 8.36 | 58 | 2.78 | 10.00 | 83 | 4.27 | 7.14 |
| Maize/Legumes, Others | 55 | 1.30 | 8.77 | 45 | 2.54 | 11.00 | 41 | 2.35 | 15.32 |
| Maize/Cassava, Others | 46 | 1.66 | 11.20 | — | — | — | 80 | 3.03 | 9.20 |
| Rice | 63 | 1.14 | 10.53 | — | — | — | 36 | .48 | 14.60 |
| Sorghum | — | — | — | 26 | 1.88 | 18.09 | — | — | — |
| Sweet potato | 76 | .58 | n.a. | 45 | .55 | 16.36 | 21 | .40 | 50.00 |

Sources: M. P. Collinson, "Usmao Area," Farm Economic Survey no. 2 (Dar es Salaam: Tanzania Dept. of Agriculture, 1962); "Maswa Area," Survey no. 3 (1963); "Lwenge Area," Survey no. 4 (1964). (All mimeographed.)

major crop acreages. The position is similar for rice. Thirty-six observations realized a standard error of 14.6 percent; eighty observations are estimated to give a standard error of 10 percent, and with 40 percent of the sample units realizing an observation on rice, again a sample size of 200 units would be required for the desired precision.

Earlier in this section on investigation we indicated the theoretical basis for this sort of calculation, in order to establish the benefits forgone because of distorted planning decisions resulting from a lack of precision in the data. We have also stress this as impractical with multivariate surveys. A subjective assessment would have to be made of the importance of the attributes; some criteria to be considered are discussed for each of these cases and, although similar in variance, have different implications for the respective systems.

### Sorghum in Maswa

Although only 42 percent of farmers grew sorghum as a crop in pure stands in Maswa 66 percent grew it mixed with maize. Planting time, methods of cultivation, and resource requirements are the same for both maize and sorghum in the system. Sorghum's place in the diet is identical with that of maize, and the two crops can be treated as substitutes. Sorghum is now a supplement to maize, although historically it dominated the whole southern area. It remains relatively prominent in this particular area because of the drought insurance it offers under the less reliable rainfall conditions. Sorghum was grown in both the other two survey areas but was observed in mixtures only on 27 percent of the farms in Geita and on 18 percent of farms in Kwimba. Maize supplanted it much more rapidly under better rainfall conditions and higher population density. The risk avoidance-role played by sorghum in Maswa is unimportant in a decision on sample size.

The key fact is its identity with maize in resource requirements and consumption role, so that, in deciding sample size, observations on the two crops can be aggregated. All farms grew either maize or sorghum in some form, and sixty-two observations on total starch grain acreage gave a standard error of 7.8 percent (forty-five observations on maize and legumes gave a standard error of 11.00 and would have required eighty-seven observations to give a 7.8 percent standard error, showing a degree of substitutability between the two crops by reducing the overall variance). Ex ante comparison of the cropping calender, methods of cultivation, and dietary substitutes would reveal the essential similarities. This sort of evaluation creates the basis for a decision on whether to increase sample size to cover ancillary crops.

## Rice in Geita

The same conclusion is reached for rice acreage, though the route to it is very different. Disregarding taste preferences, both maize and rice are substitutes as starch grain staples; their labor requirements are to some extent complementary over the critical cultivation period, allowing a spreading of peak requirements; and they are grown on distinct soil types. At the same time rice is relatively labor-intensive on a acre basis, though similar to maize in terms of pounds of grain produced per man-day of labor. It differs in using off-peak February labor in the labor-intensive operation of transplanting. This critical fact makes it an important complementary enterprise for the supply of starch grain staples and, on facts so far presented, justifies careful investigation.

However, there are other aspects. The season of the survey was particularly favorable for rice in this area, though even so it provided only 16 percent of the starch grain staple to sample farmers, while maize provided 78 percent. Improving the level of precision from a 15 percent to a 10 percent standard error on the rice acreage will not greatly upset these proportions of almost 5 to 1. This detail in analysis is possible only ex post, but in a sequential development of survey design is the kind of pointer to be followed through. More general facets of rice as a contributor to starch grain staple supplies could be established in presurvey work and would confirm its ancillary role. Initial information on the food supply pattern and the associated menu data will allow a rough weighting a main and ancillary crops. In the Geita survey area, data collected on thirty farms in 1964-65 and forty-two farms in 1965-66 revealed a very much lower proportion of food supplied by rice and shows that though rice was grown on 41 percent of farms in 1963-64, it was grown on only 10 percent of farms in 1964-65 and 8 percent in 1965-66. Thus, over the three-year period it is unlikely that rice contributed more than about 6 percent of total starch grain staple requirements. A relatively high level of error will not greatly over- or underestimate the quantitive importance of the crop to the system.

The main food crops will be indicated by the presurvey study of the area. Staples dominant through the season will be determinants of the sample size through their likely incidence in the population. Within a homogeneous system the main crops should be important on about 70 percent of the units in the population. This would allow ninety farms as a safe sample size to cover variation in crop acreages with the required degree of precision. Incidence may be disguised where mixed cropping is practiced and, unless secondary sources fully outline the mixtures found, important constituents would be missed. The Geita survey area offers an example of this with sweet potatoes. Observed growing as a pure stand on only 24 percent of farms, and averaging only .40 acres—less than 5 percent of the total cropped area—sweet potatoes were nevertheless grown, mainly mixed with maize and legumes, by 90 percent of the population and were enumerated on an average of 2.37 acres per farm.

This places considerable importance on the presurvey stage to guide the organization of the collection of crop acreage data at the level of detail needed for adequate representation in planning. Objectivity is clearly possible within this category, because it is memory-dependent only insofar as the farmer must remember to show the enumerator all his fields. Detailed collection techniques, with frequent visits from enumerators who remain in the area the whole season, can easily provide objective measurement. The practical issue is whether there are shortcuts, useful when resources are too limited to give an adequate sample size with a frequent-visit technique. This is our first manifestation of the cost/accuracy compromise which is the central problem in survey design in the field. Peasant farms are small and even objective measurement is relatively rapid, yet it is complicated by three aspects which, in addition to the definition of the productive activities requiring independent treatment, form important presurvey content. These are the patterns of land preparation and planting, the degree of physical fragmentation, and difficulties of physical access to plots which preclude the objective use of some measurement techniques.

## PRESURVEY INVESTIGATION

Presurvey investigation will be by interview locally, and agricultural officials with long service in the area will be the best respondents. Ideally, they would have a local background giving insight into the distinctive activities pursued, which constitute the first aspect to be outlined in the presurvey investigation.

### Predefining Subpopulations

Presurvey questioning under both agronomy and food supply categories will provide important pointers to detailed subpopulations in terms of food supply, insurance, and fertility objectives pursued by specific activities. It is in this category that these indications must be crystallized into a definition of those activities to be treated as independent subpopulations in the sample survey investigation.

For this purpose a sketch map of the typical layout of a local holding is useful to locate the plots of each crop, or crop activity, in relation to the homestead, the topography, and the soils straddled by the farm unit. Most plots will be designated by crop or crop association. Separate plots of the same crop or association should be probed for special objectives—timely supply of a preferred food or insurance against failure of other plots—and for the source of the difference from other plots enumerated, whether a different variety with a different maturity period, or a location in a seepage area or on a different soil type. Clues provided by presurvey enumeration in agronomy and food supply categories should be used to lead the respondent in constructing the sketch map.

For each identifiable activity discussion should cover four other aspects:

1. A rough idea of the proportion of the community engaged in the activity.

2. An idea of the importance of each activity in terms of <u>total</u> subsistence production of that type of food, i.e., starch grains, starch root, or relish.

3. Expected size of the plot bracketed, in three groups: up to half an acre, half an acre to three acres, and over three acres.

4. A note of the usual constituents on an intercropped plot.

From the information obtained the decision on groupings of minor and major activities by objectives must be taken, and from that the decision on the level of precision required for each groupings is made.

Intercropped plots have been a problem in both survey and census work because they confuse straightforward groupings on the basis of common objectives. No wholly satisfactory solution has been found. The initial problem is the identification of different intercrops. The classic intercrop is the starch staple with relishes interplanted, and often other grain or root staples. Each constituent may cover the whole plot and be planted at the same or different times, or it may cover only a part of the plot, effectively marking two plots with different mixtures. Even where the mixture is the same over the whole plot, the relative density of stand of the constituents may alter over the plot. Clearly a line must be drawn somewhere on the detail to be recorded. Two broad approaches have been adopted, one concerned to allocate the area between the constituents and the other concerned to group intercrops on some basis that will allow clarity, as well as brevity, in presentation.

## Allocating the Area

J. E. Bessel, Roberts, and Vanzetti divided the area between constituent crops but gave no basis for the division.[1] A popular but arbitrary basis is the judgement of the enumerator. M. Upton divided the area by relating the density of the population of the constituents to their density in a pure stand.[2] This "creates" an acreage figure for each crop which bears no relationship to the area of the plot cultivated. Upton does not indicate where he obtained the pure stand densities, but in the case of isolated economic trees he recommended densities from experimentation as the denominator. K. E. Hunt notes both alternatives and suggests that each constituent crop may be allocated the total area of the plot (he justifies this by comparing it with successional cropping, where the same land carries two crops in the same season, but in sequence) and that a comparison of pure-stand and mixed yields may be used to derive acreage equivalents.[3]

## Grouping of Mixtures

However, Hunt concludes that the most satisfactory approach is to group mixtures in terms of the principal constituents. M. P. Collinson provides examples of this farming systems where the most important foods are grown in mixtures.[4] The mixtures are maize grown with cassava and maize grown without cassava. The main admixtures in both classes are legumes: groundnuts, bambarra nuts, cowpeas, beans and green grains, sweet potatoes, and other, minor grains, mainly sorghum and bullrush millet. Grouping of this sort follows the idea of P. de Schlippe's field types, i.e., they are based on the major diet staple in various associations, the main subassociations being with cassava and/or legumes.[5] In the Geita survey these two associations covered 45 percent of the crop area, and together with stands of pure crops accounted for 96 percent of the acreage cultivated. Minor intercrops of great variety accounted for only 4 percent of the total area. De Schlippe makes the point that field types are probably the results of a process of simplification and codification over a long series of trials and errors. He adds that in practice "types" are few and are repeated by all members of the group. This initial grouping is essential for our application where other resources must be related to the land area of the plots. In fact, the mixture is dealt with as a composite enterprise.

To show the relative importance of the constituents as foods, Collinson reclassifies acreage on the incidence of the major foods, attributing the whole of the intercropped area to each constituent. This does not achieve the whole objective because it fails to indicate plant density, and even this is not the full story of the part the crop plays in the diet. As a bulk food, a sparsely occurring grain would be less important than a sparsely occurring legume used as a side dish. This secondary classification, supplemented by the consumption schedule, menu, and production data, will indicate the place and importance of each food. The two classifications of mixed field types for the Geita survey are given in Table 34, together with production of the constituents.

Where crops in the mixture are functional substitutes in the diet, the grouping is well justified, despite possible insurance roles. The insurance role cannot be quantified in data collection, and any distinctive labor requirements of quantity or timing will be picked up both in the labor data and in the field operational sequence.

D. Pudsey presents a list of forty-one crops and mixtures which he states is not exhaustive.[6] Not exhaustive perhaps, but certainly exhausting, and certainly obscuring the pattern of the system. The forty-one mixtures can be summarized under nine headings, as in Table 35.

The forty-one classes group well. Nine class groups account for 98 percent and 99.3 percent of the cultivated area in the respect tive rains, with only three real "composite" groups accounting for 13.6 percent and 12.6 percent of the cultivated area. The pattern of the system is brought out well by the summary, beans being the only food complementing rather than supplementing the basic staple of bananas. It is an example of a system geared to consumption needs, and a description of the consumption cycle would provide a firm underpinning to the sequence and pattern of acreages cropped.

It is concluded that any form of acreage equivalent is unsatisfactory for farm economic purposed, and that resource relationships are the basis of the analysis. The use of equivalents distorts land/labor relationships and prevents the comparative evaluation of productivity of what are effectively different enterprises. Grouping, to

## TABLE 34

Classification by Mixture based on Main Staples and Subclassification by Foods

| Crop Mixture | Maize, Cassava, and Legumes or Sw. Pot. | Maize, Legume, and/or Sw. Pot. | Cassava, Legume, or Sw. Pot. | Sorghum or Millet and Legume and/ or Sw. Pot. |
|---|---|---|---|---|
| Percentage Growing | 90 | 46 | 13 | 5 |
| Mean Acreage over Sample | 2.74 | 1.08 | .18 | .12 |

| Food type | Maize | Cassava | Sw. Pot. | Cowpea | Green Grain | Ground-nut | Bean | Bambarra | Sorghum or Millet |
|---|---|---|---|---|---|---|---|---|---|
| Percentage Growing | 97 | 92 | 90 | 62 | 53 | 40 | 30 | 29 | 27 |
| Mean Acre-age over Sample | 3.81 | 2.92 | 2.26 | 2.22 | 2.05 | 1.19 | .67 | .91 | 1.09 |
| Production (lbs) | 1207 | n.a. | n.a. | 40 | 27 | 77 | 16 | 22 | 95 |

Note: Sorghum/millet also appeared in the main intercrops.

Source: M. P. Collinson, "Lwenge Area," Farm Economic Survey no. 4 (Dar es Salaam: Tanzania Dept. of Agriculture, 1964). (Mimeographed.)

TABLE 35

Summary of Forty-one Crops and
Mixtures on an Average Farm Basis

| | Second Rains, 1964 | | First Rains, 1965 | |
|---|---|---|---|---|
| Crop or Mixture | Acres/Farm | % of Cult. Area | Acres/Farm | % of Cult. Area |
| Mature Bananas | 2.24 | 58.8 | 2.25 | 60.0 |
| Coffee (some inter-cropped) | .47 | 12.3 | .51 | 13.6 |
| Sweet Potatoes | .14 | 3.7 | .17 | 4.5 |
| Cassava | .09 | 2.4 | .05 | 1.3 |
| Beans | .10 | 2.6 | .26 | 6.9 |
| Grains | .20 | 5.2 | .01 | .3 |
| Bananas, Beans, Other | .15 | 3.9 | .24 | 6.3 |
| Roots, Beans, Other | .15 | 3.9 | .19 | 5.1 |
| Grains, Beans, Others | .23 | 6.0 | .04 | 1.1 |
| Total | 3.77 | 98.8 | 3.72 | 99.3 |
| Full Cultivated Area | 3.81 | 100.0 | 3.75 | 100.0 |

Note: It is not possible to give frequencies, since the minor intercrops often appear on only one farm, and it is not clear how many are on the same farm.

Source: D. Pudsey, "A Pilot Study of 12 Farms in Toro" (Kampala: Uganda Dept. of Agriculture, 1966) p. 45, table II. (Mimeographed.)

allow treatment of a manageable number of mixtures, is best done where constituents of the different mixtures are perfect substitutes in both dietary function and resource requirements. Once the basic "group types" have been isolated, many other mixtures will be seen as minor variants on these and can validly be grouped with them. Even where labor needs to vary, special requirements of particular group constituents will be isolated in establishing the labor pattern. Their importance will depend on the incidence of this constituent over the sample and the degree of difference in timing and intensity.

The better the framework of the system can be predetermined in this way, the more flexibility is given to survey design for other facets. Although Pudsey stresses the complexity of the system he investigated, the grouping does crystallize the basic pattern over the year and would allow shortcuts in data collection, though admittedly sacrificing considerable detail.

## TIMING OF CROP ACREAGE RECORDING

The second important task in presurvey investigation for the land category is to monitor the pattern of crop establishment, which will dictate the timing of acreage measurement and, in complex cases, the survey design. Little of the literature on census and survey work in agriculture discusses the importance of timing in recording acreage data. The simplest way to illustrate its importance is by the example of a highly seasonal system with all crops planted by the third month and harvested by the seventh. Visits before the end of the third month will miss certain crop acreages; visits at the end of the third month will catch all crops in the ground and measurement can be wholly objective; and a visit at the end of the seventh month, to collect acreage and output data together, must rely on the farmer's memory and the residues to identify particular field boundaries and their constituent crops and associations.

A highly seasonal cycle is characteristic of much of traditional agriculture, but the pattern of crop establishment is complicated by several factors: staggering the planting of crops, the need for greater intensity because of pressure on land, and the wide opportunities for planting in areas of more equable rainfall regimes. Complexity is increased either by a changed cycle length—the Ukara example shows a three-year crop sequence on the drylands, with multiple cropping and green manuring, fully utilizing available rainfall[7]—or by the overlapping of phases within the cycle. This may be double-cropping, where the same land is used, or widely differentiated planting times, where different land but the same labor is used. Survey timing in general, and crop-acreage measurement in particular, is dependent on predefinition of cycle length and the complexity of phasing within the cycle. Data collection must cover the full cycle, for important elements of consumption and/or resource use may be completely neglected by adhering to an annual season. Where double-cropping is practiced with fluid plot boundaries, single-visit surveys are ruled out and each phase of crop establishment must be covered while the plot pattern can be identified on the ground. Where crop rotation is established on fixed plots, the area measured for the second crop can be validly applied to the first. Phases are usually definable, since preparation and planting are geared to rainfall incidence, though farmers may compromise to avoid risk or to spread labor commitments. Where phases remain definable, limited visits will allow crop acreage measurement, the number of visits depending on the number of phases, although usually more than two are not feasible. Where preparation is continuous over the season, the acreage pattern evolves through the cycle, depending on the sequence of preparation and maturity of the constituent plots.

Collection under these conditions can be only by detailed recording.. It demands exceptional enumerator awareness, particularly where the same crops may be established on different plots in different phases of the cycle. It is a common failing in detailed collection techniques wrongly to record labor use against the plot of a crop which has been established for some time, because the enumerator has not checked regularly on new plots being established. Permanent crops pose less of a problem in this respect because new plots or extensions are more easily identified by the nature of the operations being performed. The plot history of biennials also requires very careful enumeration to ascertain the relation of plots in various phases to the cycle of the system. Farming systems with distinctive soil types or locational conditions may be composites of subsystems with their own phased cycles. In such cases, placing the independent cycles in juxtaposition will guide the timing of acreage recording.

As with the definition of subpopulations, the presurvey information from an enumerator of crop calendars and operational sequences in the agronomy category will help and evaluation of the possible timing of acreage measurement. The most important point is that on systems with continuous establishments of crops over the year or with overlapping seasons when a second crop follows a first on the same land), acreage measurements are difficult to record in a single visit. Under conditions of continuous establishment regular visiting is necessary, and this is thus a limitation on collection techniques and a restriction on survey design.

The final two aspects to be investigated prior to survey are both concerned with physical difficulties in enumeration: the first, the coverage of fragmented holdings and the second, access to plots of crops with high or dense growth characteristics.

## Fragmentation

Just as phasing over time creates survey problems, so does phasing over space. Fragmentation is a common feature of the structure of agriculture in areas where traditional inheritance patterns operate but population pressure has reduced the freely available land. In areas where farmers must go to the periphery of the settled area to supplement their holding, distances may be significant. Similarly, in areas where soil types are suitable for different crops but are widely separated, there may be fragmentation. As with phasing over time, the pattern should be defined in the presurvey investigation and the approximate distances to furthermost plots marked on the "typical" sketch map. The pattern forms the basis for an evaluation of the increase in cost of enumerators covering distant plots and of possible alternatives in collection. It also forms a benchmark for enumerators to question farmers who tend to underreport the number of plots on their farms. This is particularly important for limited-visit surveys. In regular-visit surveys the enumerator will pick up missing plots in the detailed collection of labor use data. Nevertheless it is important here, since detailed labor recording should begin as soon as work on a new plot is started, if it is to be attributed correctly. Other patterns established by presurvey investigation—dietary components and usual cropping pattern—also give a base for checking underreporting on particular farms. On the whole, underreporting of plots is a more common phenomenon in nationwide censuses, where enumerators are very temporary and local preparations minimal.

Table 36 illustrates how underreporting declines as the local involvement of the enumerator increases.

The average size of holding increases in all areas as data are collected by enumerators with greater levels of local involvement. Moreover, the examples above are from census-type surveys, and in farm economic surveys the enumerators are inevitably much more closely involved with the range of data to be collected. Nevertheless, fragmentation has significant implications for enumerator capacity in both detailed and limited-visit surveys.

TABLE 36

Underreporting and Enumerator Experience of Local Conditions in India

| | Survey I | Survey II | Survey III |
|---|---|---|---|
| North | 3.9 | 5.3 | 6.7 |
| Northwest | 11.9 | 12.6 | 20.6 |
| East | 3.5 | 4.5 | 4.8 |
| Central | 10.9 | 12.2 | 14.6 |
| West | 11.1 | 12.3 | 14.9 |
| South | 3.8 | 4.5 | 6.2 |
| All India | 6.1 | 7.5 | 8.9 |

Note: All figures refer to acreage.

Source: S. S. Zarkovich, The Quality of Sample Statistics (Rome: FAO, 1964).

## Access to Plots

Some methods of measurement are difficult with tall crops, or where the plot boundaries are obscured. In general, for most methods, an overview of the plot is desirable. Presurvey investigation should enumerate crops with a high growth habit and, for annuals, the time of the season they reach three-four feet high. Also, where crops are grown continuously with those of neighbors, as is often the case with seedbeds or crops located under special local conditions of soil and water, plot boundaries may be obscured.

Preenumeration of all four of these aspects will allow more informed decisions on the organization and design of acreage recording for the survey proper. In addition, the definition of the range of subpopulations to be sampled will be central to the success of investigation and planning.

## SURVEY ORGANIZATION AND METHODS OF ACREAGE RECORDING

Recording on the sampling unit benefits from the initial construction of a sketch map which roughly locates all the farmers' plots in relation to the houses and indicates the approximate distance to each. In the case of a regular visit technique, the sketch map will gradually be built up during the season, plots being entered and labeled as they are opened.

Pudsey advocates the use of a special enumerator for measurement because enumerators are prone to overlook new plots which contain the same crops as others already established, or to question the farmer in terms of plots already recorded, which leaves the initiative with the farmer to prompt him on additional plots.[8] The special enumerator covered eighty farms once every two months. As Pudsey points out, this is shorter than the growing period of beans, the fastest-maturing crops, thus avoiding any overlapping on fields. The special enumerator inspects the fields for divisions or extensions on plots, and he draws the attention of the regular enumerator to his findings. Both Pudsey and J. E. Bessell adopted this technique because their regular enumerators did not appear to be able to comprehend the measurement of areas, though they apparently were able to cope with the complexities of labor and output data collection.[9] If their regular enumerators had been specifically charged with visiting the farm fields periodically for their own inspection, the need for a special enumerator may have been avoided. It is a moot point, and would increase enumerator requirements and organizational complexity, as well as tend to supplant the work of the supervisor. Where farms are widely fragmented, however, and the enumerators must stay around the households to achieve reasonable coverage for the other data, a special acreage enumeration team of one or more may be useful to ensure that fields under cultivation are visited regularly.

With a limited-visit technique, predefinition of an expected pattern and an initial sketch map for each farm are most important, since measurement may be once and for all. The sketch map can be used to minimize the traveling over the holding in the course of measuring the acreages. Where fragmentation is general, usually as a result of inheritance, the system as a whole is likely to be too complex for limited visits. Where fragmentation is specific, i.e., one particular crop is grown away from the homestead—because of locational advantages near the river for example—this substantially disrupts the work rate of a limited-visit enumerator. He must persuade the farmer to undertake the journey, and this may be difficult. However, the necessity of this action depends on the importance of the crop. For a crop supplementary to the main pattern of the system, estimation of the area by the farmer may be adequate. There are mixed views of the ability of small farmers to estimate areas, but the problem is often misconceived because questions are frequently posed in a form which demands that the farmer adjust his usual terms of reference to those of the survey designer. It is unlikely that the problem is one of the estimation of area, which most farmers can do; it is more likely one of enumerating the local units of measure used. K. E. Hunt supports this view but complains that standards are rarely uniform over a wide area.[10] Table 37 shows the results of acreages estimated in terms of _sabenas_, the local measure of seventy by seventy paces, compared with acreage measured by the enumerators.

Of these five comparisons the differences between cassava, rice, and sweet potatoes were significant at the 5 percent level, but there is no consistent bias. Two crops show higher estimates and three lower.

For fields both estimated _and_ measured, the data were broken down into size groups, which are given in Table 38.

There were too few observations in some classes to allow statistical analysis. The general impression is that the estimates are done to the nearest half acre above the measurement. The very small plots are overestimated at .50 acre regardless of size. Above .75 acres the error is inconsistent, with five classes overestimated and four underestimated.

Acreages of ten farms which were estimated because of difficult working conditions in exceptionally heavy rain were then compared to the measured acreages on the bulk of the sample. The pattern of crops on the farms estimated is similar to that on the previously measured farms, and there are no significant differences between the measured and estimated acreages.

These tests suggest that when small plots are important in the farming system, i.e., where they are of high value or highly labor-intensive, or where fragmentation is so extensive that small plots predominate, estimates are no substitute for measurement. When plot sizes are large—over one acre—then if the pressure of fieldwork dictates, estimates can be used to supplement measurement either to save an observation otherwise lost or to keep costs down. This is particularly useful where an area is suitable for a limited-visit survey

TABLE 37

Comparison of Mean Crop Acreage from Measurement and Estimate

| Crop | Cotton | Cassava | Rice | Sweet Potato | Maize Mixture |
|---|---|---|---|---|---|
| Estimated | 1.94 | 1.17 | .94 | .70 | 1.21 |
| Measured | 1.98 | 1.38 | 1.14 | .55 | 1.11 |

Source: M. P. Collinson, "Usmao Area," Farm Economic Survey no. 2 (Dar es Salaam: Tanzania Dept. of Agriculture, 1962). (Mimeographed.)

## TABLE 38

Comparison by Size of Field, Both Estimated and Measured

| Size Group | Below .25 | .26 - .50 | .51 - .75 | .76 - 1.00 | 1.01 - 1.25 | 1.26 - 1.50 | 1.51 - 1.75 | 1.76 - 2.00 | 2.01 - 2.25 | 2.26 - 2.50 | 2.51 - 3.00 | 3.01 - 3.50 | 3.51 - 4.00 |
|---|---|---|---|---|---|---|---|---|---|---|---|---|---|
| No. Fields | 11 | 18 | 23 | 16 | 8 | 9 | 7 | 11 | 7 | 4 | 7 | 6 | 7 |
| Estimated | .40 | .62 | .71 | 1.10 | 1.00 | 1.64 | 1.43 | 1.95 | 2.24 | 2.00 | 2.22 | 4.08 | 3.68 |
| Measured | .15 | .39 | .60 | .93 | 1.10 | 1.38 | 1.65 | 1.90 | 2.15 | 2.34 | 2.68 | 3.21 | 3.68 |

Note: Observations are classed on measured area.

Source: M. P. Collinson, "Lwenge Area," Farm Economic Survey no. 4 (Dar es Salaam: Tanzania Dept. of Agriculture, 1964). (Mimeographed.)

## TABLE 39

### Significance of Differences Between Estimated and Measured Crop Acreages

| Crop | | Maize | Groundnut | Sorghum | Cotton | Sweet Potato |
|---|---|---|---|---|---|---|
| Estimated | Number | 10 | 8 | 6 | 9 | 9 |
| | Mean | 3.11 | 1.77 | 1.83 | 2.03 | .63 |
| Measured | Number | 51 | 39 | 34 | 46 | 39 |
| | Mean | 3.11 | 1.60 | 2.33 | 2.98 | .66 |

<u>Note</u>: Means relate to aggregate area of each crop per farm, not to plots.

<u>Source</u>: M. P. Collinson, "Maswa Area," Farm Economic Survey no. 3 (Dar es Salaam: Tanzania Dept. of Agriculture, 1963). (Mimeographed.)

except for soil-type fragmentation on a single crop not central to the system. Estimates of the areas of such plots can provide acceptable data for our farm management application.

## ALTERNATIVE ACREAGE MEASUREMENT TECHNIQUES

Descriptions of the various techniques of acreage measurement are available from a wide range of sources.[11] The particular aim of this section is to discuss the suitability of alternatives in meeting the different conditions for survey design found in traditional agriculture.

Regardless of the sophistication of some techniques, errors in measurement of fields in most traditional systems are inevitable. Irregular plot shapes and haphazard boundaries involve an element of subjective judgment, particularly in compromising a straight line and a crooked plot edge: an eye estimate is required to balance the bits of cultivated land left out with the bits of uncultivated land left in the area measured. Although sophisticated techniques minimize the influence of many sources of error, they are as easily abused as crude ones. Hunt refers to a test on 250 fields in which only 17 percent of the measurements came within ± 5 percent, while 8 percent outside ± 50 percent of the true figures.[12] The key to accurate measurement lies in experienced enumerators with high job morale and effective supervision, which is one reason for recommending a permanent unit for operational surveys of the farm management type.

Another general problem is the definition of crop area to be measured. Shall the canopy area of crops or their base area be taken, and should conservation works around fields or paths be included or excluded? Where a farmer has prepared a field for a crop, but left a part of it unplanted, how shall this area be recorded? The first point to make is that most of these facets can be the subject of clear enumerator instructions that will be consistent throughout the sample. For our application, economic criteria can be used to evaluate each problem. If land is scarce and the spread of canopy shade prevents the the close planting of another crop, the area covered by canopies is dictating the use of the land and therefore is properly attributed to the crop concerned. This might similarly apply to tree crops, where root poach has the same effect. Where land is plentiful, on the other hand, and labor the limiting factor in the system, the decision should be based on the related labor use, usually to the base area of the crop, though probably something beyond the boundary formed by crop trunks or stalks. In cases where a farmer has cultivated a plot and left part of it unplanted, which is a reflection of his management ability, the total plot area should be recorded and yield per acre, for this farmer, will be correspondingly and appropriately depressed. His weeding input and harvesting input will also be low. With labor the limiting factor, the proper picture will emerge, with a large cultivation input related to the large plot and a low input on postplanting operations related to the smaller crop area. If the cultivation period is limiting in the system, there will be a significant reflection on his management ability; but if postplanting operations are limiting, the phenomenon will be less serious. His decision to leave the rest unplanted may even turn out to be good management if alternatives are more productive.

Sketches of the plots are used with all methods of area measurement; a preliminary rough sketch ensures that the enumerator gets an overall view of the plot before measuring and can be used for triangulation and to draft the lines and angles to be measured. The rough sketch will be adequate with some techniques; but a scale drawing will also be made, usually in the office, from the angles and distances recorded if more elaborate details are needed.

Distances and angles are the two general aspects of measuring plot acreages. Angles may be estimated by eye or may be measured by prismatic compass. Eye estimation effectively limits the possible approaches to that of triangulation, where the irregular shape of the field is broken down into triangles which are then measured for distance either by base times perpendicular height or by perimeter formulas. Measuring angles by compass or plane table results either in a traverse around the perimeter of the plot, building up a polygon, or in building a framework for the plot by lining diagonals across the area to the boundaries. The successful eye estimation of angles is probably limited to triangulation by the base times perpendicular height formula, and lack of equipment often enforces adoption of this method of acreage measurement. In this method only a right angle is needed, though it must be located carefully relative to the apex of the triangle imagined on the plot boundary. Triangulation itself requires a clear view across the plot and is practically limited to low crops or to high annuals soon after establishment. This limitation also applies to the construction of the plot by diagonals. Where the plot shape is obscured either by very broken ground or by tall annual or perennial crops, a perimeter traverse and the use of a compass are necessary, but there are still substantial problems. Distances can be measured by pacing, measuring wheel, and fiber tape or steel and chain. Each has its advantages and disadvantages.

## Pacing

Pacing is probably the most controversial method of measurement. Unzalpi survey opted for pacing because the time required for alternatives was "about double (sometimes treble) the time needed to measure by pacing." This is a time advantage. The use of a tape or chain requires two men for the measurement of distance; and where another man must mark the angles when a traverse technique is needed, a team of three is required. Pacing requires one man less than the use of a tape or chain. On the other hand, pacing brings problems in standardizing pace lengths, both between enumerators and over varying ground conditions. Bessel, Roberts, and Vanzetti chose to try to standardize the pacing of enumerators by regular training periods.[13] A more practical alternative is to measure the paces of individual enumerators, under the varying conditions expected to be experienced during the survey, and derive coefficients to convert the pacing of each enumerator into standard units of measure. This involves the enumerator marking on his sketch maps of the plots the condition ruling: uphill, downhill, over ridges, or whatever. Formal sampling methods should be used to derive the coefficients and the level of reliability.

Zarkovich quotes an interesting comparison of pacing with chain and compass undertaken in Uganda, which shows both the possibilities and problems of measurement by pacing.[14] Table 40 sets out the data.

There is no indication of how good the experimental conditions for this comparison were. What is clear is the huge difference in pacing error between the two areas, which is attributed to differences in quality and training of the enumerators. The conclusion must be that pacing can give both useless and useful results, depending on the organization and supervision. Where these are good, it has the advantages of cheapness and the need for only one man to measure.

## Measuring Wheel

The measuring wheel also requires only one man, but like pacing it suffers from the need for correction according to the type of ground being measured. K. E. Hunt goes so far as to question its usefulness on ground "so uneven that the axle of the wheel moves up and down by, say, six inches within six feet of run."[15] Perhaps this is going too far, for corrections over the main types of ground can be systematically derived, as for pacing. Although a certain amount of

TABLE 40

The Accuracy of Pacing as an Acreage Measurement Technique in Two Areas of Uganda

| Crop | Area 1 Chain and Compass | Area 1 First Pacing | Area 1 Error in Pacing (percent) | Area 2 Chain and Compass | Area 2 First Pacing | Area 2 Error in Pacing (percent) |
|---|---|---|---|---|---|---|
| Cotton | 42,978 | 56,769 | 32.1 | 121,715 | 125,517 | 3.1 |
| Coffee | 65,198 | 97,535 | 49.6 | 80,422 | 77,580 | -3.5 |
| Matoke | 70,652 | 86,543 | 22.5 | 127,541 | 131,345 | 3.0 |
| Sweet Potatoes | 5,368 | 7,321 | 36.4 | 12,384 | 12,488 | .8 |
| Cotton, Beans | 6,693 | 9,791 | 46.3 | 23,566 | 26,443 | 12.2 |
| Cotton, Coffee | 20,658 | 25,025 | 21.1 | 40,825 | 41,067 | .6 |
| Matoke, Coffee | 35,602 | 39,757 | 11.7 | 30,905 | 30,567 | -1.1 |
| Matoke, Cotton | 3,609 | 5,465 | 51.4 | 17,283 | 19,341 | 11.9 |

Note: Measurements are in square yards.

Source: Investigation into the Measurement of Acreage Statistics (Kampala: East African Statistical Unit, 1959).

attention to its path is inevitable to minimize the deviations required by tree stumps etc., an initial bad tendency is for enumerators to watch the wheel and steer it rather than to watch the point they are moving to and merely trail the wheel to let it measure. The rate of work with the wheel and by pacing is high, though dependent on the type of crop being measured.

### Tapes and Chains

Each particular type of tape or chain has its disadvantage in the field. Cloth tapes are susceptible to wet conditions and liable to break at the ends, while steel tapes are liable to twist and be badly obstructed by obstacles lying in the fields. Chains tend to strain and are cumbersome to gather up. The main problem with all these is the need for a two-man team for measurement, plus a greater need for a third man keeping a check on the line being followed. With both pacing and a wheel the individual has little need to watch the ground and can maintain a fixed line. With tapes and chains the attention must be on the points on the ground where lengths join. The rate of work with tapes and chains, even with a team operating, can be very low, again depending on the crop being measured.

Because of the variety in conditions of work, even in the same farming area, there are few detailed accounts of the speed of different methods. Where triangulation is possible, in new crops with easily viewed fields, with either a wheel or pacing, and using a rough sketch map outlining the field and measuring and entering base times perpendicular height dimensions, two or three plots averaging 1.50 acres can be measured per hour by experienced enumerators. Detailed accounts of the time taken to traverse plots are rare. The following is taken from experience of the Central Statistical Bureau, Tanzania, using a plane table and alidade with a measuring wheel and drawing the plot to scale in the field:

> From the checks it appeared that a small field could be measured in an hour and a half and a large field in three hours, but this assumes the measurement to be correct the first time. . . . Should re-measurement be at all common then field teams are unlikely to average more than one field per day as the time getting from base camp to the fields has also to be taken into account.

The actual acreage measurements in the pilot survey described took four months, a team completing thirty-six households, usually with two fields each. The report notes:

> The method of measurement is difficult to perform accurately, particularly in the early stages of the survey when practice is being obtained. This is unfortunately the time when with greatest enthusiasm and smallest growth of plants and bush, the greatest efforts should be made to complete the major part of the work. After three months growth the crops are so high as to make accurate observations extremely difficult.

A major problem in this method is failing to close a traverse when a clear view of the plot boundaries or penetration of the crop is impossible. Reports raise the question of how large a closing error should be allowed; A Moody, using experienced enumerators on permanent plots of coffee and bananas with boundaries obscured by contiguous plots of neighboring farmers, used 3 percent of the circumference length and reported that 14 percent remeasurement was required.[16] K. E. Hunt suggests a closing error of between 5 and 10 percent will not result in more than perhaps 15 percent of the observations requiring rechecking.[17] As noted by the Central Statistical Bureau of Tanzania, which reports some remeasuring four or five times, too stringent a closing error lowers morale and encourages the team to close the final leg of the traverse freehand to avoid a repeat of the work.[18] This, rather than an increase in the accuracy of traverses, is often the most notable result of a few weeks' experience.

Under good organization most methods of distance measurement can give accurate results. Pacing is the most sensitive to poor organization because of the differentials to be controlled, both between enumerators and between types of ground condition. The use of a measuring wheel reduces the source of error to differential ground conditions. Tapes and chains, subject to their own particular idiosyncrasies, are more accurate for the measurement of distances; but from all other points of view pacing or the measuring wheel is preferable, since they allow a single enumerator to concentrate on the line to be followed while measuring. A further team member is needed for this with tapes or chains, since accuracy is reduced by diversions from the boundary being measured. This gives a cost advantage to pacing and the wheel under conditions where triangulation is possible, or where the field to be traversed is not contiguous with others. Three team members will often be needed with tape or chain, two using the tape and a third holding the line. In circumstances where triangulation is easy, with a clear view of plots and low crops, the two men on the measure will hold the line adequately. Similarly, although the example given is an extreme case, with the enumerating team drawing the plots to scale in the field, the time taken by traversing, particularly with tape or chain and compass or plane table, is greater than triangulation or traversing with pacing or the wheel.

Two main aspects affecting survey design emerge from this discussion. One is the question of the team required for measurement, and the other is the time taken. Several researchers have advocated the use of a special team for measurement over and above the regular enumerators; in these cases they were using detailed visit collection techniques. With lower frequencies of visiting and larger overheads in traveling between farms, a special team seems to have both advantages and disadvantages. D. Pudsey found it useful both to prompt the regular enumerators on new fields they had missed and to have stand-ins available who knew the farmers when a regular enumerator had to be withdrawn.[19] The prompting required on new fields might be a self-generating phenomenon, arising from withdrawing from the ordinary enumerator the responsibility to monitor the changing field pattern, since measuring the fields would be his only incentive to get out to the farm. The appointment of supplementary measuring teams might create their own work for them. Hall, using a radial compass to locate plot corners from the two points on a map of the holding, and a wheel for distances, engaged a special individual assisted by the enumerator for the area.[20] He reports two holdings completed in a day by this means, but the method does imply a view not only of the plot but also of the whole holding, which limits its application.

For both types of survey, limited and detailed visit, the use of special teams will increase costs. With limited visit surveys so much information is tied into the measurement of plots that it is imperative that the enumerators themselves do this work. This effectively restricts limited visit work to those conditions where measuring can be done by pacing or measuring wheel and visits can be timed to give easy access to plots, or where boundaries are easily determined because of the isolation of plots. For limited-visit designs it is important that calculations and scale drawings be done in the field, to allow rechecking, by the supervisor at the end of each day's work, if possible, and not on the farm, which would preclude a high rate of work in measurement. These conditions on the whole coincide with the clear seasonality in agriculture and fairly low density settlement. For detailed visits in areas where conditions allow the use of techniques which can be managed by individual enumerators, perhaps aided by household children, there seems little justification for a special team. A team will be required, however, under conditions of dense settlement with contiguous plots, particularly of the same crop, and where penetration through the fields is impractical. In these circumstances field checks should ensure that adequate measurements have been made to calculate the area. Since timeliness is of little importance, scale drawings can be done in the office, preempting any temptation to close the final vector of a traverse to avoid remeasurement; nevertheless, the allowable error before remeasurement should be fixed with an eye on the morale of the enumerators and supervisors.

The category of land use throws up important limitations on survey design. As indicated earlier, the usefulness of limited visit surveys is restricted by a flow of irregular events, in this case crop establishment, over the whole production period. Once other limitations have been synthesized from discussion of the remaining categories, the final chapter of this section will evaluate the circumstances where limited visit techniques are acceptable shortcuts in survey investigation.

## NOTES

1. J. E. Bessel, J. A. Roberts, and N. Vanzetti, "Survey Field Work," Agricultural Labour Productivity Investigation, report no. 1 (Universities of Nottingham and Zambia, 1968); "Some Determinants of Agricultural Labour Productivity," Agricultural Labour Productivity Investigation, report no. 3 (Universities of Nottingham and Zambia, 1970).

2. M. Upton, Agriculture in South-Western Nigeria, "Development Studies," 3 (University of Reading, 1967).

3. K. E. Hunt, Agricultural Statistics for Developing Countries (Oxford: Oxford University Press, 1969).

4. M. P. Collinson, "Bukumbi Area." Farm Economic Survey no. 1 (Dar es Salaam: Tanzania Dept. of Agriculture, 1961). (Mimeographed.)

5. P. De Schlippe, Shifting Cultivation in Africa (Humanities Press, 1956).

6. D. Pudsey, "A Pilot Study of 12 Farms in Toro" (Kampala: Uganda Dept. of Agriculture, 1966.) (Mimeographed.)

7. H. D. Ludwig, Ukara: A Special Case of Land Use in the Tropics, "Africa Studies" (Munich: IFO, 1967). (In German.)

8. Pudsey, *op. cit.*

9. *Ibid.*; Bessel, Roberts, and Vanzetti, *op. cit.*

10. Hunt, *op. cit.*

11. *Ibid.*

12. *Ibid.*

13. Bessel, Roberts, and Vanzetti, *op. cit.*; see also S. S. Zarkovich, *The Quality of Sample Statistics* (Rome: FAO, 1964).

14. *Investigation into the Measurement of Acreage Statistics* (Kampala: East African Statistical Unit, 1959).

15. Hunt, *op. cit.*

16. A. Moody, "A Report on a Farm Economic Survey of Tea Smallholdings in Bukoba District, Tanzania," East African Agricultural Economists Conference paper (1970).

17. Hunt, *op. cit.*

18. Tanzania Central Statistical Bureau, *Kilosa Acreage Survey* (Dar es Salaam, 1963).

19. Pudsey, *op. cit.*

20. M. Hall, "A Review of Farm Management Research in East Africa," East African Agricultural Economists Conference paper (1970).

# CHAPTER 11

# THE VARIABLE ATTRIBUTES: LABOR

The pattern of labor availability and use over the season is the key to understanding traditional African agricultural systems, and its quantification is the major objective of investigation. Both availability and use need to be treated as flows and, any quantities specified are meaningful only at points in time. Labor use presents no conceptual difficulties; availability, on the other hand, requires some initial discussion before alternative investigational techniques are outlined.

## THE CONCEPT AND TREATMENT OF LABOR AVAILABILITY

Planners have tended to overlook labor availability as a flow resource within the traditional community. Usually a theoretical level of availability is assumed as a constant constraint, or observed usage at peak periods has been accepted as a limit throughout the season. Both are inadequate, though the latter is much superior in simulating the existing situation. Where technological changes will reallocate labor and perhaps shift the seasonal peaks, it may be artificial to assume the same level of availability. The routine of the system will clearly be related to the present labor peaks, ensuring that labor will be free to be applied at these periods. Other commitments will be concentrated in relatively slack periods. Reallocation of labor through innovation may create peaks which clash with these other commitments, reducing the potential benefits of the change. It is important for planning that the timing and intensity of factors reducing availability are enumerated, and that labor availability as well as use is set out as a profile over the season. The customary pattern will be flexible over time, but there is no doubt that innovations which demand immediate changes of habit will find that these constitute a formidable obstacle to acceptance.

The family is the core of labor supply to the peasant smallholding. D. Pudsey has shown how different bases for the estimation of availability can give a 300 percent variation in total working hours available.[1] He made the point in terms of total hours available over a fifteen-month period, but the principles hold when the calculation is done on the more significant basis of supply available at any one point of time over the season. He shows the effect on supply of treating three factors as variables: the age at which children should be considered as potential labor, the number of hours to be assumed in a working day, and a full-time commitment in other work or at school.

Other factors are also important, and the following discussion covers the main aspects affecting seasonal labor availability. The point should be stressed initially that observed pack labor usage does reflect a combination of all the factors to be considered at the present peaks in the season. The use of this level as a constraint in planning has considerable justification, for it is the level reached when the community has adapted its system of life to its motivations. Nevertheless such adjustment takes time, and where reallocation disturbs traditional community patterns, the rate of adoption and change will be effectively slowed. The main information needed to quantify potential labor supplies is presented and discussed briefly below.

## Off-Farm Commitments

Outside commitments prevent family members from contributing to agricultural work because they are absent from the farm. The main possibilities are wage employment, school, and seasonal activities such as hunting or fishing which might draw a part of the labor force away from the farm. The aim in enumerating these commitments is to establish whether they are regular every season and, if so, to pinpoint their timing and extent over the season. Where absence covers the whole season, the individual is lost to the labor force, while absence for a part of the season implies a varying labor constraint in any planning model. Incidence varies with the area; R. W. M. Johnson reports that about 45 percent of householders were out of the area at certain times of the season and 23 percent were away in the critical planting period in December.[2] This has far-reaching implications, not only for the labor supply but also for decision making and for the effectiveness of extension. The family left at home is unlikely to be receptive to change without the authority of the householder. Pudsey records off-farm employment at only 7.2 percent of nonfarm activity, implying a very low frequency in the population.[3] In his sample schooling accounted for 29.1 percent of nonfarm activity; the significance of this for labor supply depends on the man-equivalent values of the missing individuals and the coincidence of school holidays and important agricultural periods. In such a case off-farm employment could be ignored in the planning model, but the high level of school attendance will vary the availability profile.

## Nonagricultural Activities

Activities which are unrelated to the farming system may limit an individual's time in work even though he is present and normally working on the farm. Domestic chores for the women are an obvious and important example, and Pudsey records them as occupying 13.7 percent of all nonagricultural activities.[4] The boundary between farm and nonfarm activities is blurred, and neither harvesting foods as required nor crop marketing is wholly agricultural. In general, activities which are cropbound, particularly those which are necessarily timely, are properly shown as operations of the enterprise concerned. Some nonagricultural activities, such as feasts and communal building, can be classed as social commitments and will follow a seasonal pattern which complements agricultural needs. These may be important barriers to increased labor supply at specific times of the year. Others are dependent on incidence in the local community; funerals or family illness requires community work, and will be random and not represent a foreseeable obstacle to labor supply.

## Specialization of Task by Sex or Age Group

An additional facet of family labor is of specialization by sex/age groups on particular crops or particular operations. It will be important to distinguish in planning where labor sources are not perfectly substitutable. Although distinct functions will fit the existing labor profile, reallocations must keep within the boundaries of the specializations to be acceptable.

## Nutritional Constraints on Availability

We noted in Chapter 2 that, among other researchers, A. T. Richards, working among the Bemba, and R. H. Fox, working in Gambia, have observed that the capacity for physical effort is reduced by undernourishment.[5] We saw Fox's particular concern with the balance of consumption and production where he superimposed an energy-intake profile on the energy-expended profile dictated by crop labor needs. He showed a reduction in the length of day worked by the men during the physically arduous cultivation operations in which expenditure exceeds intake. Even where such a clash arises during a slack labor period, enumeration is desirable to improve planning constraints. However, as Fox's work demonstrates, quantification of the phenomenon is complex, and only a descriptive outline can be established in the course of farm economic investigation.

### Use of Hired Labor

The use of hired labor adds to family supply and can be classified into regular, seasonal, and casual. Regular hired labor works on the farm throughout the year and seasonal hired labor works during a particularly demanding period, either very short or prolonged. The characteristic of both these types is a repetitive pattern from season to season, a deliberate increase in the scale of the system. Casual hired labor, on the other hand, is to meet the contingencies of the particular season caused either by weather or by family circumstances. A delayed start of the rains may create pressures to complete plantings, or unduly heavy weeding may be caused by a particular rain sequence. For example, A. Moody has noted a case where, on a research station, 35 percent of available labor time was lost by heavy rain in one important month of the season.[6] Within this definition, hiring of casual labor will not be repeated regularly from season to season and has no special place in planning constraints. It needs identifying in the course of survey enumeration to allow basic rather than contingency relationships between inputs and outputs to be derived, and is one possible indicator of the normality of the season.

Hired labor is an additional resource and is properly added to family availability. But where casual labor is substituting for family labor as a result of contingencies, this should be qualified. Where hiring is seasonal, the timing is important, for although the amount hired may represent only a small fraction of total availability over the year, it may be a significant proportion of the total over the period of hire.

## MAN-EQUIVALENT VALUES AS A BASE FOR INTERFARM COMPARISON OF LABOR CAPACITY

Two variables included in Pudsey's comparison of availability measures were length of working day and the age at which children qualify for the labor force. These are facets of the need for a common denominator to allow the comparison of labor availability and use over families with different sex and age compositions. The range of sex/age combinations active in traditional agriculture is very wide, and the need for a denominator is greater than in advanced agriculture. It is important to note that man-equivalent values are concerned to standardize measures of the rate of work of family members while at work on the farm. The values are irrelevant to the relative amount of time spent at work or in leisure by different groups, and no basis for covering variations in the quality of work by groups has been derived, so that values are useful only to standardize quantities. As with any composite measure, a degree of detail is lost; and there are variations in the efforts of individuals of the same sex and age in the same or different families. Equivalent values are used because their advantages outweigh their disadvantages, particularly in our application, with analysis and planning based on an area model made up of averages.

Various researchers, such as M. Hall, have pointed out that relative values of different sex/age groups will change with the operation.[7] Men may cultivate faster, but harvest or weed more slowly, than women. Some traditional systems in fact acknowledge the superiority of particular sex/age groups; Richards notes that among the Bemba, where only the young men tackle the work of chopping down the high tree branches in building the chitemene fires, many tribes involve only the menfolk in clearing new land.[8] Ignoring changes in these relative values can distort the constraint used in planning. When a woman works half the speed of a man on one operation and twice as fast on another, fixing her equivalent value at .50 on the basis of the first will grossly underestimate family labor capacity on the second.

Clearly, it would be impossibly complex to incorporate constraints which varied according to the timing of operations performed at varying rates by different sex/age groups. The only practical approximation is to establish relative performance on the labor-intensive operations which make an important contribution to the peaks, both existing and expected, in the system.

In communities where the pattern of production responsibilities allocates fields to individuals or sex/age groups in the household, values can be based on an analysis of the rates of work of the discrete groups on the same operation. The problem is most difficult where work is done by the whole family group, under which conditions a work study approach is feasible. Some researchers, such as T. J. Kennedy, have used work study as a means of collecting labor input data.[9] As we shall discuss later, data from small areas are susceptible to a scale effect which exaggerates the rate when multiplied up to a per acre basis. However, this upward bias is unimportant to an evaluation of relative performance.

Work study should be undertaken independently of the survey, under conditions where the sampling can be controlled by the sex/age combinations required. Government stations allow this degree of control, though it is important that the study duplicate the tools and methods of the local traditional system. As with any sampling procedure, the number of observations should be sufficient to show the variation in the population and for significance tests to confirm the differences found between groups. The period of observation should be as long as possible, perhaps a full day's work, and should be the same for each group. The focus of the investigation should be on operations which are shown to be critical to the system and those likely to prove critical, as a result of planning and innovation, for the construction of differential seasonal labor constraints where elements within the family labor force are not wholly substitutable.

The constraints on availability enumerated by the survey can be written into the planning model as variations in seasonal supply. Any social commitments or sex/age distinctions founded on the need for survival, though outdated, will still form effective barriers to change. Enumeration of these factors modifying the apparent labor capacity of the family helps quantify the gaps between use and supply more precisely, reducing speculation on underemployment and target-income motivational patterns.

## INVESTIGATION OF FAMILY LABOR AVAILABILITY AND LABOR HIRE

In a labor-limited system, general measures of labor availability and use are closely related to acreage. In the discussion of land it was shown that between sixty and seventy observations gave a standard error better than 10 percent on individual crop acreages. The same sample size will give greater precision on more general attributes. For example, sixty-two observations gave the following percent standard errors: total acreage cultivated 6.99 percent; family size, 6.04 percent; total available labor, 5.74 percent.[10] Total acreage is inevitably less variable than the component crop acreages. The variation of general labor parameters over the population will be measured within the size of sample required to give adequate precision on the main crop acreages.

Family labor availability requires little presurvey enumeration. Aspects which can usefully be outlined before the survey are the likelihood of off-farm commitments in the area and the major nonagricultural activities, such as water drawing, in the household. The outline of off-farm commitments should cover the usual type of regular employment in the area, the age/sex groups which are concerned, and the period of the year they are away. The outline underscores the need for awareness, in detailed collection, of the timing of changes in household composition which will influence labor availability on the farm. In limited-visit surveys it should be an explicit question in the enumeration of household composition whether all members have been present throughout the season. The family structure and composition is the core of farm labor supply and presents no problems in enumeration. Age may be more difficult. Absolute accuracy is unimportant, and approximations are often possible by relating births to prominent events in the community calendar.

The diagnosis of nutritional constraints on labor capacity is a difficult task and cannot be fully investigated in the course of a farm economic survey. Discussion in presurvey investigation may shed some light on the likelihood of a reduction in labor capacity. Further indication can be obtained in analysis, by relating the lengths of day worked on arduous and lighter operations at critical seasons to the pattern of food supply at these same times. A shortage of staples and a decrease in length of day worked suggests an energy problem. This can be supplemented, in both presurvey and survey questioning, by establishing the incidence of endemic diseases over the year, since resistance will be weakened by nutritional inadequacy.

Hired labor presents a more difficult problem. Since it is an attribute which may be found on a only portion of local farms, a decision is often required on whether or not to include it in the planning model and, if it is included, on the precision with which the level of usage and the cost should be measured. It is an attribute which can be grouped with family labor because its main impact is in increasing the scale of the system—of limited importance in an application, because no change in factor or factor product relationships is implied. For this reason it is included in our planning sequence, because it is an event which occurs on the majority of local farms; but there is no adjustment in sample size to increase the precision of observation.

In order to prepare the appropriate type of questionnaire, presurvey enumeration should probe the type of hired labor use—regular, seasonal, or casual—and its incidence in the population. Where it is seasonal, the time of year and the crops and operations with which it is usually associated should be established. Where it is casual, the most likely reason should be probed. Finally, estimates are needed of its incidence among the farm population, the usual cost, and method of payment (cash or kind).

Regular hired labor presents no problems in survey enumeration for any type of collection technique. Similarily, the hire of seasonal labor is an important enough event to be remembered for questioning in a limited-visit technique. Casual labor is also easily enumerated, with a possible source of error in communal labor parties, when it may be difficult for the farmer to remember the number of people present. Such groups are usually employed for a single day, and errors would be relatively insignificant in terms of total supply over the week or month concerned. Casual labor which is employed over the season irregularly but frequently does present an enumeration problem for limited-visit surveys. It is an unusual circumstance in traditional agriculture proper, since a casual labor pool which can be drawn on as required is an unlikely feature of African rural communities. It is, however, an increasing phenomenon where high-value cash crops, such as tobacco and tea, have been grafted onto the traditional system, and especially where land hunger is beginning to show, urban unemployment is growing, or estate production is declining. In areas where it is preenumerated as an important source of labor, survey design is best based on detailed collection techniques and enumerators need a constant awareness of changes in the working force on the farm.

## LABOR USE

Labor use presents no conceptual difficulties although, as with availability, the importance of the flow of use over the season can stand reemphasis. Total use will be correlated with total output, but the relationship is often secondary in labor-limited systems. Either seasonal requirements for cultivation or weeding limit the acreage which can be managed, or harvesting requirements limit the output which can be handled. The profile over the season, showing the labor supplied to meet the crop and operational requirements of the system, is the central set of data for analysis and planning. Of the four components of this profile—the acreages of crop activities, the operational sequence for each crop, the timing of the sequence, and the rates of work on each operation—the last three are enumerated in this subcategory. In a sense, work is like lunch: as a regular event we are fully aware when it occurs, but the content is variable and particulars about what we ate when are not easily recalled. The central question of survey design in much of traditional African agriculture is whether data in this critical category can be measured by limited-visit surveys in which, in the extreme case, the events to be recorded stretch back over a full season. It is the aim of this section to analyze the sources of variation in rates of work; to establish that, in many circumstances, adequate labor data for planning can be obtained from limited-visit surveys; and to specify the conditions under which detailed techniques are required.

Basically our planning task involves juggling land/labor relationships as innovations are interpolated into the existing system. The timing of operations, and thus the pattern of labor use over the season, are remolded in the planning process. As long as traditional methods are used, the rate of work, as a component of labor use, is unaffected and is thus the most important parameter to be carried from the existing to the revised system. The rates of work on each operation vary over the farm population, and identification of the sources of variation is a first step to their measurement by survey investigation.

## SOURCES OF VARIATION IN LABOR-USE DATA

Two tables are presented as examples of the levels of variation in rates of work observed in some case studies. Table 41 sets out data reocrded by R. W. M. Johnson. The standard errors have been derived by dividing the standard deviations given by Johnson by root _n_. In doing this it has been assumed that each plot gave an observation on each crop.

From the differences in the mean levels of operations, it is clear that ploughing is not common to the three crops grown. The information given is inadequate to say whether the differential in ploughing is due to increased quality of work or a different sequence or method of cultivation for the three crops. The interaction of precision and sample size suggests the operations have similarly shaped distributions. Eighty observations would allow a level of precision better than a 10 percent standard error on all operations on the three crops, though weeding groundnuts is marginal.

Table 42 illustrates a gradual improvement in precision by measuring the number of observations and the improved definition of operations over a period of three years, although the progression is frustrated to some extent by a failure to preenumerate a mixture of flat planting and ridging in the area surveyed in 1963.

TABLE 41

Examples of Precision Achieved in Collecting Labor Inputs
(expressed as rates of work in hours)

| Crop | | Ploughing per Acre | Weeding per Acre | Harvesting per Acre | Manuring per Acre |
|---|---|---|---|---|---|
| Maize (101 obs.) | Mean | 17.9 | 42.5 | 20.8 | 6.4 |
| | Percent Standard Error | 7.2 | 7.1 | 7.8 | 6.9 |
| Groundnuts (85 obs.) | Mean | 52.9 | 98.9 | 129.5 | – |
| | Percent Standard Error | 8.3 | 9.9 | 11.0 | – |
| Millet (51 obs.) | Mean | 32.6 | 94.6 | 55.5 | – |
| | Percent Standard Error | 11.7 | 11.2 | 10.0 | – |

Source: R. W. M. Johnson, "The Labour Economy of the Reserves," Occ. Paper no. 4 (Salisbury: University College of Rhodesia and Nyasaland, 1964).

TABLE 42

Improvement in Precision with Experience in Surveys in Sukumaland Cotton Areas

| Crop | Operation | 1962 No. Obs. | 1962 % S.E. | 1963 Flat No. Obs. | 1963 Flat % S.E. | 1963 Ridged No. Obs. | 1963 Ridged % S.E. | 1964 No. Obs. | 1964 % S.E. |
|---|---|---|---|---|---|---|---|---|---|
| Cotton | Plough/Plant | – | – | 20 | 12.1 | – | – | – | – |
| | Sesa | 30 | 23.1 | – | – | 6 | 21.0 | 59 | 9.1 |
| | Ridge/Plant | 35 | 15.2 | – | – | 6 | 16.6 | 61 | 9.9 |
| | Weed 1/Thin | – | – | 49 | 8.6 | 6 | 21.3 | 61 | 9.3 |
| | Weed 2 | – | – | 48 | 7.4 | 5 | 21.9 | 50 | 13.0 |
| | Weed (AU) | 3.8 | 17.1 | – | – | – | – | – | – |
| | Pick 1 | – | – | 49 | 10.6 | 6 | 38.4 | – | – |
| | 2 | – | – | 47 | 10.9 | 5 | 41.1 | – | – |
| | 3 | – | – | 28 | 15.3 | 2 | – | – | – |
| | Pick (AU) | 38 | 20.5 | – | – | – | – | – | – |
| | Grade | – | – | 47 | 11.6 | 6 | 24.5 | – | – |
| | Uproot | 35 | 19.2 | – | – | – | – | – | – |
| Maize Mixtures | Plough/Plant | – | – | 18 | 18.8 | – | – | – | – |
| | Sesa | 16 | 14.6 | – | – | 17 | 17.3 | 67 | 7.6 |
| | Ridge/Plant | 18 | 24.9 | – | – | 25 | 11.3 | 68 | 8.5 |
| | Weed | 18 | 15.4 | 19 | 22.3 | 25 | 15.2 | 67 | 12.2 |
| | Pick Maize | 18 | 33.0 | 16 | 23.7 | 24 | 17.5 | 28 | 13.6 |
| | Remove Husks | – | – | 11 | 26.9 | 19 | 21.8 | – | – |

Source: M. P. Collinson, "Usmao Area," Farm Economic Survey no. 2 (Dar es Salaam: Tanzania Dept. of Agriculture, 1962); "Maswa Area," Survey no. 3 (1963); "Lwenge Area," Survey no. 4 (1964). (All mimeographed.)

The evidence suggests that a standard error better than 10 percent can be achieved with between sixty and eighty observations on most operations, within a well-defined type of farming area. We have noted that observations on rates of work will be available only from crop activities enumerated among the population. If it is assumed that the main crops, and their associated operational sequences, should feature on 70 percent of the farms of the area, then a sample size of about 115 farms will be required to give eighty observations on the main crop operations. A detailed analysis of sources of variation in rates or work will provide guides for economizing on sample units where resources are limited and also a basis for operational classification, allowing the identification of subpopulations which need separate investigation.

Four groups covering twelve independent variables have been identified as influencing the rate of work as a dependent variable, although there would be strong multi-collinearity in a regression analysis incorporating the set. For example, the rate of work will vary with soil type, but will also vary with rainfall; and an excess of rainfall will influence soil types in different ways. The twelve are grouped for convenience, and each group and its component variables are discussed briefly.

## Crop and Operation

Crop and operation, the first two independent sources of variation, are the basis for classifying work rates in our analysis. Crop is not always a strong influence, particularly on preplanting operations which are often common to several crops. This gives an initial lead to the grouping of operations common to a range of crops as a basis for increasing the number of observations within a given sample size, and the grouping is particularly useful because preplanting operations are often the most intensive in traditional systems with hand equipment.

Many researchers have drawn attention to the dangers of recording multiple operations. Hall uses as an example the pruning and mulching of bananas during weeding; Pudsey, the digging of land while sowing beans; and others include simultaneous weeding and thining or planting new intercrops while weeding the established stand.[11] Where parts of these multiple operations are very minor (this is often true of the planting operation, which almost always immediately follows cultivation to minimize weed competition), description of the sequence is desirable. Where the component parts are significantly labor-intensive or where the sequence may be altered by the introduction of an improved technique, description is essential and every attempt should be made to record the components independently. Presurvey investigation will need to identify the components of each operational sequence.

## Soil Type, Cropping History, and Plot Size

The second group is of three independent variables which form areas of greater detail in definition, which may or may not be required in the planning model.

Soil type has an independent influence on the rate of work. Any operation which involves movement of the soil will take longer in heavier soils. Distinct productive activities by soil type have been identified and a decision will need to be taken on the desirability of their being treated as a subpopulation. The direct influence of soil type on the rate of work will weight this decison. When it is significant, it will add to the need for independent investigation of the activity as a distinct subpopulation.

Cropping history may also influence the speed of work. Certainly preparing a new seedbed out of a fallow area, or opening bush, will require not only more time on the cultivation operations but possibly its own sequence of special operations. On a less extreme level, preparing a seedbed after a crop with heavy root growth may take longer than after one with light root growth. Pudsey, in a very detailed set of labor data collected by plot, has given groupings which illustrate the need for a decision on subpopulations of labor data identified by cropping history.[12] Standard errors have been calculated from the raw plot data in Pudsey's report.

In Pudsey's grouping, shown by the first three columns of figures, small sample sizes have given good levels of precision for stable subpopulations. Other groups show relatively large variation. A comparison of means shows that the only subpopulations which justify separate investigation are opening bush before sweet potatoes and before other crops, and seedbed preparation for sweet potatoes. Grouping across all observations on opening bush seriously distorts the rate of work required for most crops, as would grouping across seedbed preparation for sweet potatoes and other crops. Cultivating out of a field which had grown sweet potatoes shows no more effort required than out of a field which had grown other crops. In fact, for this example, it is mainly cropping intentions which

## TABLE 43

### Effects of Grouping Labor Observations by Cropping History

| Operation | Cropping Sequence | No. Plots | Rate of Work Mean hrs. /acre | Rate of Work % S.E. | No. Plots | Rate of Work Mean hrs. /acre | Rate of Work % S.E. |
|---|---|---|---|---|---|---|---|
| Opening Bush | Before Sweet Potatoes | 22 | 2,098 | 10.6 | 36 | 1,391 | 10.9 |
| | Before Other Crops | 14 | 941 | 14.4 | | | |
| Seedbed Preparation for Sweet Potatoes | After Sweet Potatoes | 21 | 969 | 27.0 | 29 | 978 | 18.0 |
| | After Other Crops | 8 | 1,001 | 12.7 | | | |
| Seedbed Preparation for Other Crops | After Sweet Potatoes | 16 | 444 | 19.9 | 52 | 417 | 9.2 |
| | After Other Crops | 36 | 405 | 6.8 | | | |

Source: D. Pudsey, "Pilot Study of 12 Farms in Toro" (Kampala: Uganda Dept. of Agriculture, 1966). (Mimeographed.)

influence the rate of work, though, as might be expected, opening bush shows a clear distinction from cultivation out of last season's fields. Clearly this type of analysis is wholly dependent on having the data, but nevertheless pointers to this type of distinction can be gathered in the presurvey investigation. The key to effective grouping here would be the knowledge that farmers are concerned to prepare a particularly clean, friable seedbed for sweet potatoes, both to allow the tubers to swell properly and to reduce subsequent weeding, which is inclined to damage crops that spread.

The effect of the size of the plot on the rate of work is a particularly interesting phenomenon and is called here the "scale" effect. It is an "edge effect" similar in result to that associated with bulk harvest cuts, from small plots within a field up to per-acre yields. In this case the overheads classed as work—getting to and from the plot, getting ready to work, getting ready to leave—are just as high for a large as for a small plot. A contributory factor may be that large labor forces work faster together than individuals, or even man and wife, and will tend to work on larger plots. This higher rate of work is probably one of the benefits of community labor efforts which would otherwise be self-canceling. Often a "party spirit" exists in a large group of workers.

When bulked up to per-acre from small plots, the overhead elements distort the rate of work requirement. The effect can be removed for a sample of farms by the use of the weighted mean. Instead of calculating rates of work for each plot and averaging these, the total area covered is divided by the total time spent, giving each plot weight according to its size, large and small extremes balancing out. When variance is high, the difference can be very marked and the weighted mean is more appropriate for planning.

The weighted mean is useful in grouping subpopulations which have not been independently sampled. When it is used, the subpopulations are represented in the averages in the same incidence as they occurred in the population, thus avoiding a distortion of requirements.

It is interesting to carry the analyses on Pudsey's data a stage further, for we have now noted that scale, as well as cropping intentions and history, influences the differences in work rates relating to sweet potatoes and other crops. Pudsey presents raw data for each plot for a limited number of cultivation operations. For four of these single operations the "hours per acre used" has been correlated with the size of plot to show the amount of variation in the rate of input accounted for by scale factors. Data are presented in Table 45.

TABLE 44

Differences Between Mean of Rates, Individual Plots, and Mean of Totals, All Plots and Relation to Plot-Size Variation

| Operation | Mean of Rates | Weighted Mean | Coeff. of Var. of Plot Size (percent) | Mean as Percent of Weighted Mean |
|---|---|---|---|---|
| Seedbed Preparation, Other Crops | 405 | 354 | 73 | 114 |
| Seedbed Preparation, Sweet Potatoes | 969 | 701 | 100 | 138 |
| Opening Bush, Other Crops | 941 | 606 | 145 | 155 |
| Opening Bush Sweet Potatoes | 2,099 | 1,764 | 85 | 119 |

Source: R. W. M. Johnson, "The Labour Economy of the Reserves," Occ. paper no. 4 (Salisbury: University College of Rhodesia and Nyasaland, 1964).

## TABLE 45

Extent of Variation in Labor Input per Acre Accounted for by Plot Size Differences

| Operation | Number of Observations | Plot Size (sq. yds.) | | Rate of work (hrs., 1 acre) | | Correlations | | S.E. of r |
|---|---|---|---|---|---|---|---|---|
| | | Mean | S.D. | Mean | S.D. | r | $r^2$x 100 | |
| Second Rains, Seedbed Preparation for Crops Other Than Roots | 36 | 887 | 651 | 405 | 165 | -.40 | 16.2 | .14 |
| Second Rains, Seedbed Preparation for Sweet Potatoes | 21 | 326 | 327 | 969 | 1,095 | -.63 | 39.7 | .13 |
| Second Rains, Opening Bush Prior to Crops Other Than Roots | 22 | 1,128 | 1,634 | 941 | 640 | -.37 | 13.7 | .18 |
| Second Rains, Opening Bush Prior to Sweet Potatoes | 14 | 289 | 245 | 2,099 | 830 | -.48 | 23.0 | .17 |

Notes: $\frac{1 - r^2}{\sqrt{n}}$ has been used to estimate the standard error of r, although strictly speaking the samples are too small for the use of this formula.

All plots are very small, with only one over an acre, three more over .50 acre, and fourteen others over point twenty-five acre out of ninety-three plots measured. This may inhibit better demonstration of the relationship.

Source: D. Pudsey, "A Pilot Study of 12 Farms in Toro" (Kampala: Uganda Dept. of Agriculture, 1966). (Mimeographed.)

The sweet potato plots are about one-quarter to one-third the size of those cleared for other crops, and so the scale effect can be expected to exaggerate the differential in rates of work. Combining the data given for opening the bush and regressing rates of work on the size of plot as the independent variable give a low coefficient associated with plot size, accounting for about 17 percent of the difference in rates of work—a figure which is consistent with $\underline{r}^2$ for these operations in the table. A difference in the rate of work in opening bush for sweet potatoes is substantiated, justifying a decision to treat it as a subpopulation requiring special sampling.

Most researchers have sought to relate inputs to the whole acreage planted in a crop rather than to particular plots, and corroborative data on the scale effect is scarce.

R. W. M. Johnson examined the variation in rates of work per acre in relation to both the number of acres cropped and yield differentials, treating total hours used per acre as a variable dependent on the size of acreage cultivated and yield.[13] In a later article he modifies this, differentiating between the hours used on the main operations —ploughing, weeding, and harvesting—and more correctly indicating ploughing and weeding operations as independent variables influencing yield, and designating hours per acre used in harvesting as the only labor variable mainly dependent on yield.[14] The confusion in the earlier paper distorts the regression analysis undertaken, with total hours per acre as the dependent variable. However, in an appendix to the chapter, Johnson reports the results of analysis of variance, which he states contradict the results of the regression, which showed area cultivated responsible for only a small reduction in variance of total labor use per acre. He does not attempt to explain the contradiction but presents full details of the analysis of variance to allow its further use, and we take advantage of this here. The results demonstrate many of the important characteristics of variation in rates of work per acre.

The analysis demonstrates that for each crop the acreage cultivated is a highly significant independent source of variation in total hours worked per acre. It also demonstrates that the level of variance accounted for by the scale factor differs for each crop, from 47.8 percent of the total sums of squares in maize to 20.0 percent in millet. Part of this difference can be accounted for by the weighting of total hours used per acre by output-related labor use in harvesting. Maize used only 23.7 percent of total labor in harvesting, compared with 30.0 percent for millet and 45.9 percent for groundnuts. This leads to two other important points emerging from Johnson's analysis in addition to the confirmation of the scale phenomenon:

Results of an Analysis of Variance on Total Rates of Work per Acre for Three Crops

| Crop | | Sum Squares Raw | Sum Squares % Total Sum | D.F. | Mean Squares | F Ratio | P |
|---|---|---|---|---|---|---|---|
| Maize | | | | | | | |
| | Acres | 130,593 | 47.8 | 2 | 65,296 | 235.5 | .01 |
| | Yields | 115,992 | 42.8 | 4 | 28,998 | 104.6 | .01 |
| | Error | 26,058 | 9.7 | 94 | 277.2 | | |
| | Total | 272,643 | 100.00 | 100 | | | |
| Groundnut | | | | | | | |
| | Acres | 512,574 | 23.5 | 2 | 256,287 | 21.1 | .01 |
| | Yield | 713,965 | 32.7 | 3 | 237,988 | 19.6 | .01 |
| | Error | 959,160 | 43.8 | 79 | 12,141.2 | | |
| | Total | 2,185,699 | 100.00 | 84 | | | |
| Millet | | | | | | | |
| | Acres | 120,501 | 20.0 | 2 | 60,250 | 14.9 | .01 |
| | Yield | 295,078 | 49.2 | 3 | 98,359 | 24.4 | .01 |
| | Error | 185,310 | 46.0 | 46 | 4,029 | | |
| | Total | 600,889 | 100.00 | 50 | | | |

<u>Source</u>: R. W. M. Johnson, "The Labour Economy of The Reserves," Occ. paper no. 4 (Salisbury: University College of Rhodesia and Nyasaland, 1964).

1. Differences in the variability of rates of work stem from the particular work pattern of the operation. Operations may be land-related, as with cultivation operations, or output-related, as with certain harvesting and processing operations. Weeding may be cited as a composite, being land-related but also influenced by the growth habit of the particular crop, itself a third criterion for classification. These distinctions will be taken up again later.

2. The high level of error variance for groundnut and millet demonstrates other general sources of variation that influence work rates on individual farms.

The scale factor, our immediate preoccupation, is important when survey conditions require a smaller sample size or reduced respondent burden. Where presurvey investigation can indicate the usual plot size for a crop activity, lower variation in rates of work will be achieved by excluding observations on exceptionally small or exceptionally large plots.

This second group of variables affecting work rates, covering soil type, cropping history, and the scale effect, represents a major area of compromise between detail and cost and features prominently in presurvey content and cost/effective survey design.

## Tools, Methods, and Group Specialization

Of the third group of independent variables influencing rates of work, two have already been isolated in identifying homogeneous types of farming areas: the tools or equipment and the methods used. A further variable included within this group is the occurrence of sex/age group specialization. Differences in tools and equipment, or the method of carrying out operations, may be found within the same system, particularly where it covers distinctive soil types. For instance, in the traditional Sukuma system a flat cultivation sequence is followed on the heavy valley bottom soils, while all crops grown on the hill sands are ridged. These different sequences must be distinguished in sampling. Grouping the labor-intensive flat cultivation sequence of the heavier soils with the ridging sequence on the sands would give a distorted average.

## Residual Variance

The final group covers four variables making up what for our application is the residual variance in interfarm work rates:

motivational, managerial, and nutritional status differences between farmers and their families, and climatic variation over the area.

Motivational differences between farmers are reflected in the effort the family will make while at work on the farm. The rate of work of a fourteen-year-old in a highly motivated family may be as high as that of the wife of an idle farmer. Managerial differences will influence the conditions under which operations are done. With the timing of operations critical to productive effort, input requirements may be reduced by correct timing and increased by delay. Late cultivation in heavy soils which have become waterlogged will increase the time required. Late weeding will have the same effect. On the other hand, delayed harvesting of rice prone to shattering will reduce the effort required to pick a given acreage. The microclimate creates similar effects in terms of more or less rapid crop or weed growth and easier or harder soil conditions. These sources have no importance to our application, and their effect will be randomized over the sample and reflected in the variance of observed rate of work.

There is an important qualification on the status of climate as a source of variation in work rates. Cross-section survey data is influenced in many respects by the climate over the period of investigation, which is itself a point on an interseasonal distribution of climatic conditions. Repeated investigation over a period of years is usually advocated to control and measure the effect on attributes required in planning. In practice, funds rarely allow repeated investigations—so rarely, in fact, that it is difficult to find an example. In the course of a planning and extension exercise in part of Sukumaland, data were collected by frequent-visit techniques on eighteen farms for the two seasons 1964-65 and 1965-66. Not all farms realized an observation for every operation, and Table 47 covers only operations made on the same farms in both seasons. The column headed "Range" shows <u>the individual farm</u> with the greatest variation between the two years.

The differences between the means demonstrates a considerable interseasonal variation in work rates, with a very wide range on the individual farm. In part this is due to a failure to define subpopulations based on group 2 variables, particularly cropping history. The particular farm with the widest range will probably be reflecting a change of cover which had to be removed in the course of cultivation. Nevertheless, it does reflect the kind of variation which can arise from microclimatic factors. The differences in interseasonal averages demonstrate the importance of evaluating the effects of the climate for the period over which data are collected.

## TABLE 47

### Interseasonal Variation in Work Rates on a Small Sample of Farms

| Crop | Operation (man-days /acre) | No. of Paired Observ-ations | Mean 1964/65 | Mean 1965/66 | Range 1964/65 | Range 1965/66 |
|---|---|---|---|---|---|---|
| | Sesa | 14 | 4.5 | 3.8 | 1.8 | 6.1 |
| Cotton | Ridge/Plant | 14 | 8.4 | 5.8 | 23.4 | 6.3 |
| | Weed | 14 | 11.7 | 7.3 | 17.4 | 14.4 |
| Maize and | Sesa | 10 | 4.7 | 9.3 | 4.1 | 30.1 |
| Legumes | Ridge/Plant | 10 | 12.7 | 11.5 | 7.3 | 42.7 |
| | Weed | 8 | 6.1 | 3.6 | 9.1 | 1.7 |
| | Sesa | 13 | 4.3 | 4.8 | 9.8 | 1.6 |
| Maize and | Ridge/Plant | 14 | 9.2 | 8.4 | 29.6 | 8.3 |
| Cassava | Weed | 10 | 8.1 | 9.9 | 4.8 | 20.5 |

Source: Compiled by the author.

## CULTIVATION, WEEDING, HARVESTING, AND PROCESSING

It is useful to examine the three main sets of operations—cultivation of the seedbed, weeding and cultivation, and harvesting and processing—and to summarize the main sources of variation operating on each.

### Cultivation

The cultivation operation is land-related, and most of the sources of variation in work rates in it stem from the condition of the land:

1. Soil differences. Distinctive soil types and even different levels of fertility of the same basic type will alter work rates in preparing the seedbed.

2. Cover differences. Opening land from bush or grass may require extra operations in cultivation, such as cutting down the trees or burning the grass. The difference in sequence should be preenumerated. The intensity of off-season weed cover or the weight of residues will vary with the crop and the time between crops, and will affect the rate of work in land preparation the following season.

3. Local rainfall conditions. Delayed rains result in hard, dusty lands, and excessive rainfall leads to waterlogging.

In most parts of Sukumaland the same operational sequence for preparation of the seedbed holds for 90 percent of the crop acreage. "Sesa" is an operation which scrapes the cover from the old ridges into lines in the old furrows. This cover is subsequently buried by the new ridges. "Sesa" is carried out regardless of the density of the cover and may require only one or two man-days per acre on infertile soils where weed regeneration and crop residues are negligible, but upwards of twenty man-days if heavier land is being taken out of dry-season grazing.

### Weeding

Possible confusion between the rate of weeding and the number of weedings should be avoided. From a management point of view these are alternatives. The timing in weeding a particular crop will interact with the number of weedings to decide the total weeding input. Early weeding may reduce the input required at one point of time but increase total input if a further weeding is required later. The relative merits of these alternatives depend on the reaction of the particular crop and their timing in relation to the labor requirements of the system as a whole. It is important to preenumerate the usual number of weedings on each crop in the presurvey investigation. Aggregating weeding inputs into a single rate of work greatly increase the variance of the measure and distort the choices of the farmer.

Three sources of variation influence the rate of work in weeding:

1. The growth habit of the particular crop and the planting practice followed. A crop quickly establishing a dense cover will require lettle weeding because it shades out weed competition. The degree will be affected by the spacing practice, and although marked differences can be expected between spreading plants and those of upright growth, grouping across crops of a similar growth type and density is valid.

2. Since weeding is a matter of shifting soil, the work rate is land-related; and so the same land-based sources of variation will influence it as do the cultivation operations.

3. The rainfall pattern during the season may encourage or deter weed growth and directly affect the rate of work required to clear a given crop acreage. Since there is no basis for isolating this, it will be randomized over the area. Because it is an important source of variation in labor requirements, an estimate of the season as a typical or atypical is important in evaluating the usefulness of the cross-section data obtained by survey.

## Harvesting and Processing

While crop processing operations are usually directly output-related, harvesting itself is a true composite operation, its relation dependent on the growth habit of the crop and the harvesting methods used in the system. Johnson's analysis of variance showed acreage accountable for variation in rates of total labor use per acre on three crops, and the level of explanation was reduced on crops with output-related harvesting operations. His data demonstrate the reduction in variance achieved when rates of work are expressed per unit of output. The reduction is greatest on the groundnut crop, where a higher proportion of the harvesting operations are output-related.

TABLE 48

Comparison of Variances on Harvesting Inputs
(per acre unit and per output unit)

| Crop | | Harvesting Labor Hours per Acre | Harvesting Labor Hours per 100 lbs |
|---|---|---|---|
| Maize | Mean | 20.8 | 5.4 |
| | Coefficient Variation (%) | 77.8 | 58.4 |
| Groundnuts | Mean | 129.5 | 44.0 |
| | Coefficient Variation (%) | 80.0 | 42.2 |
| Millet | Mean | 55.5 | 25.1 |
| | Coefficient Variation (%) | 73.1 | 41.7 |

Source: R. W. M. Johnson, "The Labour Economy of the Reserves," Occ. paper no. 4 (Salisbury: University College of Rhodesia and Nyasaland, 1964).

Precision is clearly increased, though no generalization is possible. The degree of improvement depends on the growth habit of the crop and the pattern of work for harvesting. For example, maize and millet require the worker to move from stalk to stalk; this is an "overhead" operation which may be influenced by stand density but not by the number of fruits or fruit weight. Various levels of compromise arise. On groundnuts many factors contribute to yields: density of stand is important and, where uprooting is done selectively by hand, inputs will be related to density; where it is done by ox equipment, it will be acreage-related. At the same time all haulms must be lifted, regardless of the nut yield per plant. Similarly, a high proportion of empty pods will affect final yield, and so there may still be a significant level of variation in shelling requirements for a given number of kernels. Overhead elements dominate the harvesting of many tree crops which must be picked by climbing the tree—coconuts are an obvious example. Operations with a high proportion of such overheads are logically acreage-related. Where a part of the harvesting operation is concerned only with the fruit produced, input per unit of output is more relevant to precision for planning needs. Where such work is labor-intensive and dominates a harvesting sequence which cannot be broken down into discrete operations, input per unit of output is the correct basis for decision on sample size.

The gradual harvesting of food for consumption fresh from the field influences the variation in per-acre inputs between farms. Some crops, such as green maize, are picked from the field as required, and the residual is harvested in the normal way when dried off. The level of yield and the timing of maturity will influence the proportion of the crop remaining for harvest and thus the input needed. The main decision to be made is whether the picking of green maize is to be attributed to harvesting or domestic chores. Normally the residual crop should be considered and the harvest input required to pick it; but whereas the production eaten fresh is important output data, the inputs required to gather it are best classed as a chore, and the relationship between residual output and input should be used to estimate the effects of changing yield levels on labor needs. This, of course, breaks down where high overheads of labor are needed and picking the actual fruit is a small part of the total requirement.

Often the harvesting operation can be broken down into its acreage and output-related components, thus giving a better basis for survey and planning from a limited sample. Any such division underlines the importance of careful presurvey description of all the field operations on major crop activities. Operations such as groundnut harvesting, which might involve up to ten stages, require particularly careful description. Stages may be grouped together when performed together, but the components of the group must be clearly defined before the survey and in the mind of the respondent during the survey. If definition of the operation is vague, respondents may answer on different bases, thereby creating a bias in the data. Local usage, often embodied in tribal language, may offer a useful basis both for grouping sequences of stages and for grouping operations common to several productive activities.

A second type of classification also helps to identify operations and, therefore, work rates critical to proper simulation of the existing system: the distinction between necessarily timely and postponable operations, first drawn in an economic planning context by J. Heyer.[15] Heyer perhaps went too far in excluding all postponable operations from farm work. There is a need to cover these requirements, regardless of their timing, within the labor supply available for the season. The division is a very useful one, particularly where there is a marked slack season in the system. In principle, if the aggregated timely and postponable requirements can be covered by total labor supply over the season, then necessary timely operations can always take priority and the postponable group can be treated as a residual. The timely operations thus become critical to defining the system and replanning it, and a focus for precision in data collection. Certainly where underemployment in seasonal troughs is marked, precision requirements on postponable operations can be relaxed. However, the conditions which demand timeliness are probably wider than seems apparent at first sight and can be divided into exogenous and endogenous factors.

## Exogenous Factors

Key operations, usually land preparation and planting, but also others, are climatically determined beyond the control of the farmer and form the framework of the seasonal pattern. Flexibility in these operations has associated costs, and delay usually means a reduction in yields; harvesting rice before it sheds, and processing tea or sisal before they dehydrate demand a pattern of timely action from the farmer. Other operations are much more flexible but nevertheless have a place in the sequence of the cropping calendar; other work is dependent on their timely completion; but their commencement may be fairly open-ended. W. Scheffler cites felling timber for tobacco fuel and barn-building for tobacco curing as postponable, but flexibility is clearly limited to the preprocessing part of the season.[16]

## Endogenous Factors

Less obvious are some of the factors within the system. Weeding is flexible, but only at a cost to the farmer; food crop processing is postponable until the food is required. Conditions of work, such as very dusty soil because of poor rainfall, may inhibit flexibility in land preparation operations. Storage space is often very scarce and threshing or shelling may be necessary. All these factors reduce the postponability of apparently nontimely operations.

The need for careful description of operations has already been stressed, and the conversion of an operation from necessarily timely to postponable may be a valuable contribution to dissipating work peaks. Groundnuts are a good example. In areas where they are a main crop, methods of stocking to keep them dry and prevent sprouting have been devised. In areas where they are a supplementary crop, it is still seen as necessary to pick the nuts off the haulms soon after they are uprooted—that is, there has as yet been insufficient pressure for a solution to the problem. Farmers themselves clearly realize some of the values of postponability in operations. Prolonged extension efforts have been made in Sukumaland to promote the practice of picking and simultaneously grading cotton into separate bags. This would keep the workers out in the fields doing work which is both postponable, providing there is storage space, and pleasant, when sitting under a tree in company. Efforts in this case have been in vain.

## THE PRESURVEY INVESTIGATION

Presurvey investigation breaks down into two parts: a general content centered on identifying the sources of variation in the system which create subpopulations requiring independent sampling, and a remaining content that depends on the collection technique adopted. The general content falls into three parts:

1. Describing the tasks involved in the main operations and the main sources of variation influencing them.

2. Defining common operations done on more than one crop activity, so as to assess the possibilities for grouping and to pinpoint specific operations peculiar to particular crops.

3. Identifying necessarily timely operations, particularly the labor-intensive ones.

Various classifications of crop operations have been described as useful aids in predefining the subpopulations of operations which need independent sampling. It is important to predetermine plots of the same subpopulation by confirming the homogeneity in the end use of the product and by investigating variable characteristics which might affect the rates of work, even though the end use of the product is the same. The discussion has particularly emphasized the need for adequate precision on the necessarily timely operations, particularly the labor-intensive ones in the peak periods, for these underpin the shape of the labor profile.

As indicated earlier, the question of whether limited-visit surveys can cope with labor data in the form of a continuous flow over the season is central to this chapter. By constructing a representative farm model from components which halt the flow at points defined by operation and time, many problems are avoided. In order to present the case clearly, the alternatives of frequent- and limited-visit techniques are discussed separately.

## Frequent-Visit Techniques

We have already mentioned the use of work-study techniques and their main failing of an upward bias in per-acre rates, because of the small plots measured. In addition, the feat of organization required to cover a range of crop/operation combinations or a number of farms, as well as visiting all farms to record other components of timing and sequence, means that the technique is not a practical possibility within a survey. Consideration here is limited to what is usually thought of as a frequent-visit procedure, that is, listing the work done by the farm labor force each day. The required planning components—operational sequence, operation timing, and rates of work—can all be synthesized from this kind of list. A good deal of other planning information is also enumerated in the course of this listing; the lengths of day worked by different groups and at various parts of the season are important qualifications of labor-supply constraints for planning.

Various approaches have been used for detailed collection. The method recommended by Pudsey and supported by Hunt requires a listing of all activities of each individual during the daylight hours.[17] This is probably the best technique, allowing a full enumeration of other nonfarm activities undertaken by family members. It may, however, be precluded by cost conditions, for it adds significantly to both enumerator and respondent burden. Some researchers have been satisfied to list the agricultural work done by each family member, while others have enumerated by enterprise, bringing up to date the work done on each crop since the last visit and recording any new plots or crops on which work has begun. The problem of this last approach lies in covering new enterprises started where the farmer has not decided the crop he will plant when clearing bush or grass fallow, or where he changes his mind as a result of seasonal contingencies. New plots of crops already established on one plot are particularly prone to being overlooked.

### Problems in Detailed Collection

Except for the work-study technique, all other approaches are memory-dependent to a greater or lesser degree, a point often overlooked by critics of limited-visit techniques. The degree of dependence is determined by the frequency of the visits. The use of labor is frequent but the content is irregular, and this irregularity must decide the feasible frequency of visits. Where a family working as a group will complete the whole operation on the major crops before starting new work, continuing for a period of several days or even weeks, recall will be better than where individuals work partly in the family force and partly for themselves, or where small plots predominate and the labor force switches work within the day.

Evidence on the recall of labor use is flimsy. Hunt has stipulated daily visiting as necessary for all accurate information, implying a visit to collect yesterday's data.[18] This is clearly ideal but in practice rarely achievable. It is also an oversimplification of the problem. Underlying the question of complexity of work organisation just illustrated is the demarcation of the recall period on each farm.

We have described the problem of demarcating the recall period in the mind of the respondent in order to avoid the phenomenon of inward transfer, which creates an upward bias where the period is open-ended. "Yesterday" forms a tight recall period which slackens as the period is increased to two and three days previously. Where work organization is complex and highly irregular even from day to day, historical information for two or three days previous may be too much to ask and liable to heavy transfers. No hard-and-fast rule can be given and preenumeration of the work habits of the community must form a basis for decision. Some tasks in all systems will be regular from day to day. Where there is no habit of continuous effort in fieldwork until tasks are completed, a plot approach, supplemented by a check through family individuals not questioned on the plots, may allow some extension of the recall period and a reduction in visit frequency by prompting the memory of the respondent. Because of limited resources, the need to reduce visit frequency to give better control of sampling errors usually requires a compromise on the accuracy of data on the single farm unit. Observational errors will rise when recall is stretched over a badly defined reference period. Two further respondent questions compound the problem of accuracy and visit frequency.

## Need for Multiple Respondents

Hall has raised the problem of the ability of the farmer to answer questions on the activities of all family members, but other researchers make apparently contradictory statements.[19] We have identified the source of the problem as the type of work organization traditional to the community and family. The ability of the decision maker to answer for the whole family will depend on the proportion done communally. Where decisions as to daily labor allocation by individuals are taken by the head of the family, he will still be able to answer for the whole family. Where the deployment of effort is coordinated by individual obligations to the family, the possibility of a single respondent becomes questionable. No doubt the day's activities will form a major topic of conversation at meals and in the evening, and the household head will keep himself informed of progress to hold individuals to their responsibilities. As Bessel, Roberts, and Vanzetti have noted, in large families and also in poorly motivated households, accurate answering for every member by a single respondent may be difficult, particularly where individuals will switch between crops from day to day.[20] In these extreme situations a recall period of no more than a day, and the use of both the household head and senior female member as respondents, may be necessary. This radically complicates the enumerator's work, and his coverage may be reduced drastically to two or three farms when families are at work during the day and away from the home.

Respondent Burden

Hunt has expressed concern with the respondent burden of daily visits on farms throughout the season.[21] He estimates a month as the longest period to which a household should be subjected to questioning with this intensity, allowing that the full season may be reasonable if visit frequency is reduced to two per week. As we have seen, the feasibility of maintaining accuracy with reduced frequency depends on the form of work organization (communal or individual) and the work habit (continual on a single crop operation until complete, or switching from crop to crop in a variety of tasks within the day).

The fact that the planning model is area-based, and thus dependent on average relationships and not comparative relationships between farms, creates alternative solutions to the respondent burden problem through variations in sampling techniques. Each has its problems, and a decision to follow an alternative must be made after preenumerating the characteristics of the labor economy of the particular area.

Hunt recommends the use of subsamples. Recording on twelve different farm groups for a month each and six different groups for two (separated) months each are two possible examples where daily visiting is required. Two problems arising here are continuity and completeness in the operational sequence and adequacy of the sample size. The subsamples chosen may be variable in timing as well as in work rates. Thus double-recording of the same operations and omissions may be experienced on transfer to a new subgroup. The outoff of recording in one subsample may leave a field half-weeded but fully measured, the rate of work being distorted unless a new measurement of the completed area is undertaken. The complexity of area remeasurement with each transfer of subsample seems prohibitive. More important, monthly transfers will allow collection on only 8 percent of the chosen sample at any one time. Raising subsample size by bringing other farmers in from the population would meet this criticism but would greatly complicate field organization of the survey.

Recording on a limited number of plots on the farm is an alternative which reduces the length of interview needed. It can be done where close definition of subpopulations gives an expectation of very low variability for work rates on particular activities, and so the number of farms sampled for these can be reduced. Alternatively, where farms grow several plots of what has been confirmed as the same crop, only one plot need be recorded. Both these possibilities involve enumeration on a plot basis, and the second alternative is susceptible to transfers of work done on other plots with the same crop. In systems where plots are worked by individuals, bias will arise unless all sex/age groups involved are recorded. This almost inevitably rules out any saving, for the number of plots of the same activity will usually be fewer than the number of desirable sex/age groupings. If sample plots are selected, they should be about average size in order to inhibit distortion by the scale effect.

Limiting coverage to critical periods of the season will reduce respondent burden and may be feasible where necessarily timely operations constitute the major seasonal peaks. In the case of Sukumaland, for example, two periods of three months—November to January and May to July—would cover the constraints and coefficients critical in the planning model.

The problems of labor data collection by frequent-visit techniques spring from the level of complexity in the labor economy of the system. Complexity increases as householders delegate decisions about crops to be grown to family members, creating a need for more than one respondent and increasing the time required on each farm by the enumerator. Complexity also increases as work habits become more flexible, partly because of a larger variety of crop activities or a proliferation of smaller plots. Workers may transfer within the day between crops, plots, and operations. A most important presurvey requirement will be the description of the way family labor is organized and applied, in order to allow selection of the proper collection technique.

## Limited-Visit Techniques

In discussing memory performance in Chapter 7 we noted that events occurring in a pattern aided recall, though the mechanism altered from remembering the historical details of each particular event to the use of experience to estimate an average and to apply this to the pattern formed by the frequency of incidence of the event concerned. We have noted examples from Zarkovich in which longer reference periods, which fit natural cycles, give better precision than artificially bounded shorter references periods, which suffer badly from end effects of one sort or another. It is assumed that the same type of pattern is formed for components of the labor profile by repetition from season to season. Limited-visit techniques deliberately seek to exploit this pattern in eliciting answers based on experience rather than on historical recall of labor use for the particular season. Differences in managerial, motivational, and microclimatic conditions will still vary the interfarm patterns, and their effects will be randomly distributed over the population. Using historical recall for labor recording, interseasonal variation can be controlled only by effective sampling from the population of seasons. This is usually prohibitively expensive when the resources for investigation are limited. Cross-sectional data for one season represent a single observation in this interseasonal population and may be subject to large errors. Before the data can be used for planning, they must be evaluated, and, when necessary, normalized. At the same time, the existence of this underlying seasonal pattern provides a basis for answering questions on rates of work and other components, which are experience- rather than historically oriented; and this basis is free of distortion from the climatic characteristics of a particular season.

That the farmer has an accumulation of experience with labor use is clear; the question which is difficult to answer unequivocally is whether he retains this experience in a form which the investigator can exploit to establish measures of labor input. Decisions as to what crops, how much of each, where and when to grow them, imply a measurement of his needs, of his available land resources and their productivity, and of the capability of his family labor force. He may have two maize stores, and if both are full he will have enough grain to last through to the new crop. The quantity contained by these stores must bear a relation to an area cultivated in the past and to the results achieved. He needs other foods and perhaps some cash. The requirements of the range of crops in terms of climate, soil, and timing must be integrated into his calendar, within the capacity of his available labor force. Anthropologists and other researchers have demonstrated the detailed knowledge of the peasant farmer in the definition of local soil characteristics, the discrimination of crop subspecies by peculiarities of growth habit, and the use of fertility and phenological indicators. It is now accepted that traditional farming systems and the practices followed in them create a balance, often subtle and delicate, between the community and its environment. Such acceptance acknowledges that traditional farming is systematic. The decision task in a labor-limited system is that of balance between family needs and family labor resources. Land and the ecological regime it supports are only intermediary to this vital balance.

The predominance of evidence of traditional farmers' familiarity with land/plant complexes is perhaps due to the "land"-mindedness of interested observers. Even in areas of western Europe, where land is a major constraint on farm scale, most farmers would readily estimate the size of labor gang required to complete a task on a field of a particular crop they were growing. The basis for such estimates is accumulated experience, which is equally available and more critically important to the traditional farmer. Indeed, given labor supply as the main constraint on the farming system, it would be more logical to assume that traditional farmers have no concept of land as a factor in their production process. Certainly the farmer cannot articulate his experience in terms of formal units of measure, which are a feature of a type of education he will not have had. Questioning phrased in his own terms is the key to tapping his experience. Defining these terms is critical to the construction of an effective questionnaire for limited-visit surveys. This applies equally to daily visits. Certainly the farmer will know how long he worked yesterday, but he may not be able to answer in terms of hours. Specific evidence of a pool of knowledge on labor use has been lacking. The results of limited-visit surveys measuring rates of work will be presented in this study as evidence of the feasibility of tapping the accumulated experience of traditional farmers to furnish adequately precise planning data on labor use.

## Collection Methods

The approach adopted is enterprise-oriented to give the respondent a clearly defined focus for his experience. With a single-visit survey at the end of the season, the reference period is well defined, being the whole season; and with two or three visits, timing is arranged so that operational subsequences are complete. Information gathered in one visit will close the reference period for subsequent visits. Although the main labor use component to be measured quantitatively is the rate of work, the operational sequence and its timing are both required for proper representation of the profile. The operational sequence on a field represents a series of important events which is repeated from season to season. It is well established in the experience of the farmer and can be enumerated from either experience or memory. It will be important to preenumerate operations susceptible to seasonal sources of variation, for example, a second or third weeding. When the area as a whole is in an extraordinary season, such operations should be investigated for incidence over seasons to give some basis for normalizing their importance. Operational timing is susceptible to the same sources of variation, and the approach to enumeration should be the same. Just as the labor profile can be reduced to components, so the rate of work itself has three facets: the labour force involved on the job, the length of the workday, and the number of such days required to complete the operation. Where labor force grouping is common to specific crop operations and the workday length to the time of season and sex/age grouping, both these facets can be enumerated generally, leaving only the number of days to be measured. Enumeration is more difficult where individuals are responsible for their own fields and are merely under obligation to help the household food supplies. The three facets are discussed in turn.

Labor Force. The labor force is easiest to enumerate where the whole family labor force works together on whatever crop/operation combination is given priority by the decision maker. This straightforward organization is complicated by crop or operational specialization. The decision maker will be aware of specializations and is a suitable respondent for their enumeration. When he is excluded from certain work groups, particularly where the specialization is based on a whole crop, the number of respondents is increased for the enumeration of the workday length and the number of days required.

The pattern is further complicated when individuals merely have obligations to supply or have minimum communal work obligations, with their own fields over and above these. Again the head of the household will be the appropriate respondent to describe this pattern, but respondents for the two other components will have proliferated. The time required to enumerate the fields of a household where individuals grow their own crops is increased only if some respondents are not available. Ensuring their availability will be an important part of survey preparation. A difference from frequent-visit techniques, is that respondents will not be antagonized so much by being available for a limited number of visits.

Workday Length. The main differences in workday length are likely to be less time spent in work by the young and very old, and time out for the domestic commitments of the women. The other main source of variation will be seasonal urgency, with longer days worked at critical times of the year. Again, this may be modified where operations require efforts beyond the nutritional capacity of the food supply, which may effectively reduce the length of the workday. The daily routine is obviously familiar to all families. It is unlikely to be directly quantifiable in terms of hours. Experience has shown the best approach to be in terms of when the work is started and finished. Farmers will indicate the time of day very accurately, and family routines may be geared to the position of the sun. Sex/age group differentials can be probed by establishing whether all family members go out to and return from work together. Quantification of the lengths of days worked into hours rests with the enumerator.

With a communal labor force the decision maker will quantify the differences in working day, and both sex/age group variations and seasonal urgency variations should be investigated by the questionnaire, either on a general basis or when each enterprise is covered. Where specialization occurs by sex or age group, the length of workday may best be enumerated through a senior member of the group concerned when the particular enterprise is covered, and the sources of variation will be the same. In a system with individual plots, the workday will be much more flexible and also outside the family control, so that each respondent should be enumerated in relation to his own plot.

Days Required. The main source of variation in work rates within a community is the number of days required to complete an operation. It is best enumerated in relation to a clearly defined enterprise on the ground, a plot of the particular crop or crop mixture. The aim in enumeration is to place the respondent in a decision context as near as possible to that with which he is familiar.

## Enumeration of Labor Input Data

The approach in the field is vital to successful enumeration. The following description presumes that the appropriate respondent has been selected and investigates all three components in relation to a specific plot.

It is important that the enumerator and respondent go to a plot of the crop or mixture for which labor use is to be recorded. The respondent's visual impression of the size of the plot will be the key to accuracy in the replies he makes, and being physically present on the plot creates the reference point for the interview and the context for both memory-dependent and experience-dependent questions.

For memory-dependent questions the crop on the field is a key to operational sequence and timing; for experience-dependent ones, the size of the field. While at the field the period of reference is clarified; either it is the whole season or, when it is the last recorded work on the field, its timing is confirmed with the respondent to close the old and open the new reference period.

The operational sequence and its timing are enumerated on a mixed experience/memory basis. Presurvey investigation will have outlined the normal sequence and timing of operations and these, set out in the questionnaire, will form a basis for checking omissions or interpolations which are out of the ordinary. The enumerator must be particularly careful to use open-ended questions to confirm or elicit information of this type. Both the start and the finish of those main operations which are continuous over a significant period should be recorded. This information forms a check when subsequently enumerating the days required, as well as allowing the calculation of a center date.

To enumerate the number of days' work required, questioning switches to each operation and uses experience rather than memory in response. The interviewer should explicitly adopt an <u>ex ante</u> outlook on the crop to be grown and must clearly dissociate further questions from the particular experiences of last season. There are three rate-of-work components:

1. The <u>usual</u> labor force for this operation on this crop, prompting the respondent as to the crop, operation, and time of season.

2. The <u>usual</u> workday for the members of this labor force, prompting the respondent as to time of season and seeking sex/age differentials, particularly the domestic commitment of the women.

3. The number of days required to complete the operation with the labor force described on the field under enumeration. The respondent can usefully be prompted as to the crop, operation, and labor force involved.

Answers are quantified by reducing the labor force to selected man-equivalent values and relating the number of days of so many hours to the size of the field.

<u>Effectiveness of the Technique</u>

The evidence to be presented in support of limited visits as an effective technique for the collection of labor data was accumulated between 1962 and 1966. The method and approach described have evolved out of this experience, but none of the data presented have been obtained by the use of the completed technique as described. Facets of the technique arose at different stages in the research. The original work sought to tap the farmers' memory over a period up to nine months long, and it was not until 1965 that it became clear that farmers were using their know-how and experience to build their answers rather than actually remembering the work done months previously. It was this realization that accounted for the apparently consistent results that had been obtained, and allowed a rationalizing of the approach to focus explicitly on experience rather than memory.

Four sets of evidence of the accuracy of data collected by limited-visit techniques are presented in turn.

Comparison of Survey Results. Farm economic surveys were carried out within Sukumaland, an area of broadly homogeneous farming. Each survey was done in a different subarea and a different season, and all data were collected by either one visit at the end of the season (1962) or two visits (1963-64). Table 49 presents the arithmetic mean of operations common in two or more of these areas from these surveys.

In these surveys the number of observations was often smaller than desirable, though the position had been improved by 1964. All these data were collected by limited-visit techniques, necessitated by the scarcity of enumerators. Since there is no comparison of methods here, the example cannot test the reliability of limited visits. The data do demonstrate a consistency in general magnitudes and also in interrelationships within and between areas:

1. The relationship between sesa, ridging, and weeding as operations is consistent for cotton and maize over the three areas.

2. The relationship between sesa and ridging in cotton and maize, with greater efforts being expended on maize, is consistent in the three areas.

3. Rice is shown to be a more labor-intensive crop than either cotton or maize in both the areas in which it is grown. Other researchers, notably N. V. Rounce in 1948 and D. Rotenhan in 1963, have confirmed the general magnitude of these rates on the main operations.[22]

TABLE 45

Labor Input Data from Three Farm Economic Surveys in Sukumaland
(man-days/acre)

| Crop | Operation | 1962 Number Observed | 1962 Average | 1963 Number Observed | 1963 Average | 1964 Number Observed | 1964 Average |
|---|---|---|---|---|---|---|---|
| Rice | Dig, Transplant | 14 | 41.0 | — | — | 27 | 51.8 |
| | Weed | 14 | 48.9 | — | — | 26 | 33.9 |
| | Pick | 14 | 37.2 | — | — | 15 | 44.1 |
| Ridged Cotton | Sesa | 30 | 8.7 | 6 | 7.3 | 59 | 8.8 |
| | Ridge, Plant | 35 | 13.5 | 6 | 12.5 | 61 | 10.6 |
| | Weed 1 | — | — | 6 | 8.3 | 61 | 7.7 |
| | Weed 2 | — | — | 5 | 6.8 | 50 | 5.8 |
| | Weed (all) | 38 | 14.5 | — | — | — | — |
| | Harvest (all) | 38 | 17.0 | 6 | 21.5 | — | — |
| Maize Mixtures, Ridged | Sesa | 16 | 10.1 | 17 | 11.6 | 67 | 10.8 |
| | Ridge, Plant | 18 | 14.5 | 25 | 17.6 | 68 | 14.1 |
| | Weed | 18 | 5.9 | 25 | 8.6 | 67 | 9.6 |
| | Harvest | 18 | 4.7 | 24 | 6.7 | 28 | 3.6 |

Source: M. P. Collinson, "Maswa Area," Farm Economic Survey no. 3 (Dar es Salaam: Tanzania Dept. of Agriculture, 1963). (Mimeographed.)

Comparison of the Accuracy Achieved in Planning with Single-Visit Labor Data and Actual Farm Records. The data used in planning the first season's program for a trial farm were general survey data collected by a single-visit technique. Those used in planning the following two seasons were recorded in detail on the farm itself. Table 50 summarizes planned and actual labor use over the three seasons and gives an error index computed by summing the monthly deviation of actual from planned usage, expressing it as a percentage of total planned use for the three seasons.

While this single instance is inconclusive, it is evidence in support of single-visit data as adequate for farm planning under conditions of uncertainty. In each season, actual total use was close to planned total use. Similarly, the level of deviation demonstrated by the "error index" is consistent over the three seasons. Reference to the profiles set out in Chapter 8 showed the deviations due to shifts in timing, caused mainly by weather and family contingencies in the particular seasons, while rates of work were relatively accurately predicted.

Direct Comparison of Techniques in the Same Population: Aromatic Tobacco. In 1964 a detailed data collection technique, using weekly visits, was adopted for a survey on aromatic tobacco.[23] Lack of trained enumerators led to use of local field staff of the Ministry of Agriculture supervised from the research center and, in the field, by the district agricultural staff. At the end of the season a single-visit survey was mounted by the permanent team of the survey unit at the research center. A second sample of farmers was selected from the same frame and visited once only. The objective was a direct comparison of the two methods within the same farming population, as well as a check on the work of the temporary enumerators.

Three groups of data were collected: a control group to establish the comparability of the two subsamples, a group of labor input data to test the possibility of collection by a single visit, and a group of food crop production data. The data on control attributes are presented in Table 51. Data in the control group were collected by the same techniques of enumeration in both single- and frequent-visit surveys. It is noteworthy that precision is greater for all attributes with the single-visit survey, partly because of the larger number of observations. The fairly high incidence of hired labor, on 23 percent of the farms in the single-visit sample, suggests a lack of awareness on the part of the temporary enumerators used for the detailed collection and a lack of presurvey investigation for survey design. Both criticisms are fair comment.

TABLE 50

Comparative Accuracy Achieved in Planning by Use of Single-Visit Survey Data

| | Single-Visit Data | | | Farm Records | | |
|---|---|---|---|---|---|---|
| | 1962-63 | | 1963-64 | | 1964-65 | |
| | Man-Days % | Index | Man-Days % | Index | Man-Days % | Index |
| Planned Total Labor Use | 290 | 100.0 | 508 | 100.0 | 442 | 100.0 |
| Realized Total Labor Use | 309 | 106.6 | 505 | 99.4 | 486 | 110.0 |
| Aggregated Monthly Deviation of Realized from Planned Use | 139 | 47.9 | 263 | 51.8 | 198 | 44.8 |

<u>Source</u>: Compiled by the author.

TABLE 51

Comparison of Data from the Same Population, Collected by Different Techniques: Control Data to Establish Comparability

| Attribute | One Visit | | | Weekly Visit | | | |
|---|---|---|---|---|---|---|---|
| | No. Obs. | Arith. Mean | % S.E. | No. Obs. | Arith. Mean | % S.E. | Significance of Levels of Differences |
| Family Size | 51 | 5.6 | 7.6 | 43 | 5.5 | 10.2 | n.s. |
| Family Labor Available (man-equiv.) | 51 | 2.41 | 5.9 | 43 | 2.74 | 9.9 | n.s. |
| Hired Labor Used (man-equiv.) | 12 | .05 | — | 0 | — | — | — |
| Acreage of Aromatic Tobacco | 48 | .20 | 6.1 | 33 | .15 | 12.7 | * |
| Production of Aromatic Tobacco (lbs.) | 46 | 53.8 | 9.5 | 30 | 49.2 | 14.0 | n.s. |
| Acreage of Maize | 51 | 2.05 | 7.7 | 43 | 2.26 | 10.2 | n.s. |
| Acreage of Groundnuts | 47 | 1.62 | 8.9 | 42 | 1.76 | 11.9 | n.s. |

*The only significant difference on aromatic acreage had $t = 2.29$, significant at the 5 percent level.

Notes: Data on the production of aromatic tobacco were collected at marketing points. It is more useful as a control.

Acreages of maize and groundnuts are food acreages covering all areas of the crop, whether pure or interplanted.

Source: Compiled by the author.

Comparison of Data from the Same Population, Collected by Different Techniques: Labor Input Data Compared Between Samples (man-days/acre)

| | One Visit | | | Weekly Visit | | | |
|---|---|---|---|---|---|---|---|
| | No. Obs. | Arith. Mean | % S.E. | No. Obs. | Arith. Mean | % S.E. | Significance of Levels of Difference |
| Aromatic Tobacco: Seed-bed Prep. | 46 | 2.8 | 14.3 | 34 | 3.4 | 12.4 | n.s. |
| Aromatic Tobacco: Ridging | 45 | 34.2 | 8.7 | 20 | 37.2 | 11.6 | n.s. |
| Aromatic Tobacco: Planting | 45 | 32.1 | 12.7 | 17 | 34.7 | 18.2 | n.s. |
| Aromatic Tobacco: Weeding | 46 | 13.8 | 10.2 | 11 | 24.0 | 27.1 | * |
| Aromatic Tobacco: Picking and Stringing | 43 | 398.2 | 9.7 | 24 | 282.2 | 15.6 | n.s. |
| Maize Mixture: Ridging and Planting | 51 | 23.2 | 10.6 | 37 | 24.2 | 12.5 | n.s. |
| Maize Mixture: Weeding | 51 | 9.1 | 13.0 | 21 | 4.2 | 12.5 | n.s. |
| Maize Mixture: Picking Maize | 51 | 5.4 | 12.4 | 21 | 8.8 | 15.7 | n.s. |
| Maize Mixture: Picking Nuts | 47 | 6.5 | 21.2 | 23 | 14.2 | 17.3 | * |

*Differences are significant at the 5 percent level.

Source: Compiled by the author.

Significant differences occur on two operations, weeding aromatic tobacco and picking groundnuts from maize mixtures. The difference in picking and stringing aromatic tobacco is virtually significant at the 5 percent level, with $\underline{t}$ = 1.87. These and other operations were susceptible to definitional problems. Groundnut harvesting was not broken down for the weekly-visit enumerators, and stages might well have been disaggregated by some respondents. Aromatic tobacco weeding is a joint operation combined with pulling the soil up around the plants and remaking the ridges to support the crop, once the weeds have been removed. Picking and stringing is made up of several stages and is interesting because corroborative evidence is available to show its very intensive labor requirement. Working among large farmers in Rhodesia, Morrow breaks the operation down into four stages: picking and putting the booked leaf into a box, needling the leaves, stringing the leaves, and hanging the strings on racks.[24] For six reapings Morrow gives a per-acre requirement of 207.6, which he defines as 100 percent efficiency. He comments that large-scale farmers should aim to organize their work gangs to achieve 85 percent efficiency. This would give them a per-acre requirement on the order of 260 man-days. By comparison, the work rate achieved under family farming conditions, with a relatively unfamiliar crop and on minute acreages susceptible to a heavy "scale effect," would seem spectacular at 282 man-days per acre. This is further evidence to suggest that the single-visit figure may be more realistic; and it is added to by the fact that hired labor, apparently missed by weekly enumeration, was used predominantly for picking and stringing the tobacco. A degree of the difference in the intensity of labor use will be accounted for by the significant difference in the size of acreage grown over the two samples, the limited-visit farmers having a higher scale effect with smaller plots. The easily defined cultivation and planting operations gave almost identical answers by both techniques.

Direct Comparison of Techniques in the Same Population: Cotton. In the course of an investigation, planning, and extension project in the area of one cooperative society (about twenty-five square miles) in Gelta district, Tanzania, from 1963 to 1966, four collections of data were made on samples of farmers from within the same population. The methods of collection differed from a single-visit survey to visiting on alternate days, and unfortunately comparison is confounded by two factors: collections were made over three different seasons, and detailed visit techniques were used only on nonrandom samples of farmers selected for their labor efficiency. Before presenting the data the four samples are briefly described:

In the 1963-64 season eighty-nine farmers were surveyed by a two-visit technique to provide data to allow construction of a typical farm unit for the area. Farmers were randomly selected from the registered list of the Livenge Cooperative Society.

In the 1964-65 season thirty farmers were selected for farm planning and extension; of these twenty-seven were from the original sample of eighty-nine. Selection was made on acres cropped per unit of available labor during the 1963-64 season as a reflection of vigor in farming. In the course of an extension program on the thirty selected farms, data were collected by visits on alternate days.

At the end of the 1964-65 season a one-visit survey was carried out on thirty-five farms randomly selected from the registered lists used as a frame for the original sample. Only family size and crop labor input data were collected.

The extension project was continued into 1965-66 to assess the turnover of farmers, since some were lost to the program and others joined. Extension was carried out on forty-two farms for the season and detailed data were collected by twice weekly visits. The forty-two were a core of eighteen previously selected farmers plus volunteers but were not randomly selected.

Rate of work data for the four groups is set out in Table 53.

The only direct comparison is of the data collected in 1964-65 and is distorted because twenty-seven of the thirty farmers in the detailed survey were selected to participate in an extension program on the basis of efficiency in labor use. Only limited data were collected in the single-visit survey, whose sample was selected at random from the same population, and no direct comparison of labor efficiency between the two samples is possible for 1964-65. Evidence from 1963-64, when the twenty-seven farmers were part of the original sample of eighty-nine, does demonstrate their improved efficiency and a higher rate of work index for the cultivation operations. Table 54 compares the selected twenty-seven farmers, the full sample of eighty-nine, and the sixty-two who were unselected.

The selected farmers had a faster rate of work than the whole sample, which would contribute to the increased area cultivated per unit of available labor.

Overall, however, this comparison suggests an upward bias in single-visit data. It is believed that this is due to three shortcomings in enumeration, which are now explicitly covered by the approach which has been described.

1. A failure to set up the particular field as the point of reference for the farmer.

## TABLE 53

**Rates of Work on Main Crop Operations from Four Samples of Farmers from the Same Population**
**(man days-per acre)**

| | | Two-Visit (1963-64) | Detailed (1964-65) | Single-Visit (1964-65) | Detailed (1965-66) |
|---|---|---|---|---|---|
| Number in Sample | | 89 | 30 | 35 | 42 |
| Family size | | 6.6 | 6.7 | 7.4 | 9.6 |
| Acreage cropped (per man-equiv.) | | 2.92 | 4.04 | n.a. | 2.85 |
| Cotton | Sesa | 8.8 | 4.3 | 9.8 | 6.1 |
| | Ridge/Plant | 10.6 | 8.2 | 12.0 | 8.1 |
| | Weed All | 12.4 | 11.1 | 23.5 | 8.9 |
| | Pick/Grade | n.a. | 28.4 | 34.8 | 27.3 |
| | Uproot | n.a. | 1.7 | 2.0 | 1.6 |
| Maize Mixtures | Sesa | 10.8 | 3.9 | 8.5 | 7.2 |
| | Ridge/Plant | 14.1 | 8.2 | 11.9 | 12.5 |
| | Weed | 9.6 | 4.5 | 7.9 | 5.4 |
| | Harvest Maize | 3.6 | 1.8 | 2.5 | 4.2 |

Source: Compiled by the author.

TABLE 54

Labor Efficiency Characteristics of Subsamples of Farmers

| | Original 1963-64 Sample, 89 farmers | Selected 27 Farmers | Unselected 62 Farmers |
|---|---|---|---|
| Acres Cropped per Man-Equivalent | 2.92 | 4.04 | 2.43 |
| Rate of Work Index in Cultivation | 100.0 | 82.8 | 107.5 |

Source: Compiled by the author.

2. A failure to enumerate length of workday for all members of the labor group for each operation, thus tending to attribute a day length to the group as a whole.

3. Having enumerated the labor group for the operation, the farmer estimates the number of days required to complete the operation on the field in question. It may be necessary to remind him who is in the labor force in order to be clear that the number of days refers to this group, or to supplement the question as to the period required with a further question asking whether all members of the group would appear for work every day over the period.

It is believed that the evidence confirms the usefulness of limited visit survey techniques in collecting labour input data. Other workers have adopted the technique from time to time, usually because results are urgently required. J. D. MacArthur reported apparent success in a study of Mwea Tebere rice farmers; and D. Rotenhan, working in Sukumaland, obtained consistent results from three small samples.[25] H. A. Luning expressed reservations in the more complex farming systems, including both coffee and tea enterprises, in the Rungwe area of Tanzania.[26]

Two factors seem to prevent the effective use of limited-visit techniques. Where workers switch from task to task, there is less basis for a general framework of experience which the method can exploit. Where continuous cropping is practiced, the timing of limited-visit surveys may fail to cover the period when some crops are in the ground. Use of the established crop as a reference point for quantification is an important feature of enumeration; and so where the sequence of crop establishment cannot provide these points, a frequent-visit technique will be required.

## NOTES

1. D. Pudsey, "Pilot Study of 12 Farms in Toro" (Kampala: Uganda Dept. of Agriculture, 1966). (Mimeographed.)

2. R. W. M. Johnson, The Labour Economy of the Reserves, Occ. paper No. 4 (Salisbury: University College of Rhodesia and Nyasaland, 1964).

3. Pudsey, op. cit.

4. Ibid.

5. R. H. Fox, "Studies of the Energy Intake and Expenditure Balance Among African Farmers in the Gambia," Ph.D. thesis (London: Medical Research Council, 1953).

6. A. Moody, "A Report on a Farm Economic Survey of Tea Smallholdings in Bukoba District, Tanzania," East African Agricultural Economists Conference paper (1970).

7. M. Hall, "A Review of Farm Management Research in East Africa," East African Agricultural Economists Conference paper (1970).

8. A. T. Richards, Land, Labour and Diet in Northern Rhodesia (Oxford: Oxford University Press, 1939).

9. T. J. Kennedy, "Cotton Farmers' Motivations in Kisumi," East Africa Economic Review, new ser. II.

10. M. P. Collinson, "Usmao Area," Farm Economic Survey no. 2 (Dar es Salaam: Tanzania Dept. of Agriculture, 1962). (Mimeographed.)

11. Pudsey, op. cit.

12. *Ibid*.

13. Johnson, *op. cit*.

14. R. W. M. Johnson, "The African Village Economy, an Analytical Model," *Farm Economist*, XI, 9 (1968).

15. J. Heyer, "Seasonal Labour Inputs in Peasant Agriculture," East African Agricultural Economists Conference paper (1965).

16. W. Scheffler, *Smallholder Production Under Close Supervision: Tobacco Growing in Tanzania*, "Africa Studies," XXVII (Munich: IFO, 1968). (In German.)

17. K. E. Hunt, *Agricultural Statistics for Developing Countries*, rev. ed. (Oxford: Oxford University Press, 1970).

18. Hunt, *op. cit*.

19. Hall, *op. cit*.

20. J. E. Bessel, J. A. Roberts, and N. Vanzetti, "Survey Field Work," Agricultural Labour Productivity Report no. 1 (Universities of Nottingham and Zambia, 1968).

21. Hunt, *op. cit*.

22. N. V. Rounce, *The Agriculture of the Cultivation Steppe* (Longmans, 1948); D. von Rotenhan, *Land Use and Animal Husbandry* in Sukumaland, "Africa Studies," XI (Munich: IFO, 1966). (In German.)

23. M. P. Collinson, "Maswa Area," Farm Economic Survey no. 3 (Dar es Salaam: Tanzania Dept. of Agriculture, 1963). (Mimeographed.)

24. Morrow, *The Study of Labour Use on Aromatic Tobacco Growing on Large Scale Farms* (Salisbury). (Mimeographed.)

25. J. D. MacArthur, "Labour Costs and Utilisation in Rice Production of the Mwea/Tebere Irrigation Scheme," *East African Agricultural and Forestry Journal*, XXXIII, 4 (1968).

26. H. A. Luning, "Patterns of Choice Behaviour on Peasant Farmers in Northern Nigeria," *Netherlands Journal of Agricultural Science*, XV (1967).

CHAPTER

# 12

# THE VARIABLE ATTRIBUTES: CAPITAL AND LIVESTOCK

Among traditional herding communities, livestock might validly be included with capital in a single category. Because of its wider importance as an enterprise, with objectives ranging from subsistence production to fertility maintenance, it is covered as an independent category. Capital is discussed first.

## CAPITAL

Capital is subordinate to labor and land as a factor of production in traditional agriculture. Farming systems in Africa are never capital-intensive in the sense in which the term is used of advanced agriculture. Family labor generates most of the capital content in the system. Yet, as in any agricultural system, what capital elements there are have great importance in development. Assets which are specific to a particular enterprise are a cost of undertaking it and, once made, the investment creates an inflexibility in the farming pattern which inhibits changes in both methods and enterprises. In a market economy the possibility of resale reduces this inflexibility, but it makes little contribution where assets have no market—and this is the case with land in many traditional agricultural communities. It is important in planning improvements to identify investments both as potential barriers to innovation and as a necessary cost of further expansion in the enterprises concerned. At the same time, because assets are "labor-created" and labor-maintained, they will represent an important element in the labor profile of the system, which it is important to record. This is particularly true in seasonal systems with definite troughs and peaks in the profile, where the pattern is liable to be altered by changes in methods. Shifts in the profile may change the opportunity costs of using labor for capital projects, either jeopardizing the long-term balance of the system or reducing the attraction of the changes being mooted.

The asset structure of the traditional holding covers the fixed and movable equipment and, particularly important, improvements to the land. In addition, working capital is included in this category. These three facets can be divided into specific and general assets, specific assets being the investments associated with a particular enterprise and general assets being those useful for all enterprises.

## General Assets

General assets include general-purpose buildings and equipment, usually limited to cultivation tools. Storage facilities are perhaps the only general-purpose buildings likely to be improved by innovation. Low-cost alternatives will be rare, and the difficulty of valuation by opportunity cost is immediately highlighted. Benefits to be had from the release of labor by the provision of central storage or water supply have rarely been quantified. Characteristically, traditional agriculture has a limited range of general capital goods; and with the intimate relationship between subsistence production and domestic consumption, the demarcation of productive farm capital may be difficult. However, since purchased capital goods are unusual, this factor will not seriously distort planning.

Bush-clearing and permanent land improvements may be either general or specific. The clearing of cover from soil types used for a range of farm crops creates a general asset, as do general soil conservation measures. On the whole, bush clearing and general permanent improvements are at the opposite end of the evolutionary path for traditional systems. Clearing is a feature of land-extensive systems in areas of high population density. Specific improvements, on the other hand, may be a feature of any stage of development. Certain working capital items, including some maintenance costs, are also properly classed as general assets.

## Specific Assets

Assets linked to a particular enterprise are also few and far between in traditional agriculture. Coffee fermentation tanks, groundnut shellers, and all processing equipment tend to be specific. Unless there is a land market locally, specific assets which are fixed cannot readily be mobilized as capital for alternative uses. On the other hand, the mobility and relatively short working life of machinery creates less inflexibility in the system. Bush-clearing of soils for a particular crop and permanent improvements such as rice bunding are similarly specific to enterprises. Permanent crops represent a major asset in many traditional farming systems; and investigation to determine establishment costs, in terms of the labor required up to maturity, will be important for planning. Most working capital items are specific, and recording these is particularly important. Outlay on purchased inputs added to outlay on hired labor measures the costs farmers are willing to incur out of existing income levels. It constitutes a limitation on the rate of adoption of innovations and therefore is a particularly important constraint in the planning sequence. Seed is a working capital item which is usually provided from the previous crop and is often ignored in costing traditional farms. For certain crops with high seed requirements, low yields, and poor storage qualities, seed supply may be an important constraint on development, particularly when improvements incorporate very high plant populations.

Planning requires the investigation of the cost and productive life of all assets and the capacity of equipment. Capacity is important to define the points of expansion at which new capital investment is required. Equipment may be acreage- or output-related, and the direct measurement of capacity is complicated by the spread of its use over time. It is best calculated from labor-use data which cover the operations associated with the use of the item concerned, for the capacity of any item will be unique to the conditions of the particular system. The cost and productive life of assets needs direct investigation in this category.

All assets are either purchased or constructed by the use of family labor: there are few purchased assets in traditional agriculture. The central problem in the investigation of capital is in the case of labor-generated assets where recall of the time required to carry out the constituent operations of capital projects may be poor for assets with a long life. For such assets the incidence of new investment in the current season will be limited among the population and thus also in the sample. The organization of enumeration constantly meets with this problem of ensuring enough observations to get reliable data. Similarly an important part of presurvey work is to identify such assets in advance.

### Presurvey Investigation

With asset structure an important criterion in identifying

homogeneous type of farming areas, the main presurvey need is to outline the typical structure. Assets should be itemized under fixed equipment, land improvements, and working capital and should be classified as specific or general, and purchased or built by family labor. The keys to eliciting information on land improvement will be provided by the preenumeration of the rotational cycle and conservation practice.

Where assets are constructed from a variety of materials with different productive lives, particular items which form a significant proportion of the total cost should be treated as independent assets; otherwise the fixed-cost element of an enterprise may be considerably biased. For example, a barn for flue-curing Virginia tobacco may use zinc-dipped flue pipes which must be purchased and have a life expectancy about half that of the structure itself. Since they form perhaps 85 percent of the capital outlay required for this type of barn, they justify independent treatment. These "partial" assets should be identified while drawing up the inventory for the typical farm.

For each item an estimate of its productive life and, if purchased, its cost should be obtained. The purchase of capital items is an important event, and recall of the cost and timing of purchase presents no problem for the farmer. An estimate of productive life is particularly important for an idea of the likely incidence of labor-based investments among a given sample within the period of investigation. Where asset life is prolonged, coverage in the sample will be poor. Some farmers in the sample will be replacing the asset in the current period; enumeration for these is no problem, though with frequent-visit techniques a special recording sheet is helpful—treating the event as an enterprise, as it may be spread over a long period. Where the capital items concerned are specific to important enterprises and significant amounts of labor are likely to be involved, it is important that incidence of visits be increased—particularly when construction is customarily done at important or potentially important periods in the seasonal profile. An alternative to a larger sample is the use of the investigational technique described in the labor category, relying on the experience of the farmer of the time needed for the work rather than on his memory. As with other labor investigations it will be useful to describe the operational sequence involved in the presurvey stage, as a basis for questionnaire construction. An example of the kind of confusion which can arise again refers to a curing barn for Virginia tobacco. M. P. Collinson reported that eleven out of forty-nine growers built barns during the season, with an average of forty-two man-days required for construction.[1] However, there was no presurvey awareness of the need to distinguish construction and maintenance, and it is almost certain that the two are confused, with forty-two man-days an underestimate of the labor required for gathering materials and constructing a barn.

The establishment of permanent crops and other specific investments present a special case which is outlined here in relation to permanent crops. These follow two types of cycles. A production cycle, which describes the changes in yield over the life of the crop, can be superimposed on an enterprise cycle, which is market-related. Given a viable market opportunity, there will be establishment up to the level of resource availabilities on individual farms. There may be reestablishment once its productive cycle is complete, if the marketing opportunity remains attractive. It will be useful to identify in the presurvey stage the likely phase of any local permanent crops, or enterprises requiring specific capital investment, within either cycle. There will be little purpose, and some difficulty due to a scarcity of observations, in recording the labor required for establishment of a crop which is well into its production cycle and unlikely to be reestablished. On the other hand, if the potential from improved practice is high, recording may be important.

## Survey

In the survey a questionnaire can directly enumerate all assets which are purchased, including working capital items, and those specific to an enterprise should be noted. Most of the capital elements will be derived from family efforts, and these will be enumerated in the same way as the labor data in the survey. Where incidence is low, observations are increased by using limited-visit labor techniques tapping the experience of the farmer. While a special questionnaire will be needed for general assets, specific items can be enumerated within the labor questionnaire for the enterprise concerned.

Owing to the ease of recall associated with capital assets, there are three approaches to the enumeration of the productive life of each asset; the use of two of these allows a cross-check on responses.

1. The age of existing assets is enumerated, and the sample data will show a decreasing frequency toward the time of replacement.

2. The proportion of the sample reporting new investment in the current season allows a second estimate of asset life.

3. Finally, the farmer can be asked either how long his

previous structure lasted or how much longer he expects his current one to last.

Where enterprises are new and the original asset is not yet exhausted, there will be added difficulty in estimating productive life of the asset. This was experienced in the tobacco example quoted. Farmers who had recently started growing were optimistic about barn life. While the proportion of farmers replacing their barns suggested an asset life of between four and five years, farmer estimates averaged 6.5 years.

The final question in this capital category relates to the valuation of land and permanent improvements. Valuation is relevant only where there is an established market for either land or standing crops. It is at its most problematical in areas with an emerging market. The incidence of transactions will be few, and value differentials for land improvements will be irregular. Presurvey investigation in the agronomy category will elicit local practices and survey will confirm their incidence in the population.

The category imposes no limitations on survey design. Difficult events can be covered by flexibility in collection techniques to ensure adequate precision in planning data, rather than raising the cost of investigation by an increase in sample size.

## LIVESTOCK

Livestock is treated as a self-contained category which will cover the information required for a full description of livestock enterprises. Aspects already mentioned are manuring, important for fertility maintenance, and herding communities, where cattle are a foundation of social custom. Few farm economic surveys have covered livestock; Collinson, D. Rotenhan, and H. D. Ludwig have touched on aspects, but comprehensive coverage is difficult because of the special nature of livestock as an enterprise and as a data collection problem.[2]

Treatment is complicated by the motivational priorities manifested in livestock in many communities, bound up with intergeneration transfer of wealth by inheritance and with marriage and dowry rights. Multiple rights in animals create ownership identification problems and complicate disposal possibilities. Primitive herding and grazing practice, often involving communal action and seasonal migration, add to the same identification problems and complicate survey design.

K. E. Hunt has advised careful consideration of the usefulness of livestock data before investigational surveys are undertaken.[3] As with cropping activities, a major criterion will be the proportion of the farm population who are livestock owners, an important point to be estimated in presurvey investigation. Preliminary evaluation should also decide whether cattle can have a role in short- and medium-term development, or whether basic reform of structure in the land tenure pattern and market outlets must precede management innovations. At one extreme might be a cattle-keeping community with marriage and inheritance customs tied to livestock, an area with no established market outlets for meat or milk products and with communal grazing supervised by the children, thus absorbing only minor quantities of labor and occurring on land not fit for cultivation. In such circumstances investigation is unlikely to contribute significantly to subsequent extension strategy. Justification for investigation in this type of situation would be to describe present livestock management as a basis for a technical research program, useful only if the required manpower and facilities for such a program are available.

With evidence of real potential, however, enumeration becomes important. An extreme case would be in pastoral nomadism, where <u>any</u> advance of the community is dependent on changes in their animal husbandry. Less extreme but still vital circumstances would be where technical improvements are available and stock appears to have a comparatively high potential. This is also true where stock is absorbing both land and labor in competition with crop enterprises, including cases where their social role is accelerating fertility loss in conditions of increasing population density. A dual classification of livestock enterprises has been made, and certain features of organization will tend to follow from the dominance of one role or the other, although the two are by no means mutually exclusive.

## Extensive Livestock Enterprises

Extensively run cattle are characteristic of traditional cattle-keeping communities which use the animals as a store of wealth for reasons of family security. With an advance in marketing, the glimmerings of a meat enterprise may emerge, particularly where increasing population density begins to pressure traditional grazing management practices. The most difficult enumeration problems fall into this class which is characterized by communal grazing and herding and by the division of ownership even of single animals. Investigation is complicated by the cattle being the traditional tax basis for nomadic or seminomadic farmers, so that any counting of animals is likely to be treated with suspicion and evasion. Where cattle in these circumstances are not a serious competitor for resources, there are no signs of an emergent enterprise, and no improved technology is available, enumeration is not worthwhile. It is important where population pressure is causing a clash between arable and grazing land, the two being substitutes, or where some diversion is occurring, albeit into a poorly differentiated meat market.

## Intensive Livestock Enterprises

Intensive livestock enterprises are characteristic of densely populated areas, i.e., urban vicinities and rural areas with a high-value cash crop giving a market for milk, or where fertility problems have stimulated stock-keeping for manure. Characteristically, farms in these areas have one or two animals living on the holding. Milk production demands closer control of breeding than is found in extensive systems; and manure requires closer control over housing and feeding and is properly a cost against the farm generally or against a particular crop. Where population density creates grazing problems, fodder production results; and this can be highly labor-intensive and in close competition for resources with crop enterprises. Intensive systems are relatively easy to enumerate, since the social complexities are reduced or even lost altogether and animals are in situ on the homestead.

Four subcategories of data are distinguished: management, herd composition and structure, fecundity, and production. Their relative importance and the presurvey and survey content will be weighted by the role of livestock in the system under investigation. An initial classification of the role of local livestock—extensive or intensive; for milk, meat, manure, or social purposes—will provide a guide to limit the content and to design the investigation. Unless livestock is the sole means of community support or a major potential cash enterprise, its attributes will not dictate survey design or sample size. It will be a question of fitting in with a design for the arable enterprises. Where stock is dominant, an independent investigation may well be justified.

## Management Practice

As with cropping practice, livestock management practices can be expected to be common to stock owners throughout the area and as such can be enumerated in the presurvey investigation, by local interview. Over the range of possible roles being covered, however, it provides a good deal of survey content. Management practice is further divided into four subcategories: feeding, housing, watering, and breeding.

### Feeding

Feeding practice is itself subdivided into grazing and the use of fodder.

Grazing. Rotational practice, the stocking rate, and length of grazing day are the most important facets of grazing practice. Details will vary, depending on whether there is communal or private organization. Grazing practice under private control implies an enterprise integrated with the arable system, and enumeration presents fewer difficulties. However, the onus in collection is thrown onto the survey proper, since practice will tend to vary with the management of the individual farmer. Objective measurement of grasslands and the enumeration of fencing as a capital asset will significantly increase survey content.

Under communal organization, rotational practice may be much more difficult to enumerate and the stocking rate well nigh impossible to assess, except by broad aggregate measures for the area as a whole. Presurvey investigation through the local grazing authority will be important. The basis of rotational changes should be investigated, to determine whether practice revolves around differences in soil type or location, including the possibility of seasonal migration, and whether the timing is on a seasonal basis or over longer periods. The rights of stock owners to overgraze fallow lands, including crop residues, should be stabilized.

The length of the grazing day is important under both types of organization. Estimates can be obtained in the presurvey stage and confirmed by enumeration during the survey. Herding practice will be more complex in communal organization, with the possibility that family herds are grouped to minimize the labor required in supervision. Where migratory transfers are practiced, herding may be done by "tribal relations" in the migration areas for some sort of fee—a further complication to the enterprise. Custom can be reliably established by local interview.

The presurvey enumeration of grazing and herding practice is particularly important under communally organized systems. Identification of the possibility of seasonal migrations to new grazing areas, and of communal herding, will influence both the timing of the survey and the approach to the enumeration of herd numbers. There is only a limited content under communal organization—confirmation of length of grazing day and the labor commitment required in herding. There is, as we have noted, a good deal more when grazing is under individual control within the farmlands.

Fodder. The use of fodder is associated with private grazing organization and intensive stock keeping. Whether fodder is collected or grown on the farm, there are immediate implications for family labor use for which data are required. Practices associated with fodder can be outlined in the presurvey stage. If use is seasonal, the timing and extent of the period are important. Other points to consider are the frequency of collection and whether it is the responsibility of particular family members. Both the source of fodder and the operational sequence in either collecting or growing it should be established. To quantify fodder labor requirements in the survey, it should be treated as an independent enterprise and enumerated in the same way as other crop activities.

## Housing

Stock housing will be enumerated as a specific asset, and investigation will involve the procedure outlined in the capital category. A distinction is needed between purchased and labor-generated components, with an operational sequence outlined for labor use. Survey will enumerate the cost, productive life, and capacity of the unit. Extensive herding systems will be limited to a simple night barn, but in intensive systems some description of the method of construction and the layout will be necessary as a context into which improvements for manure and milk production must be introduced. It will also be important to outline usual practice in the conservation and storage of manure, and there will be survey content covering the labor routines involved in both milk and manure production. For manure this will include the use and collection of litter for the housing. Any special housing arrangements for young stock are outlined in the presurvey stage.

## Watering

Watering practice in the area is enumerated in the presurvey investigation. The main distinction is whether the animals are brought to the water, characteristic of an extensive system, or water to the animals. Both alternatives imply the use of labor; and the pattern of watering outlined prior to the survey should be filled out, like any other operation, in a labor-input questionnaire in the course of the survey. The frequency of watering, distance to water, and seasonal variations in sources of supply are the main points for preenumeration. Important management implications arise from the distance animals must be moved to water. As S. Groeneveld has pointed out, as the distance increases, the grazing area is more and more restricted, thus limiting the size of herd which can be carried.[4] Water sources and seasonal variations in the area can be checked during the survey and are particularly important where tenurial reform is seen as a prerequisite to development, with consequent problems of access and the possibility of public investment in watering points.

### Breeding

Breeding management has two facets: the degree of control over calving—particularly the time of calving—and calf rearing practice.

Practice will tend to be the same over areas in which animals play the same role within the farming system and can be described in presurvey investigation. Information should include whether farmers control servicing and, if so, the timing which is followed in the area. In extensive systems, where animals run together, the methods of control should be described. The main points on calving practice are whether both sexes are retained, how long they are allowed to suckle, or whether they are hand-reared, and at what age they are weaned. Survey content will be limited to confirming weaning age on the current crop of calves, though with an individual-based organization it may be valuable to confirm the incidence of the practices described in the presurvey investigation.

The survey content of these four data groups is dominated by labor items which can be handled by the techniques discussed in the labor category and which are central to survey design. There are no special problems for limited-visit surveys, since all the events for enumeration are regular, often daily, events and have a pattern which is readily established by presurvey investigation. Problems may arise in frequent-visit techniques where supervisors and enumerators must be aware of the likely onset of seasonal events: calving, fodder use, and migratory grazing. Few livestock systems have been thoroughly investigated, and a thorough presurvey outline of the management routines followed is essential to the construction of an effective questionnaires for all collection techniques.

## Herd Structure and Composition

Herd structure and composition is wholly quantitative and

enumeration is by survey. The main information needed is on herd size. Data collected in the form of an inventory can also give useful data on herd dynamics, current levels of removal, and transactions within the community and with other areas. Observation of the herds and counting the sex/age groups are the techniques for enumeration. Where presurvey investigation can establish sex/age groupings traditionally distinguished by the community, it is useful to exploit the same classification in the survey. For frequent-visit techniques the inventory can be an open one at the beginning of the investigational period and closed at the end. For limited-visit surveys, one visit to enumerate this subcategory should be taken at the end of an annual period. Timing the visit after the completion of, or before, calving for the season will create a better reference period for the farmer. Presurvey investigation of the seasonal cycle will indicate the most appropriate timing. An example of data collected by a single visit from which an inventory can be constructed is given in Table 55.

The major problems in collection arise in extensive systems, where communal herding and migratory grazing imply difficulty in identifying ownership. However, except for large herds, perhaps over fifty animals, there will be few response problems, since cattle are an intimate part of family life. Frequent-visit techniques, forewarned on migratory grazing and communal ownership, will have the full year to observe and will be able to choose the appropriate times for enumeration over the investigation period. Enumeration should be infrequent in order to minimize response burden, since events are well remembered and, usually, few and far between. Limited visits, on the other hand are at a disadvantage unless the livestock grazing cycle is allowed to dictate the timing of one of the visits. Where this is impossible, visiting is susceptible to the absence of animals or unknown interfamily groupings. Where such practices are common and survey timing is irreconcilable with a clear picture of livestock holdings, and where livestock is important as an enterprise or a large resource user, an extra visit is necessary for effective enumeration. As we shall see, this is also true for sales of livestock products. Limited-visit enumeration without the more complex communal or migratory aspects presents few problems. An opening inventory should be established by working backward over the year for each subgroup of the herd, covering births, sales, deaths, purchases, and slaughterings; such transactions will have a very limited incidence in most households and are important and memorable events.

This subcategory is particularly susceptible to local prejudices against herd counting, created by past tax experiences. Great care is needed during survey preparation to reassure not only the community

## TABLE 55

### Numbers of Stock on Sample Farms at Date of Survey, and Changes During the Year

| Age Group | | Cows +2 | Cows 1-2 | Calves | Bulls 1-2 | Bulls +2 | Goats | Kids | Sheep | Lambs |
|---|---|---|---|---|---|---|---|---|---|---|
| Numbers Present | | 835 | 332 | 503 | 217 | 165 | 429 | 102 | 592 | 175 |
| Numbers Consumed | | 85 | 40 | 77 | 27 | 13 | 132 | 122 | 49 | 58 |
| Purchases | No. | 7 | 9 | – | 4 | 1 | 14 | 2 | 15 | – |
| | Price (E.A. shillings) | 137 | 129 | – | 100 | 160 | 20 | 7 | 20 | – |
| Sales | No. | 1 | 2 | – | 2 | 6 | 1 | – | – | – |
| | Price (E.A. shillings) | 100 | 240 | – | 100 | 160 | 20 | – | – | – |

Note: No distinction is drawn between transactions within or outside the survey area.

Source: M. P. Collinson, "Usmao Area," Farm Economic Survey no. 2 (Dar es Salaam: Tanzania Dept. of Agriculture, 1962). (Mimeographed.)

leaders but also the farmers themselves that no use will be made of the information for tax purposes.

## Fecundity

Further investigation of herd dynamics can be justified in areas where beef projects are to be introduced, but it is of limited relevance under intensive systems where a very small number of animals are kept for milk or manure. Of course, with more sophisticated enterprises, with calf sales as a source of income, it reasserts its importance. Table 56 shows Rotenhan's estimates of breeding dynamics from cross-section data in three surveys in Sukumaland. A particular point of interest is the inhibition of dynamics in Ukerewe, which has the very high population density of 415 persons per square mile.

To complete the picture, rather more information is needed on the mortality rate among calves and on the frequency and extent of herd decimation due to shortage of food and water or to the incidence of disease. This is important in establishing whether continuous buildup of herd size is accelerating a fertility problem or whether the problem is caused solely by increasing arable requirements.

All the information is historically oriented and therefore essentially memory-dependent. An initial important point to be established in presurvey investigation is whether the farmer can be expected to know the calving history of his female animals. Even where he does have this knowledge, enumeration will be limited to animals which have been with the present owner throughout their life. This is unlikely to rule out many animals in traditional herding communities, where transactions are few. Presurvey enumeration will also give the incidence of serious decimations to serve as a reference point for questioning on individual farms.

Survey questioning will be limited to mature animals, and area averages will be derived by grouping the answers for each attribute. Where herds are relatively large, a subsample of females will be sufficient to cover variation within the survey area. The farmer's probable knowledge of the history of a particular animal should be ascertained by asking how long he has owned it, and how many calves it has had, before more detailed questioning on calving history. Age at first calf, the number of calves, calving interval, and calf mortality, for a sample over the area, will give a picture of herd structure and growth. The incidence of herd decimation is easy to record for the individual farm and will normally parallel the history for the community

TABLE 56

Example of the Type of Information Needed to Show Herd Dynamics

| Area | Shinyanga | Kwimba | Ukerewe | Average |
|---|---|---|---|---|
| Average Age of Cows over One Year | 7.6 | 9.1 | 5.9 | 7.5 |
| Average Age of Mother Animals | 9.2 | 10.9 | 10.3 | 10.0 |
| Age at First Calving (months) | 38 | 37 | 54 | 39 |
| Period Between Calves (months) | 16 | 18 | 26 | 18 |
| Number of Calves per 100 Females | 39.3 | 37.4 | 12.0 | 36.5 |

Notes: Average age of first calving is from animals with more than two calves.

The number of calves is of live births.

Source: D. von Rotenhan, Land Use and Animal Husbandry in Sukumaland, "Africa Studies," XI (Munich: IFO, 1960).

as a whole. The severity of particular decimations may be more problematical, though they will be well remembered by the farmers.

## Consumption and Sales

Enumeration of this subcategory aims to measure the use being made of livestock by the household, and hence their value as a capital asset. Results may be surprising. Collinson, in a tentative evaluation of extensively run livestock in central Sukumaland, estimated a return on capital value of the order of 25 percent from produce in the form of meat and milk, including the increase in inventory over the season.[5] He estimated the productivity of labor used in cropping to be slightly higher than stock: E. A. Shillings 5/15 versus E. A.

Shillings 4/65. However, since efficiency in the use of available labor was only 33 percent for cropping, the productivity of available labor in cropping was E.A. Shillings 1/41, whereas with stock, requiring labor over the whole year, the return remained at 4/65.

Data to be collected are products utilized or sold from the herd; these cover animals, hides and skins, milk and milk products, and manure. Animal transactions are relatively few and are well-remembered events, while the use or sale of hides and skins is closely associated with domestic consumption of animals. These two items present no enumeration problems over the range of collection techniques, though, being quantitative, they are properly content for the survey itself. The main problems in collection are in milk, milk products, and manure, and are similar to the problems in collecting some crop output data. These items form flows over the season; and where there are sales, they are often to a poorly structured and uncontrolled market, particularly among extensive cattle herders.

The initial information required in the presurvey stage is on the range of livestock products in the community. Further questioning should try to identify the market outlets and stages of processing of dairy produce, or the timing and sequence of work in manure spreading. Irregular sales of milk and milk products to an undifferentiated market within the community are beyond the scope of a limited-visit survey design, for neither memory nor experience can cope effectively with the events concerned. Where it is an important farm enterprise, a frequent-visit technique will be necessary. There are possibilities for a limited-visit design if the produce is sold to a marketing channel which records purchases made from individual farmers. Sales of sample farmers can be monitored through this as a secondary source of information.

## Milk

In frequent-visit surveys, production of milk should be recorded at each visit. Having preenumerated the containers used for holding the milk, the number of these filled daily should be recorded and the number of cows milked noted. This should be supplemented at monthly intervals by the enumeration of any new calves. Where milk is processed, each stage should be followed through. The capital equipment involved will be a specific asset and enumerated in the same way as other assets, while the labor requirements of both milking and processing will be recorded against the stock as an enterprise. Both of these aspects can be covered by the technique being used within the survey design. Detailed collection techniques require care to avoid double-counting where milk is accumulated before processing, since a reference period longer than a day or two is difficult for the farmer to isolate and thus is susceptible to end effect. The pattern of processing over the season will depend on the shape of the herd-lactation curve. It may be seasonal during periods when supply exceeds domestic needs. Presurvey investigation should describe the shape of the curve, which will depend mainly on whether animals are calved at one particular time. Where calving is not specific to a particular time, processing will be more closely dependent on herd size and production over the season will be less variable. Incidence of processing among stock owners should provide some estimates of the size of herd at which processing becomes profitable as well as the amount of domestic consumption.

Limited-visit surveys can give an estimate of total milk production. If the shape of the lactation curve of local animals can be found from local experiments comprising present and improved stock, the level of production enumerated during survey visits can be tied into the curve to give the expected quantity produced over the lactation. Survey procedure will depend on whether the herds are calved at a specific time of the season. Where they are, the number of calves is enumerated and estimates are made of the amount of milk produced in a day at three points of the curve: one month after calving, the date of survey, and the end of the lactation period. The aggregated results for the sample will give area estimates at three points on the lactation curve. Response will be based partly on recall of the level of production at the beginning of the milk "season," partly on current production, and partly on the farmers' experience of the seasonal pattern of production repeated year by year. Where there is no specific calving period, the sample of individual cows enumerated for fecundity can provide the same estimates for their last calving. In this case results will be grouped at the points located on the "usual" lactation curve and similarly allow estimates of average production. This data on total milk production, supplemented by conversion ratios where it is processed, will allow an estimate of milk products as a potential cash enterprise. Comparison of farmers who do process, and those who do not, as subsamples will show the herd size necessary to cover processing overheads under current management practice.

## Manure

Manure presents few problems. Actual labor used in carrying and applying manure will be enumerated in the course of the labor-input questionaires. The important point to enumerate here is the quantity which will be moved in a day. The containers used should be measured and the number of loads carried by family members in a working day established. The routine will be firmly fixed in farmer experience, and an estimate of the number of loads carried will be obtained by any collection technique. Family members likely to be involved in the work can be preenumerated before the survey.

Livestock, then, gives a closely integrated data category isolated from the main categories because extensive stock keeping in particular is often virtually independent of the crop enterprise and requires almost a special survey investigation. It should be emphasized again that survey design may be limited by the need to accommodate livestock. Where this is apparent from the presurvey investigation, careful consideration should be given to short- and medium-term potential either as enterprise or as resource competitor before a decision is taken to include full livestock coverage. In communities keeping stock extensively, where there is little immediate prospect of an enterprise potential, coverage may usefully be limited to aspects of resource competition. This is particularly true where periodic decimation precludes long-term accumulation of larger herds and where a fertility problem is caused by increasing arable requirements.

## NOTES

1. M. P. Collinson, "Aromatic Tobacco and Virginia Tobacco: A Comparative Survey of Two Tobaccos on Family Farms in the Tabora Region of Western Tanzania," East African Agricultural Economists Conference paper (1970).

2. M. P. Collinson, "Usmao Area," Farm Economic Survey no. 2 (Dar es Salaam: Tanzania Dept. of Agriculture, 1962) (Mimeographed); D. von Rotenhan, Land Use and Animal Husbandry in Sukumaland, "Africa Studies," XI (Munich: IFO, 1966); H. D. Ludwig, Ukara: A Special Case of Land Use in the Tropics, "Africa Studies," XXII (Munich: IFO, 1967).

3. K. E. Hunt, Agricultural Statistics for Developing Countries (1969).

4. S. Groeneveld, Problems of Agricultural Development in the Coastal Region of East Africa, "Africa Studies," XIX (Munich: IFO, 1967).

5. Collinson, "Usmao Area."

CHAPTER

# 13

# THE VARIABLE ATTRIBUTES: OUTPUT

Output is a measure of the performance of the system and quantifies the results of resource use at present levels of technology and management ability. It forms a benchmark for the evaluation of alternative improvements to be tested in the planning sequence. Enumerating the quantities of foods produced also gives a basis for assessing the importance of each subsistence activity to the family.

Self-evidently the content of this category is the amount produced of each crop grown and the value realized for any produce sold. (Livestock products were discussed separately in the livestock category.)

It is useful to reiterate that average yields per acre are required for the area as a whole, and that the planning model places no importance on establishing the input/output relationships for each particular farm. There are two approaches to output measurement. In the first, the total production of each crop is reduced to a per-acre basis. This method is subject to severe limitations in systems where the same crop is grown by different techniques requiring separate representation in the planning model. Unless grouping across all activities growing the same crop can be justified, the method of yield measurement must distinguish between different activities. Total output divided by total acreage cannot do this. The second approach is to raise sample results to a per-acre base, which has the advantage that a specific plot can be treated as a sample unit and every identified activity covered. It gives added flexibility to the sampling scheme.

Sampling techniques involve either crop cutting or crop measurement; cutting locates and harvests small plots within the field, and measurement uses relationships between physical plant characteristics and final yield. Virtually all the work on objective yield measurement techniques has been done with single crops, with their maturity habits dictating the timing of the survey. The major difficulty in our application is the need to measure output for all the crop activities in the system over the same production period. This complicates the design, for crops in the same system may have very different maturity habits. Tea, with a picking season which lasts most of the year and with changing yield expectations over the life of the bush, may be grown alongside annuals which require picking only once during their life. In between these are crops like cotton and fruit, which matures as the plant develops, requiring several pickings over a longer period. The matter is complicated further in farming systems with no distinct season and the possibility of a crop maturing for harvest in any month of the year. Even in a highly seasonal system harvests may be prolonged; in Sukumaland the new legumes and sweet potatoes become available in March or April, and the final cotton is not picked until August, a spread of five or six months.

A final complication is the gradual harvesting of produce for immediate consumption, widely characteristic of traditional agricultural systems. Crops such as green maize, sweet potatoes, and cassava roots and leaves are picked or uprooted from the fields as required. They present measurement problems of both quantity and standardization of product, since the crop may be consumed at varying stages of maturity. D. Pudsey shows the flow of output from sweet potato plantings over eleven farms.[1] Over a fifteen-month period a total of 13,450 pounds was harvested; 432 pounds was the lowest monthly total and 1,336 pounds the highest, with maturity varying between four and thirteen months and a spread of up to seven months in the maturity of potatoes taken from the same plot. The pattern demonstrates the value placed on a supply of fresh sweet potatoes by these farmers. So complex a pattern cannot be simulated in an area model, and an approximation is required to show the cost, in terms of utilities forgone, if the pattern of supply is compromised by resource reallocation as a result of innovation.

## VARIATION IN OUTPUT DATA

The survey will measure interfarm variation in yield levels, but exaggerated precision in sampling is spurious if interseasonal variation is very large. The key to the precision required is the use of the data as a benchmark in evaluating changes. The greater the increments in productivity expected from innovations, the lower the precision needed in enumerating existing output levels. For example, where fertilizer use on maize will increase the yield from 400 to 1,600 pounds per acre, a 15 percent standard error on existing maize yields may be adequate and the extra costs required to reduce it to 10 percent may not be justified. Initially survey design must cater to possible eventualities, but there will be indicators of local yield levels to allow a comparison of existing and potential performance.

The emphasis placed throughout on the identification of independent production activities is fundamental for output data. Differences in yield levels and in harvest timing will be the result of growing the same crop as different activities and, together with the importance of the contribution to total production, will confirm the need for independent representation in the planning model. Pudsey gives results for differentiated bean-growing activities with yields recorded by plot, which are set out in Table 57.

The significance of each of these four subpopulations and the basis of the decision as to whether each should be investigated independently, would rest on either the contribution each made to the total supply of relishes in the system or the use of peak labor required by each activity. Aggregating the four subpopulations together gives a coefficient of variation of 11 percent, greater than for the smallest and most variable of the four and confirming the need for independent investigation.

## PRESURVEY REQUIREMENTS

Presurvey investigation is mainly concerned to establish characteristics of the output data which will limit the choice of techniques for measurement.

For some idea of whether high precision is required, estimates of local and improved yield levels should be obtained. This is a difficult point, since yield per acre is an inappropriate criterion for comparison in a labor-limited system; and to compare the labor productivity of the alternatives begs the major task of the survey. The same standards of accuracy—a 7-10 percent standard error—should be sought for output data. But on an activity for which the required number of observations seems unlikely to be obtained, the test of expected differentials is an aid to a decision. It can be qualified by a subjective evaluation of the amount of extra labor which will be absorbed by the improvements, and by the shifts in timing required, which might well bring labor requirements into peak periods. A similar preliminary enumeration of the relative importance of different

TABLE 57

Differences in Yield/Acre from Beans Planted
Under Varying Management Conditions

| Management Characteristic | | Number of Plots | Mean Yield (lbs./acre) | % S.E. |
|---|---|---|---|---|
| Pure Stand | 1st Rains | 14 | 826 | 18.3 |
| | 2nd Rains | 9 | 520 | 26.7 |
| Intercrop | 1st Rains | 7 | 668 | 23.2 |
| | 2nd Rains | 13 | 249 | 28.1 |

Source: D. Pudsey, "A Pilot Study of 12 Farms in Toro" (Kampala: Uganda Dept. of Agriculture, 1966). (Mimeographed.)

types of starch staple and relish will aid the decision on whether an enlarged sample is needed.

Characteristics of output data which will influence the type of collection technique used are summarized under two main heads.

### Identification of Harvest and Postharvest Practice on the Same Crops Grown as Distinct Crop Activities

The same crops grown as independent activities may be harvested and stored differently, in which case there is no problem in measurement. Where the produce is bulked after picking, after processing, or in storage, recording must be done before this occurs. The following points should be investigated:

1. Whether each method of growing the crop gives the same type of good at the same time of the year.

2. Whether produce is processed before storage and in what state and place it is stored, with particular note of when produce from different activities is stored together in the same state.

3. Whether the store is accessible and easily measurable, either by volume or by a physical count.

4. For crops consumed directly out of the field, the timing of the consumption period and of the final harvest of the residual crop.

5. For crops requiring more than one picking, the number of, and intervals between, pickings.

## Identification of Market Outlets for Produce

The market outlets for any produce which is sold should be enumerated. Monopoly outlets, which may be a useful source of sales data, should be discussed, particularly for likely loopholes in the records of produce purchase.

Techniques for yield measurement range from estimation to crop cutting. Most of the necessary conditions for accurate measurement by crop cutting are known, many derived from the work of Mahalanobis at the Indian Statistical Institute and Sukhatme at the Indian Council of Agricultural Research.

Zarkovich notes of crop cutting:

> At first sight it appears to be a simple, objective procedure leading to unquestionably accurate results. In fact however, it is a complex procedure with many elements that may be subject to error. The application of the procedure requires extreme care or its results may be more inaccurate than eye estimates.[2]

The main advantage of crop cutting is its basis on the plot, so that all independent subpopulations can be measured. The statistical and procedural requirements for crop cutting have been well-documented, and the steps are outlined here.

### Random Selection of Fields as Sample Units

Our application requires only an area average and may use all plots of each crop on the farms selected or may select a plot on each farm, or as many as are required, to give the necessary level of precision. There is a good deal of flexibility in sampling procedure. Subsampling of plots on the farm is valuable in reducing enumerator time, thereby potentially increasing his coverage of farms, keeping down costs, and reducing respondent burden.

Relating resource use and productivity for the single farm, on the other hand, implies a much more comprehensive procedure to cover variation within the plot.

## Location of Sample Plots on Selected Farm Plots

The plot should be randomly located in the field. A rigid and clearcut procedure forestalls any initiative of the enumerator in favoring areas of better growth, an important source of bias in the method.

## Size and Shape of Sample Plots

The best size and shape of plots is still a subject for discussion in the literature. However, it is clear that accuracy in marking out and weighing depends most on the experience and enthusiasm of the enumerators. Small plots are more susceptible to abuse from poor enumeration, and large circular plots appear to give the most reliable results. With large plots, fewer cuts are needed to achieve the same sampling error, and thus the extra time required for cutting is balanced out to some extent but will vary according to the type of crop. Table 58 illustrates the balance between sampling error and time taken in cutting.

Although the general relationship is shown by the data, plot sizes between 1/50 and 1/100 of an acre, at the higher end of this scale, are more appropriate for survey work with the standard of enumerator usually available. These would give an index of between 150 and 200 in terms of the final columns of the table. The smaller plot size would be used on crops grown in small fields.

## Method of Harvest

The crop is cut and processed, preferably being weighed at each stage when drying, shelling, or threshing is to be done. It is accepted that the methods used in harvesting and processing should simulate those of the farmer as closely as possible; any neglect of this forms an important source of bias. The importance of the gap between biological yield, the weight of the whole crop from the sample plot, and the useful yield—that which the farmer has available for disposal—has been repeatedly stressed. Zarkovich gives the example of three methods of potato lifting and the different levels of

TABLE 58

Relative Efficiency of Sample Plots of Different Sizes

| Size of Plot (sq. ft.) | No. of Cuts Needed for the Same Sampling Error | Average No. of Hours of Work to Cut Crop | Total Time Needed | Total Time Needed When 100.9 sq.ft. = 100 |
|---|---|---|---|---|
| 12.5 | 3.47 | 0.3 | 1.04 | 67 |
| 50.3 | 2.19 | 0.6 | 1.31 | 84 |
| 100.9 | 1.74 | 0.9 | 1.56 | 100 |
| 201.1 | 1.39 | 1.4 | 1.95 | 125 |
| 544.5 | 1.00 | 2.6 | 2.60 | 167 |

Source: S. S. Zarkovich, The Quality of Sample Statistics (Rome: FAO, 1964).

damage caused: wooden ploughs and manual lifting, 4.6 percent damage; throw wheel machines, 1.1 percent damage; elevators and sieve-down machines, 12.6 percent damage.

In addition, crop losses greater than 11 percent have been observed; and when subsequent storage, handling, and transport losses are added, storage losses being particularly prevalent in developing countries, the useful yield is significantly lower than the biological yield. Traditional practice of harvesting the crop as required for consumption is impossible to simulate for sample plots, there being no possible basis for relating the choice of the householder to the fruits on the plots. An alternative approach is essential for this type of crop; and its occurrence should be identified before the survey, since the need for additional techniques will be important to survey design.

Early demarcation of sample plots is necessary to avoid biases from early crop failures. This creates problems of control and the farmers must be prepared to protect the plots from casual picking, although this is difficult where other members of the family have disposal rights over the produce from the crop concerned.

Our farm economic application is concerned with all the crops on the farm, and the problem of organizing yield cuts according to the different planting and maturity permutations limits the use of crop cutting to detailed-visit surveys. Only by frequent visits to the farmers can the enumerators keep in touch with the state of the crops. Prolonged harvesting of such crops as tea and coffee, and crops like cotton, requiring several picks, confirm this limitation. In surveys organized on a frequent-visit basis, the extra work load involved in crop cutting is likely to limit the coverage of the enumerator, particularly where the harvest period is prolonged; this is also true of systems characterized by continuous cropping and no marked season. Also, the more plots to be covered per farm, the heavier the enumerator's work load.

Table 59 shows some examples of variation and precision in crop cutting samples in Sukumaland, Tanzania.

Here grouping across maize mixtures, justified by the homogeneity in labor requirements with or without cassava, improves the precision. Overall, for the farming system concerned, a sample size of between fifty and sixty observations would realize a standard error of less than 10 percent on the maize crop. Where grouping across mixtures with and without cassava is justified, this number of observations could be realized on thrity-five to forty farms. Thirty observations would give the same level of precision for the cotton crop.

## RECORDING PRODUCTION

Household budget surveys have increased our experience in recording consumption data. By recording and weighing the foods consumed over the year, comprehensive and accurate data on both quantities and timing of supplies can be obtained. However, the response burden is so heavy that surveys are usually made over short sample periods tied into the food year of the community.[3] A pure consumption approach to the estimation of total output used in the home would be too onerous for both respondent and enumerator. Subsamples might be used for consecutive periods tied to the food year, but the enumerator burden would remain. Further, the consumption period is not synonymous with the production period; and unless the farmer could be persuaded to store his harvests separately for subsequent enumeration as consumed, there could be no conclusive relation between the two. To do this would prolong the survey period, with consequences for both costs and coverage. The problems associated with a pure consumption approach seem intractable.

TABLE 59

Precision Achieved in Crop Cutting by Plot in Detailed-Visit Farm Surveys, Sukumaland (1/50th acre plots)

| | Cotton | | Maize from Mixtures | | | | | |
|---|---|---|---|---|---|---|---|---|
| | | | With Cassava | | No Cassava | | All | |
| | 1964-65 | 1965-66 | 1964-65 | 1965-66 | 1964-65 | 1965-66 | 1964-65 | 1965-66 |
| Number of Plot Cuts | 19 | 37 | 21 | 35 | 22 | 20 | 43 | 55 |
| Mean Yield (lbs./acre) | 819 | 644 | 516 | 447 | 665 | 317 | 592 | 404 |
| Standard Deviation | 440 | 287 | 376 | 262 | 483 | 134 | 440 | 232 |
| Coefficient Variation (%) | 53.7 | 44.6 | 72.8 | 58.6 | 72.6 | 42.3 | 74.3 | 57.4 |
| % Standard Error | 12.2 | 7.3 | 15.8 | 10.0 | 15.5 | 9.5 | 11.3 | 7.7 |

Source: Compiled by the author.

Researchers have used a modification of the consumption approach, tackling measurement at an earlier stage but carrying out the measurement in much the same range of ways as is available for consumption surveys. M. Hall evolved the technique of measurement just after harvest to cope with crops of bananas and other foods in a system of continuous cropping with small amounts being produced throughout the year.[4] As with crop cutting, this technique is practical only within a survey designed for frequent visits to the farm throughout the year. It can cope with problems of prolonged or continuous harvest and of gradual consumption.

In the extreme case, enumerators would be required to weigh the amounts of crop harvested each day. This is clearly impractical, since the respondent burden would be very large, with the farmer having to wait for the enumerator before storing or processing the produce. In communities where individuals are responsible for particular crops or plots, several respondents may be needed. The farm coverage of the enumerator would be considerably reduced, both by the need for daily visits and by the increase in complexity and physical work on each farm. In practice, collection is memory-based, supplemented by the careful evaluation of units of measure and conversion ratios for the same crop in various states of maturity and processing. Some researchers, such as Bessel, Roberts, and Vanzetti and D. Pudsey, have gone so far as to provide special containers—forty-gallon drums and four-gallon tins—to their sample farmers as basic units of measure which they retain after the survey as an incentive to participate.[5] All containers used to collect and carry the range of crops grown must be enumerated for volume/weight ratios of the crops in the states in which they are transported. This can be done concurrently with survey enumeration, new combinations being investigated as they arise. Where the containers are standard throughout the community, evaluation of volume/weight ratios should be done on a formal sampling basis to the required level of precision. Some products can be recorded in their natural units, such as bunches of bananas or number of coconuts. Careful subsampling is required to establish the weights involved in both containers and natural units. Random observations over all the farms in the sample can provide the conversion ratios for an area average.

Plot differentiation presents a difficult problem for recording by this technique, not so much in recording the number of units coming off a particular plot as in the way the produce is kept once it is picked. Pudsey notes the case of beans harvested in bundles and hung up to dry. Once they are hung up, plot identity is lost and threshing yield must be averaged for the whole crop. Where variation in shelling percentages, threshing, drying, or grading results, or unit weights, is a function of the basis of plot differentiation, the comparison is distorted and the method cannot be used unless grouping the activities can be justified. The problem does not arise only between plots of a crop grown in different seasons. Groundnut plots of distinctive times of planting might well realize the same volume of unshelled nuts, but the shelling percentages may differ radically, giving contrasting weights of nuts. The proper organization of sampling for container measurement and conversion ratios is central to effective use of this technique. It should be based on the principles that only area, not farm, averages are required, considerably reducing the work required on each farm, and that the necessary degree of differentiation of crop source will depend on the proposed complexity of the planning model. Finer divisions than this are superfluous. Table 60 shows the kind of data needed for each crop.

The technique is unsuitable for limited-visit surveys, though a consumption approach may be a useful supplement in communities with a food year of a regular pattern. Enumeration of the quantities of home-grown produce eaten on the day's visit, or the previous day, gives a basis for aggregation on the food calendar prepared in the food economy category. The feasibility of this approach will be restricted to one or two main crops, either basic to the diet for the whole year or with a clearcut seasonal incidence, but it is a useful check on other methods being used.

Crops recorded by Pudsey using this technique show a modest sample size for a 10 percent standard error.

The decision whether to increase sample size to cover the more variable bean activities recorded by Pudsey and set out in Table 57 would depend on the relative importance of their contribution to food supplies.

## MARKETING OUTLETS

The use of marketing outlets for recording total output is a very easy alternative to any other collection technique, particularly where outlets are restricted and sellers are registered. Sales data are an important check on other techniques; and where the response burden is high due to a large number of crops on the farm, it can be reduced by dependence on market data for those which are sold. Sales data are particularly valuable for crops with continuous or prolonged harvest periods.

## TABLE 60

Container Weights and Conversion Ratios for Some Crops

| | Maize | | Rice | Millet Sorghum | | Groundnut Bambarra | | Cowpeas Bean Grams | |
|---|---|---|---|---|---|---|---|---|---|
| | Cobs | Shelled | Paddy | Heads | Threshed | Pods | Shelled | Pods | Shelled |
| Dry weight per bag (lbs.) | 120 | 200 | 175 | 140 | 100 | 84 | 180 | 40 | — |
| Dry weight per 4-gal. tin (lbs.) | 20 | 36 | 29 | 24 | 35 | 13 | 28 | 7 | 39 |
| % Conversion to Shelled Weight | 80 | — | 80 | 75 | — | 65 | — | 66 | — |

Source: M. P. Collinson, "Usmao Area," Farm Economic Survey No. 2 (Dar es Salaam: Tanzania Dept. of Agriculture, 1962). (Mimeographed.)

TABLE 61

Example of the Precision Achieved by a Daily Recording Approach to Output Data and Observations Needed for a 10% S.E.

| | Pure Beans, 1st Rains | Pure Sweet Potato | Millet |
|---|---|---|---|
| Number of Observations | 14 | 24 | 17 |
| Mean yield (lbs./acre) | 826 | 4,723 | 507 |
| Standard Deviation (lbs./acre) | 566 | 2,908 | 239 |
| % Standard Error | 18.3 | 12.6 | 11.4 |
| Number of Observations for 10% Standard Error | 47 | 38 | 22 |

Source: D. Pudsey, "A Pilot Study of 12 Farms in Toro" (Kampala: Uganda Dept. of Agriculture, 1966). (Mimeographed.)

The main shortcoming with market data is their aggregated nature. Unless there is a clear timing difference between deliveries from various plots, production cannot be attributed to particular soil and management situations. The data imply a total output approach.

Even closely controlled market outlets are by no means foolproof. Very often the operation of controls indicates that alternative outlets may be more attractive but are judged to be against the interest of the economy. Individuals who are not registered growers may sell through those who are in order to avoid a registration fee, perhaps forfeiting the right to a second payment at the end of the season. Data on cotton collected from the same farms by both crop cutting and the sole market outlet allow a comparison of results which is set out in Table 59. In 1964-65 three farmers were known to have sold cotton for others, which inflated their per-acre results when calculated on deliveries and distorted the average for the relatively small sample of eighteen farmers. The table gives a comparison of results from sales and crop cutting data with and without the three offenders. Added to this, cotton was differentiated between plots grown with and without fertilizer; and the results of one crop cut in each subgroup were weighted by acreage to give an average yield per farm for the comparison with market data. A similar comparison is made for a larger sample of farmers in the following year, without any known distortion of the sales data.

When market sources are used exclusively, this type of distortion is difficult to detect. Usually when farmers do sell others' crops, the amount is quite large, since their profit per unit is inevitably small. Unrealistic yield figures which are turned up in data analysis—e.g., greater than three standard deviations above the mean—may be rejected. Usually the phenomenon is limited to transactions from non-registered to registered sellers and the bias is upward; however, where individuals have debts to the marketing organization, particularly if credit is allowed off-season, it may also occur between registered sellers, and in this case the errors cancel out. Judgment is required to evaluate the market outlets for shortcomings of this sort. Local extension staff or marketing officials will often be able to pinpoint individuals in the sample susceptible to the types of pressures implied by the tactics adopted.

Table 63 gives some examples of the level of precision achieved in five farm surveys, where yield estimation was wholly dependent on sales information from cooperative societies.

Precision here is good, with fifty observations giving a 10 per-cent standard error. It must be noted, however, that these are special cases in which the crops introduced into the traditional system are grown on a homogeneous basis throughout the community, with no mixtures, on the same soil type, and planted within a well-defined period and in highly seasonal areas. Where the marketed crops are more closely integrated into traditional husbandry, or where the system is less markedly seasonal, wider variation can be expected, especially where produce marketed is surplus food requirements—characteristic particularly of maize and bananas as cash crops.

## YIELD ESTIMATION

Estimation techniques are the only alternative to market outlets for enumerating output data in limited-visit surveys. Since market sources are limited to cash crops, estimation procedures are inevitably required for subsistence production in limited-visit techniques. Several approaches to estimation are possible, but they are divided into two groups: to establish total farm output by enumerating each

Comparison of the Precision in Yield Estimate
Through Crop Cutting and Sales Data on Cotton

| Year | Basis of Estimate | Group | No. | Mean lbs./ acre | Standard Deviation | % Standard Error |
|---|---|---|---|---|---|---|
| 1964-65 | Cutting | Fertilizer | 25 | 1,269 | 423 | 6.7 |
| | | No Fertilizer | 19 | 819 | 440 | 12.2 |
| | | Full sample, Weighted Yields | 18 | 1,024 | 485 | 11.1 |
| | Market | Full Sample, Farm Yields | 18 | 1,450 | 1,008 | 16.5 |
| | Cutting | Selected Sample | 15 | 1,005 | 502 | 13.0 |
| | Market | Selected Sample | 15 | 1,068 | 468 | 11.4 |
| 1965-66 | Cutting | Full Sample, Weighted Yields | 38 | 762 | 267 | 5.7 |
| | Market | Full Sample, Farm Yields | 38 | 767 | 445 | 9.4 |

Note: In 1964-65, for the full and selected samples, only those farms have been used with data available from both sources.

Source: Compiled by the author.

## TABLE 63

**Precision Achieved in Yield Data by Relating Sales Data to Cropped Acreage on the Farm**

| | Cotton | | | Aromatic Tobacco | |
|---|---|---|---|---|---|
| | 1961-62 | 1962-63 | 1963-64 | 1963-64 | 1963-64 |
| Number Observations | 82 | 54 | 83 | 46 | 30 |
| Mean Yield (lbs./acre) | 586 | 577 | 461 | 297 | 353 |
| Standard Deviation | 399 | 353 | 315 | 183 | 222 |
| % Standard Error | 7.5 | 8.3 | 7.5 | 9.1 | 11.5 |

Source: Compiled by the author.

possible crop use, a technique which does not allow plot differentiation, or to estimate yields on a field basis, by either eye estimation or plant measurement, which do allow plot differentiation.

## Enumeration by Crop Use

With crop production and food consumption as the central theme of his livelihood, the traditional farmer inevitably has a very clear idea of his needs and of his current and likely future situation. The basis for this knowledge, however, is purely domestic and is impossible to break down into components for enumeration: the components themselves, the capacity of his family, and their food preferences present equally complex enumeration problems. The farmer will know that if his grain stores are full after harvest, there will be little problem in feeding his family until the following harvest; this conclusion involves a quantification balancing needs and supplies. Estimation can be approached by volume measurement of stored crops, supplemented by consumption and sales up to the date of enumeration; and questioning should be taken crop by crop to give continuity in response over the stages of consumption, storage, and sales.

### Consumption

Timing of visits is important in order to minimize the consumption element in the sum as the most probable source of error—and this factor immediately rules out the use of the technique in continuous cropping systems. There is every likelihood of a pattern in consumption which will have been outlined while exploring the community food year in the presurvey investigation. The pattern is exploited in the questionnaire sequence, emphasis being placed on the quantity used each day since consumption of the food concerned started for the season. The person responsible for collecting the food for preparation prior to eating is the best respondent, and three points should be followed up:

1. Whether the crop was collected from the field or from the store.

2. The state of the crop when collected and the receptacle in which the crop is carried—its volume the usual level to which it is filled.

3. The frequency of collection. This sequence can be checked occasionally by reenumerating on the basis of eating. The frequency of eating and quantity eaten should be consistent with the frequency of collection and the quantity collected.

## Storage

Although there is a clear basis for enumeration of quantities in storage, there are problems of access, since stores are sometimes an integral part of the living accommodation and are often closed with mud or other materials to protect the crop. Likely access problems require presurvey enumeration and may need special attention when the community is being prepared for the team's visits. Certain storage methods—for example the tying of maize cobs to tall racks outside the house—lend themselves to easy counting. Others, such as a storage facility consisting of a large cylinder of plaited wood in the roof of the house, where the smoke will protect it from insects, present problems. Usually quantities of the basic staples are so large as to preclude weighing, though this may be feasible with the supplementary relish crops and stored seeds, which are often in self-contained lots, and may be as quick as volume measurement.

First, all storage facilities containing the specified crop must be enumerated, omitting any containing last year's produce. Where the crops from activities which must be enumerated independently are stored together, as is possible with continuous cropping, separate enumeration will be impossible and this method is impractical. Then the volume of storage containers is measured or the number of units estimated if the crop is tied up in bundles or on a rack. Finally, the state of the crop is recorded.

On-farm investigations must be supplemented by independent sample studies of threshing, shelling, drying, and grading percentages and of the volume/weight ratios of crops in different states. Local material should be used, if necessary from farmers in the area.

## Sales

Information on sales is readily enumerated through a controlled outlet, where the pattern of sales is frequent and regular or sufficiently infrequent to make each sale an important event. Where sales are frequent and irregular, with no market control, enumeration is very difficult; and if such sales make up a significant proportion of total output or a significant part of cash income, the method is ruled out.

Total output is the sum of these uses and allows the calculation of yield per acre over the whole area of each of the crops grown on the farm. There is no differentiation possible by varying management or natural conditions.

No direct comparison of this type of estimate with crop cutting has been made. Table 64 presents data collected in surveys in which this technique was used over the period 1961-65: three in different seasons in Sukumaland and two in the same season on different sample farms from the same population in Tabora. Data on plot cuts made in Sukumaland in 1964-65 and 1965-66 are given in the last two rows as an indirect comparison of the similarity between the results of both methods.

## Crop Measurement

Eye estimation of the yield of standing crops is still the traditional basis for crop production statistics in many developed countries, though the trend is toward more objective treatment. Farmers' estimates of the yields of standing crops in traditional agriculture is problematical because of the difficulty in finding a unit of measure for estimation. No doubt the farmer has clear ideas of whether he is about to have a good or poor harvest; but quantifying this, except in terms of his storage capacity, is difficult. Training enumerators in the task is a possibility; but unless they see the results of familiar stands bagged and weighed, they have little chance to accumulate the experience required. On the whole, because of the large range of crops usually involved, training is impractical. The use of experimental or demonstration plots as examples is often precluded by the wide differentials between these and local yields, so that they provide no useful common denominator. Nor do mixtures, giving an overview only of the tallest constituent, lend themselves to eye estimates.

A good deal of work has been done on the relationship between the physiological development of crops and final yield, particularly in the United States. Although the principles of the relationship are increasingly understood, the techniques of measurement are still in their infancy; and the conditions of crop growth in particular areas require careful local experimentation to derive the appropriate coefficients. The density of stand, the number and condition of fruits, and their probability of survival to maturity form the core of the information needed. Houseman and H. F. Huddleston estimate that it requires three years' research to establish coefficients for a crop grown in a particular environment.[6] In the conditions of the developing countries, bearing in mind the high level of manpower required, the

## TABLE 64

**Yield Data Collected by Summing Estimates of Consumption, Stored Crops, and Sales**

| | Number Farms | Maize Mean Pounds/ Acre | % S.E. | Number Farms | Groundnuts Mean Pounds/ Acre | % S.E. |
|---|---|---|---|---|---|---|
| Sukumaland | | | | | | |
| 1962-63 | 62 | 565 | 8.9 | 49 | 196 | 13.5 |
| 1963-64 | 86 | 452 | 10.1 | 31 | 123 | 23.6 |
| Tabora | | | | | | |
| 1963-64 | 51 | 459 | 11.4 | 47 | 185 | 20.6 |
| 1963-64 | 43 | 471 | 11.9 | 47 | 176 | 17.2 |
| Sukumaland | | | | | | |
| 1964-65 | 43 | 592 | 11.3 | 16 | 43 | n.a. |
| 1965-66 | 55 | 404 | 7.7 | 15 | 129 | 21.4 |

Note: The yield cuts made in 1964-66 are from the same farm population as the estimated Sukumaland data for 1963-64.

Source: Compiled by the author.

development of such techniques must have a low priority. It might, however, form a useful supplement to present experimental work.

Crop measurement requires all the paraphernalia of crop cutting, and sampling and organization are equally elaborate. Its advantage lies in seasonal agriculture, where measurement can be made on all crops at once during later maturity and formulas can be used to aggregate varying maturity states of the fruit counted.

One further problem deserves some comment, particularly in the light of the planning sequence adopted: the valuation of output, and particularly of intercrops. Whenever crop sales are enumerated, the price, units of sale, and the time of sale in the season should be recorded as a guide to the commonality and seasonal stability of markets in the area. It is worthwhile including a supplementary question on the purchase of foodstuffs usually grown, or of seeds, to add to the number of observations on local price levels. In areas or on crops where transactions are predominantly between farms, this will double the number of data obtained.

A good deal of discussion has centered on the valuation of subsistence crops. The yields of mixtures can be recorded by any of the methods described, subject to the limitations of the methods themselves. Combining the outputs of constituents to measure the productivity of the resource use involved may present a serious difficulty. When the constituents are all marketed there is no problem; but when various nonmarket priorities are satisfied by the constituents, comparisons with other activities satisfying a different priority mix are difficult. Retail food prices form a usual basis for valuing subsistence crops, as a measure of the opportunity cost of changing to enterprises with marketed output; but these fail to cover preference and insurance criteria, and the uncertainty or seasonality of the local retail food outlets. The problem is explored a little further, but not solved, in the planning sequence.

The conclusion must be that available techniques for the collection of output data severely limit survey design. Limited-visit designs are infeasible when the same crop is grown under different conditions in the system which are important enough for independent representation in the planning model. The exception to this is where the output of the discrete activities is kept separate through processing and storage, allowing estimates of total production by each activity. Even frequent-visit surveys may require a combination of collection techniques, with crop cutting supplemented by harvest recording on crops which are picked continuously over the season. Limited-visit techniques are feasible only where total output of each crop in the system can be related to the total acreage it covered, and this may group several plots if the sources of variation between them do not justify their independent representation in the planning model.

## NOTES

1. D. Pudsey, "A Pilot Study of 12 Farms in Toro" (Kampala: Uganda Dept. of Agriculture, 1966). (Mimeographed.)

2. S. S. Zarkovich, The Quality of Sample Statistics (Rome: FAO, 1964).

3. E. Reh, A Manual on Household Consumption Surveys, "Nutritional Studies," 18 (Rome: FAO, 1962).

4. M. Hall, "A Review of Farm Management Research in East Africa," East African Agricultural Economists Conference paper (1970).

5. J. E. Bessel, J. A. Roberts, and N. Vanzetti, "Survey Field Work," Agricultural Labour Productivity Investigation Report no. 1 (Universities of Nottingham and Zambia, 1968).

6. E. E. Houseman and H. F. Huddleston, "Forecasting and Estimating Crop Yields from Plant Measurements," FAO Monthly Bulletin of Agricultural Economics and Statistics, XV, 10 (1966).

CHAPTER

# 14

# CONCLUSIONS ON SURVEY ORGANIZATION AND DESIGN

This chapter summarizes the limitations on designs which have emerged from the discussion of the characteristics of the attributes which must be measured by farm survey. It concludes that most programs will be composite, with the survey unit varying the design according to the conditions of the particular farming system under investigation. The first part of the chapter provides some pointers to survey organization in the field, emphasizing the selection of and conditions of work for enumerators and on thorough preparations in the community to be investigated.

The appropriate location for a survey organization is the local research center, though it cannot be overemphasized that units must be coordinated nationally as well as with local research and extension personnel. In many ways, setting up the survey units is the most difficult phase. The purchase of equipment and the selection and training of enumerators can be done during the period in which the economist in charge is working on the identification of homogeneous type of farming areas.

A four-wheel-drive vehicle will normally be essential for each survey team. Transport within the survey area will depend on the nature of the terrain, though bicycles are normally the most suitable, with supervisors either on motorcycles or with the survey vehicle. Other equipment will depend on the predominant type of survey design. Portable field accommodation is always useful, particularly for limited-visit surveys when the staff is in the field for only a two or three months. Where the ministry concerned has vacant houses in the area, these are an obvious first choice. Field equipment will depend on the particular methods of measurement adopted, though certain items, such as balances and measuring wheels, boards and clips, will be generally useful. Care should be taken in choosing the calibrations for balances, which will vary with the type of crop and the particular containers to be measured. Whatever specialized equipment is required should be ordered as early as possible, since supply is unlikely in less than three or four months in many developing areas.

The selection of enumerators who are likely to succeed in what is essentially a public relation is difficult. A period of probation is strongly advised. The most one can say is that enumerators must be capable of writing and of arithmetic to the level of calculations required in the fieldwork. The unit should be structured as a career opportunity, with the best of the initially selected enumerators having the chance of further training for the formation of a supervisor cadre. A fairly high turnover must be expected in the early periods of fieldwork, and the planning of commitments should anticipate losses. This of course presumes a permanent team, which is preferable from many points of view. Almost inevitable the situation of a farm economist setting up a survey program in African agriculture is that of a newcomer trying to break into a technically orientated establishment. He must fight for funds and facilities and the key to success is staff, for with staff come the associated rights to allowances, transport, and accommodation.

There is a great deal of support for permanent enumerators (by Catt and MacArthur, for example) and the distinction between census work and an ongoing, operationally orientated survey program should be emphasized.[1] The understanding of the enumerator grows while he is on the job, and reliability and consistency depend on experience. A permanent team reduces the supervision requirement, though failings associated with low morale, poor conditions of service, and idleness still require control. Temporary workers introduce a new dimension of error arising from inexperience. In a way this is a dangerous comment, since levels of supervision, even of permanent enumerators, are usually lower than is desirable. The importance of systematic supervision for the reliability of the data and the morale of a survey team cannot be overestimated. A supervisor must share field conditions with the enumerators, in order to maintain satisfaction in the unit and prevent a fall in morale, as well as to judge if conditions are adequate.

Under certain conditions, when unit resources are stretched, temporary enumerators are a necessity.[2] Where the investigation is straightforward, useful data can be obtained by them. Two points, however, deserve emphasis:

1. Detailed supervision is a prerequisite for the use of temporary enumerators, for an experienced supervisor is conscious of errors of inexperience which must be controlled. Usually temporary enumerators in survey work are staff already resident in the area, a factor which can be good, in that they are accepted by the community, or bad, because they will have an initial advantage over their supervisor. The local senior officer must ally himself closely to the supervisor in order to establish the authority needed for effective control.

2. It is dangerous to employ part-time temporary enumerators. Adding a further item to the work schedule of local staff will cause resentment. If they are responsible to another authority for other work, a harmonious team is unlikely. Temporary enumerators should be used only in expediency, and the expediency should be important enough to divert staff completely from other duties.

The failings of poorly supervised temporary enumerators are exemplified by the results of a frequent-visit survey later checked by resurvey.[3] Both surveys covered samples of fifty growers drawn from the same population. The single-visit survey by the permanent survey team recorded the use of hired labor on 23 percent of the farms at a particular peak period of the season. Detailed collection, with inexperienced enumerators, recorded no use of hired labor on any farms.

There has been a good deal of discussion on the desirable level of education for enumerators. A consensus would suggest that between six and ten years is sufficient. The quality required varies with the task. A more perceptive staff is needed for single-visit surveys involving coverage of the farm for a whole season and requiring an insight capable of cross-checking to give internal consistency to the questionnaire. Similarly, with a complex system of agriculture, an understanding of the reasons for questions allows a development in the process of the interview. These qualities are not always commensurate with longer schooling, but in general a better-educated staff helps in establishing a survey program because they will absorb the work faster, can train less qualified men as the unit expands, and will provide the cadre of supervisors.

Besides perception and the ability to read and write, D. Pudsey has stressed the need for the enumerator to fit into the local community.[4] There is a tendency among the educated—who are, after all, an elite—to grow away from their roots. A certain stigma attaches to agriculture and the traditional way of life. While this feeling is understandable, and disappears as sympathy with the rural masses develops, it causes arrogance and impatience on the part of the enumerator which destroys the rapport required for any kind of reliability in response. It is important that the enumerator be prepared to fit into his appropriate level in the community hierarchy, asking no distinction by his position as an official.

Survey organization can be broken into three phases: area preparation, data collection, and data tabulation and analysis.

## AREA PREPARATION

The preparations required before the survey can be carried out include three stages: a conditioning excercise in public relations, a sampling design, and the presurvey investigation. We have already dealt with presurvey investigation, and the first two stages now deserve some comment.

### Area Conditioning

The farm population under investigation should fully accept the implications of the survey before data collection begins. This acceptance is particularly important to limited-visit surveys, when there is no time, as there is with frequent visits, to establish a rapport with farmers over the survey period. With detailed collection techniques there can be a month of trial collection before the season begins to test out farmer/enumerator relationships. Even with this type of design, the basis for success is one of good relationships with the community before visits to individual farms begin.

K. E. Hunt makes the very valid point that an institutional hierarchy which leads down to the individual farmer is a useful vehicle in preparing a population for investigation.[5] The institutional chain may be of several types, but the overheads of preparation are reduced if it can form sampling strata at the same time. Cooperative movements, political parties, and marketing or administrative organizations, which need to penetrate to the individual household to perform their particular function, are all possibilities. But it is important that the hierarchy used be one with which the community freely identify itself. In particular the community leaders must be in favor since their cooperation is the sine qua non of an effective survey.

An example of preparation for two surveys in Sukumaland will illustrate some of the more important points. Both followed the same path through the cooperative movement. At the start discussions were held with the regional agricultural officers and the Victoria Federation of Cooperative Unions, a tertiary-level organization in the movement. The political arm of the administration was included in early discussions, for failure to do this may precipitate disaster at a later stage in the work, when local political officials resent penetration of their areas without the approval of their seniors. The area having been selected, letters from the regional agricultural office and from the cooperative apex organization to their local representatives cleared the route into the area and local preparation began. Initially local meetings with government departments and with the cooperative union, the secondary-level organization in the movement, were held for information. Cooperative primary societies were selected at the union from the society population of the area.

The manager of the cooperative union, the district agricultural officer, and the survey team held meetings with the committee of each primary society selected as a first stage unit. The purpose and procedure of the survey was explained and, if the society leaders were agreeable—as indeed all were—the registered lists of the society were used as a frame to select a sample of farmers. A date was arranged and an extraordinary general meeting of the society was called, to which the selected farmers were explicitly invited. At the subsequent meetings the purpose and procedure of the survey were explained to the farmers, as society members, and the list of selected farmers was read out to the meeting. The fact that the body of the meeting was in favor of the work created considerable pressure on the selected individuals for their cooperation, not only then but also during the interviews. The active participation of primary society officials in the random selection procedure and in the subsequent use of society records was extremely valuable. Individual respondents clearly felt they had an obligation to cooperate to the community, rather than to an outside, unknown body. Indeed, at the meetings the only objections raised were that some of the selected farmers were unsuitable to represent the community. It often took a good deal of explanation to convince members that bad farmers needed to help just as much as good ones. There can be no substitute for an articulate but modest local member of the survey team conducting the case for the survey at these meetings. The presence of the unit leader, an expatriate in t this case, may give the meetings "occasion," but the nuances of discussion in the tribal tongue must be understood.

The major need in public meetings is for a peg on which to hang the survey. There can be little immediate benefit to cooperators or the community as a whole. Where the survey can be tied to the availability of a credit scheme or improved seed at the primary society in the following season, it is helpful. Survey benefits of a medium and long-term nature have little meaning to farmers.

The local institutional officials will often be those who can provide presurvey information in the main data categories.

## Sampling

We have already noted that sampling techniques are well documented. The central problem in sample design is usually the lack of a comprehensive and up-to-date frame, and preenumeration to draw up this frame is often required. This and the need to confine enumerator areas in order to minimize transport costs and optimize coverage of sample units favor the use of a stratified sample. Any clearly demarcated administrative divisions can be used as first-stage units (or second-stage, where the type of farming area is seen as a purposively selected first stage), and preenumeration is reduced to the population of these units. The balance between the sampling fractions of the two strata should be struck to give the enumerator a fair work load within the first-stage unit, confining his area of activity and optimizing his coverage. M. P. Collinson selected twenty-three cooperative primary societies as first-stage units with five farmers from each society.[6] This proved a poor design; not only were the number of preparatory meetings increased drastically but, with the limited-visit collection technique used, enumerators had to be shifted between primary societies every third day, considerably increasing transport cost. In the next survey five first-stage units were selected with fifteen farmers from each. The lesson of enumerator capacity as a criterion in sample design had been learned.

Stratification may be useful where there are distinctive subpopulations, perhaps cattle owners using ox plows for cultivation Where subgroups are small but a representative sample is desirable, and if the frame identifies each unit as a member of the subpopulation, stratification by subgroups using a variable sampling fraction allows the enumerator's commitment to be kept in balance with the required coverage of each population. Where the existence of subgroups is known, preenumeration will often allow not only updating of the frame but also identification of the two populations.

Various types of frames have been used for farm survey work, many out of expediency. While preenumeration is superior to any existing frame, the cost and time involved need careful consideration.

Registered lists of cooperative members were used by researchers in Tanzania, though only a proportion of the farm population were members.[7] A. Larsen has investigated the reliability of these lists compared with the political cell organization into which the whole country is now divided.[8] His findings demonstrate a degree of bias in the lists, related primarily to scale differences, with larger families tending to be cooperative members. His data are presented in Table 65.

Sample sizes were large enough to contain variances even though a very wide area was covered. Standard errors on variables on which most sampling units offered an observation were below 10 percent. Of the significant differences only three, one at the 1 percent level and two at the 5 percent level, give concern.

A scale difference is demonstrated by expressing gross farm income and cultivated area on a per capita basis, which reduces the level of difference.

| | Cooperative Members | Ten Cell Members |
|---|---|---|
| Gross Farm Income per Head (shs.) | 285 | 264 |
| Hectares Cropped per Head | .375 | .355 |

That all the cooperative members grew cotton for sale, and 15 percent of the ten cell members did not, is reflected in the wider difference between crop returns, significant at the 1 percent level. Cotton sales, shown separately, are similar in both samples once the scale difference is removed.

| | Cooperative Members | Ten Cell Members |
|---|---|---|
| Cotton Sales per Head (shs.) | 110 | 101 |

Nevertheless, a scale bias is present, as is a small efficiency bias; and the comparison illustrates how this can arise from a biased frame. Where the survey seeks to investigate cotton growers, the registered lists may be justified as a frame, while where the terms of

TABLE 65

Comparison of Statistics Obtained from Random Samples of Cooperative and Cell Members, Sukumaland, 1968-69

| Variable | | Sample of Cooperative Members (n = 114) | Sample of Cell Members (n = 219) | Significance of Level of Deviation |
|---|---|---|---|---|
| Gross Farm Income | (shs.) | 1,807 | 1,525 | .05 |
| Crop Returns | (shs.) | 1,341 | 1,098 | .01 |
| Food Crop Returns | (shs.) | 644 | 601 | n.s. |
| Livestock Returns | (shs.) | 443 | 406 | n.s. |
| Farm Costs | (shs.) | 64 | 51 | .10 |
| Net Farm Income (per man-equiv.) | (shs.) | 686 | 618 | .15 |
| Hectares Cultivated | | 2.38 | 2.05 | .10 |
| Cotton Sales | (shs.) | 696 | 582 | .10 |
| Food Crop Sales | (shs.) | 12.6 | 21.4 | .10 |
| Household Size | | 6.35 | 5.77 | .05 |

Note: Cotton sales are averaged and tested only for cotton growers.

Source: A. Larsen, "A Choice of Sampling Population in Rural Sukumaland," East African

reference are to the whole population, the ten cell units offer a more representative base for sample selection. Where frames of this type are available, the cost of preenumeration is not justified. There are going to be far greater inaccuracies arising than the differences from this source. Mere thought can often improve the frame, with little effort on the part of the investigator. For example, although each cooperative society has its registered list in Sukumaland, these are rarely updated. However, paralleling the list, each society has a card for each farmer delivering his cotton to the society in the last season, whether he was a member or not. These cards form an automatically updated frame. When considerable updating is required, as is often the case with taxpayers' lists or an electoral roll, preenumeration within selected first-stage units may be equally efficient. Area preparation, including conditioning, sampling, presurvey investigation, and the field testing of the questionnaire (when a new one is to be used) should be possible within a two-month period. Initial contact should be well in advance, however, and much of the early work can be spread over the preceding six months. Area preparation requires the intermittment commitment of both the economist, as head of the unit, and the field supervisor(s). The final month requires a full commitment from both supervisors and enumerators.

## DATA COLLECTION

During both the presurvey preparation and the data collection, enthusiasm on the part of the unit in the field cannot override bad living conditions. This is particularly so in detailed collection designs, where the period is very long, but is also true of limited-visit surveys. Field staff should be as comfortable as possible, and can usefully participate in deciding what camp equipment should be purchased. Often a hard start to life in the field is useful as a sorting mechanism for unsuitable staff. Once the initial turnover is made, it is imperative that the nucleus of the unit be as settled as possible. The chain of authority within the unit for the different aspects of work, accommodation, and terms of service should be clear to all. The burden of decision should be with the head of the unit, although responsiblities can be delegated to supervisors as the unit expands and as they gain experience. Initially the supervisor should have authority only in clearly defined areas of the work itself, and unit members should have direct access to the head on other matters.

Completed questionnaires should be checked in the field by the supervisor. For limited-visit surveys these will cover a whole farm, and for detailed visits arrangements must be made for getting them to the office for processing. Questionnaire design, the spaces allowed, and the contents of each sheet will reflect the frequency of collection for checking and forwarding to the office. Some researchers, such as Bessel, Roberts, and Vanzetti, have built systems of control, including a second visit by a senior enumerator, into the organization, with bonuses for correctly completed questionnaires.[9] This is particularly desirable in teams where personal control may be difficult because of the large numbers. Small teams may rely on the morale of the unit, but success depends on the personalities involved. Where morale is difficult because of antagonism between cadres in the unit, a formal approach is necessary. Even with a feeling of purpose within the unit, checking remains important. With detailed visits where supervisors are few and must cover a large number of enumerators, a random basis for checking is often useful; the supervisor is given a randomized selection of enumerators under his care and a randomized selection of their farmers, in order to prevent a formal pattern of checking, which might be abused. The length of time in the field will vary according to the collection technique adopted, ranging from a full season for frequent-visit techniques down to a month for single-visit techniques.

## DATA TABULATION AND ANALYSIS

Where, as in our application, there is a predetermined objective of analysis and planning, it is estimated to require the full commitment of the economist for a three-month period. Clearly, where research into aspects of the traditional system is an investigation of objectives, using the data to test relationships would require a significantly greater commitment by qualified manpower.

Data tabulation, on the other hand, depends on the collection technique being used. Frequent-visit techniques, giving daily records, require a much greater effort in tabulation, since the data must be aggregated prior to tabulation. For such techniques an additional clerical staff is required, working parallel to the enumerators in the field. With limited-visit techniques the enumerators themselves will tabulate the data after the completion of fieldwork.

The later stages of area preparation and data collection itself, as necessarily timely stages dictated by the pattern of the season, will rule the organization of the program. Although both tabulation and analysis involve full commitments by both economist and clerical staff, they are flexible and can be fitted into the pattern imposed by the fieldwork in the area concerned.

The fact that both the planning and the extension of the survey are dependent on local conditions give further emphasis to the importance of the presurvey investigation. We have already belabored the point that the range of collection techniques centers on the frequency of visits to the farm. All frequencies adopt as many objective measurement techniques as possible; but with limited visits, as we have seem, conditions impose certain restrictions on their usefulness. Before comparing the costs of alternative designs at the opposite ends of the spectrum of available techniques, it is useful to summarize the main conditions of traditional systems which preclude the use of limited-visit surveys.

Acreage measurement, counting of livestock, and crop cutting for yield estimation are the only wholly objective techniques; and the first two of these demand an element of memory effort by the farmer which may be prejudiced by poor area preparation. All other data, and labor data under all circumstances, are memory-dependent. Recall is governed by the nature of the attribute and by the respondent's ability to demarcate clearly the reference period. Three sources of complexity create badly defined events and open-ended reference periods, any of which may rule out limited-visit techniques and may raise the cost of detailed techniques by increasing the frequency of visits required or by increasing the enumeration load on each farm, both reducing the coverage of the individual enumerator.

## Cropping Calendar

The climate governs the characteristics of the cropping calendar, and three levels of complexity can be identified.

1. Continuous, often interpenetrating cropping. Areas with rainfall all the year round may have no distinct seasons, so planting is relatively haphazard and crop sequence may be dictated by labor supply rather than climatic exigencies. A characteristically flat labor profile results. These areas attracted the early practice of sedentary agriculture and tend to be densely populated, often with semipermanent crops. Similarly, the overlapping of crops, often the same crops, or different crops on the same land, removes the reference points for a limited-visit approach, confusing the reference period. Frequent visits are prerequisite for survey design in such areas.

2. Areas of bimodal rainfall and two distinct planting seasons. Limited-visit techniques are possible where the seasons are separate;

but where they overlap, particularly with crops on the same land, or the same crop featuring in both seasons, three visits will be needed. Productivity of the same crop in the two seasons is likely to differ, and it will be important to estimate output of the two activities independently. Labor use must be enumerated while the crop area is visible on the ground, or the reference point is lost. The three visits will be to establish the acreages of the first season; to estimate first-season production and additional acreages; and to measure production of the second season and the acreages of later established crops. A limited-visit design may be impossible where the overlapping occurs with a variety of crops at significantly different times. The second visit must be able to be timed after the harvest of crops to be repeated, but before new plantings of the crop are established, if these touch on the same land. In farming systems where permanent fields have evolved, and crops conform to plot size rather than vice versa, the problem is unlikely to arise.

3. Seasonal cropping. Where there is a clearly defined period of the year in which all lands used for arable cropping stand fallow, all types of survey design are feasible, subject to the complexities to be discussed below.

## Variation in Soil Type or Husbandry Practice on the Same Crop

In systems in which the same crop is grown on different soil types, under different fertility conditions, or under different husbandry practices on individual farms, yield levels may differ. Unless a system of yield estimation consistent with limited visits can be used, detailed collection techniques are required to record yields by plot. Circumstances which allow of limited visits are the following:

1. Different uses are made of the respective crops and an estimation technique can discriminate between their output.

2. The present productivity of both activities, although different, is very much lower than potential results at higher levels of management. Then grouping the two activities is justifiable, particularly when they are of limited importance in the system.

## Important but Irregular and Unrecorded Output

Where the output of important enterprises is irregularly sold on an unrecorded market, limited-visit surveys cannot cover such production unless total output and other outlets can be enumerated by estimation.

Where design conditions other than accuracy are important, particularly urgency, the second and third sources of complexity may be ignored. This will depend on the judgment of the planner on the importance of the enterprises concerned in terms of the resource use and production of the system as a whole. Diagnosis of their importance is the core of presurvey investigation.

For farming systems where conditions allow the use of either limited- or frequent-visit techniques, the limited techniques offer advantages of speed and cost. Extremes of the spectrum are compared here; a single-visit survey, recording all the required information at the end of the season, is compared with daily visiting. The resource sets and organization of the alternatives are designed to exploit the particular advantages of the two methods of collection; the single visit to exploit mobility and fast coverage where collection time in the field is a low proportion of total time, and the daily visit to minimize observational errors on the single farm and to allow more detailed simulation of attribute subpopulations in the system.

The two extremes are compared under conditions where a sample of 100 units meets precision requirements, where farmers live on their farms, and where the survey is carried out by permanent staff. The hypothetical programs presented are related to the dominant conditions in survey design: accuracy, cost of coverage, and speed. This presentation shows the unit as a going concern, although in practice a period of establishment would be needed to build both types of organization up to the level of cohesion shown. Emphasis is placed on the whole program of investigation and planning, with the economist responsible for analysis, planning, and overall supervision. Criteria of unit costs and coverage are not meaningful in relation to data collection alone, which is only one facet of the program.

## EXAMPLE OF A SINGLE-VISIT SURVEY AND PLANNING PROGRAM

### The Resource Set

Table 66 sets out the manpower and costs involved in the program based on data collection by single visit.

Overall cost of this unit, on the basis of these estimates, would be £5,960, roughly personal emoluments plus 33 percent for associated

TABLE 66

Resource Set and Costs for a Program Based on Single-Visit Collection (£ sterling)

| | 4 Enumerators (each) | 1 Supervisor | 1 Economist |
|---|---|---|---|
| Personal Emoluments | 425 | 800 | 2,000 |
| Allowances | 45 | 130 | 50 |
| Transport and Traveling | 5 | 660 | 180 |
| Office and General | 20 | 20 | 140 |
| Total Cost | 1,980 | 1,610 | 2,370 |

Notes: The enumerators have eight-ten years' schooling, and the supervisor is at diplomat level.

Depreciation on a four-wheel-drive vehicle is included under transport and traveling for the supervisor, with 1,500 miles per month.

Source: Compiled by the author.

costs. Allowances and transport and travel are a major element where personnel will be in temporary field accommodation for from three to six months of the year, moving between different areas. The enumerators are of slightly higher quality than for daily visiting; the need for internal consistency in a questionnaire ranging over the whole of farm business requires greater skill than the repetitive daily questioning, though the experience and perception required for both types of work should not be underestimated. The gearing between the economist and the less qualified manpower is low, and an important feature of the programme is the flexibility in function of the supervisor and enumerators, again demanding slightly better qualifications in the enumerators to deal with a diversity of work. The result is a fairly high cost team of personnel, to be justified by its capacity to meet important design conditions.

### Organization of the Unit

Figure 4 sets out the organization of the unit in an ongoing program of work. The format of this figure and that of Figure 5, showing the survey flow for daily visiting, are different, arising from the variation in resource sets and organizations, but both have common features: both are two-way flows by staff level and time in months, showing how personnel are drawn into the various stages of the program over its duration.

Where the symbols are formed by a broken line, the stage requires only the intermittent attention of the personnel concerned and can overlap with other activities. A solid line indicates a stage to which the personnel concerned are continually committed. The figures show the number of months for each stage, which vary for data collection and tabulation between the two techniques. Data collection for the single-visit program requires a period of one month and assumes a rate of work of one farm per enumerator per working day.[10] Tabulation tion of single-visit data requires a two-month period and is done by the enumerators who are fully committed to the work.

Figure 4 shows the stages for a single-visit survey, giving a nine-month period for the investigation and planning sequence. The annual program shows three surveys possible in a year. To complete three surveys, timing is very critical for the enumerator whose year is fully committed with nonpostponable work. The capacity of the economist, fully committed in the four-month sequence from data tabulation to planning, is also a constraint on the capacity of unit. This limits the flexibility of the data tabulation task for the other levels of personnel, who must keep up the flow of processed information to make full use of the economist. The capacity of this organizational unit is 300 farmers or, more importantly, three areas per year. It should be noted, however, that a change in timing of data collections running sequentially would mean a once-and-for-all loss of output for the unit.

## EXAMPLE OF A DAILY-VISIT SURVEY AND PLANNING PROGRAM

### The Resource Set

Table 67 sets out the manpower and costs of a program based on daily visits.

FIGURE 4

One-Visit Survey Design: Organization and Annual Program Flow

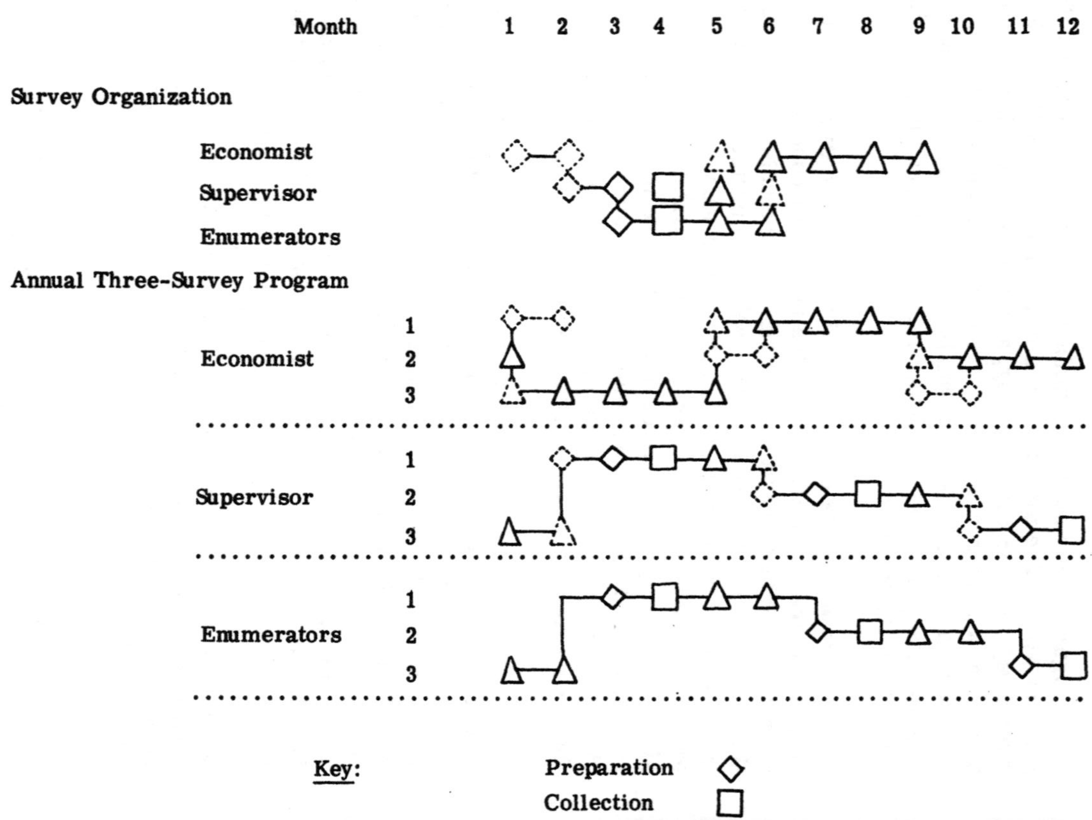

TABLE 67

Manpower and Associated Costs of a Unit for a Program
Based on Daily Visits
( £sterling)

| | 10 Enumerators (each) | 3 Statistical Clerks (each) | 1 Office Supvr. | 1 Field Supvr. | 1/2 Economist (each) |
|---|---|---|---|---|---|
| Personal Emoluments | 350 | 350 | 630 | 800 | 2,000 |
| Allowances | 10 | — | — | 30 | 50 |
| Transport and Traveling | 10 | — | — | 660 | 180 |
| Office and General | 30 | 20 | 70 | 60 | 110 |
| Total Cost | 4,000 | 1,110 | 700 | 1,550 | 1,170 |

Note: The enumerators are assumed to have six-eight years education, with the field supervisor at diplomat level.

Source: Compiled by the author.

The aggregated total cost of the unit is £8,530 personal emoluments plus 23 percent associated costs. The lower level of allowances is partly offset by the higher office and general expenses. Enumerators living in an area will be using rented or government accommodation, representing usually between 7.5 and 10 percent of salaries. The slightly lower quality of staff required in enumeration and tabulation is reflected in lower salary levels. The staff gearing is very different from the program based on single visits, with a large number of lower-level staff to each farm economist.

## Organization of the Unit

Figure 5 shows the stages of the program and the participation by the four levels of staff involved. The duration of the program increases to eighteen months, requiring a ten-month period in the field to cover the full agricultural calendar. The rate of work of the enumerators does not affect the time in the field but does affect the number required to cover a sample of 100 farmers. Coverage has been assumed at ten farmers per enumerator. This is a compromise between varying experiences in the field, since in practice researchers have rarely been able to manage daily visiting. Pudsey used an enumerator for six farmers, visiting three times a week with farmers living on their holdings.[11] Collinson and Upton report ratios of about 12.1 with one or two visits per week.[12] Bessel, Roberts, and Vanzetti aimed at twenty-thirty farmers per enumerator, with farmers living in a village, and reports achieving seventeen-eighteen.[13]

The physical pattern of settlement has an important effect on coverage for daily visiting. Extremes reported are three and eighteen farmers a day. Collinson found 10.1 very practical with enumerator areas of about eleven miles' radius and three visits each week. The area had a density of about seven farms per square kilometer, with the family living on the farm. The same ratio was achieved with three visits per fortnight in an area with one farm per three square kilometres. Tabulation is lagged one month behind collection, with one clerk to every three enumerators.

There is limited slack at each level, but the lack of flexibility due to the need to cover the full crop calendar prevents any use being made of this at lower staff levels. The economist has enough slack to cover two teams of this type, and consequently only half his time is costed to this unit. The element of general administration in his work will be very much larger because of the number of staff under him. With two teams his subordinate staff will be thirty strong. Staff

FIGURE 5

Frequent-Visit Tecnique: Survey Flow

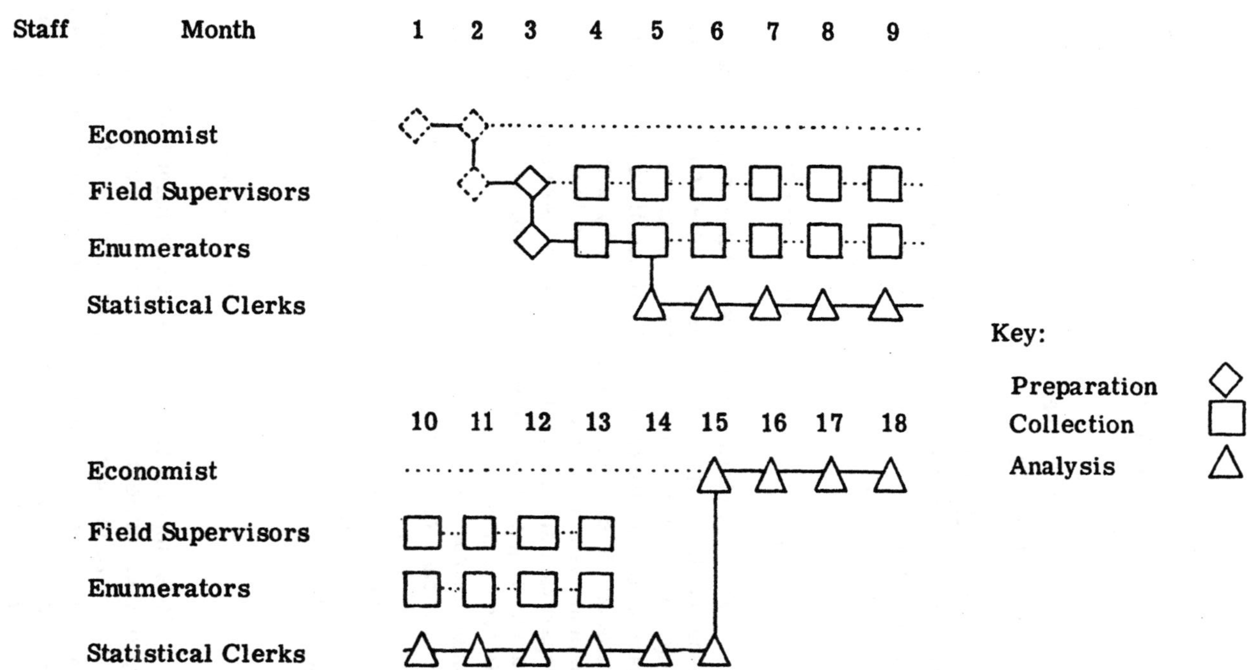

gearing is very high but capacity is reduced by the long period required in the field.

## ADVANTAGES OF THE SINGLE-VISIT TECHNIQUE

Table 68 compares the two units on the basis of four criteria of cost and capacity.

The single-visit survey gives superiority in areas covered per year as a measure of capacity, cost per area covered as one measure of efficiency, and speed of area coverage. These three criteria are discussed below.

### Area Covered per Year

With a potential coverage of three areas each year, the program based on single visits gives much better coverage than a daily-visit program. How much better depends on two factors: the relative scarcity of the different levels of qualified manpower and the possibility of finding three areas for collection by a one-visit technique in which the seasons are staggered sufficiently to allow a staggering of the area preparation and collection stages. Where diplomat-level staff is scarce, single-visit collection has a premium in the efficient use of this type of manpower. Where economists are few and the season homogeneous over larger areas, the detailed-visit program may equal the coverage capacity of a single-visit program. In practice, the limited number of farm economists working in the context of a largely technical ministry will rarely have the establishment authority to command the level of subordinate staff required by the daily-visit program outlined.

### Cost per Area Covered

It is important to stress at this point that area covered is the appropriate criterion rather than cost per farm covered. Given a sample size dictated by the interfarm variation in important data and by the level of precision required, there is no value in spreading overheads by increasing the number of farmers covered. Total program costs will increase by the extra recurrent costs involved.

Information on costs is very limited. Many farm economists have worked within an establishment, such as a research center, with

TABLE 68

Comparison of Programs Based on Single- and Daily-Visit Collection Techniques

| | Single | Daily |
|---|---|---|
| Total Cost of Annual Program (£) | 5,960 | 8,530 |
| Areas Covered per Year | 3 | 1 |
| Cost per Area Coverage (£per area) | 1,987 | 8,530 |
| Time for One Area | 9 months | 18 months |

Source: Compiled by the author.

many costs, including transport and accommodation, carried on centralized station overheads. It is particularly difficult to isolate those overheads attributable to the economics section. Most others have covered periods of establishment (inevitably with higher costs), have failed to carry the program through the stages of analysis and planning, or have operated under highly specific conditions. Nor has there been any common objective in the investigations which have taken place.

J. D. MacArthur has reported £9 per farm as an estimated cost of full business figures, but the survey design giving this cost level is not described.[14] The implications of his article suggest that this cost was for collection only and related to a relatively low visit frequency, varying from once a week to once a month. Also, the sampling techniques used placed emphasis on large numbers to give reliability, partly because of the lack of a clearly defined objective in data collection and partly because of a failure to appreciate the importance of describing the characteristics of data distributions to show the useful sample size.

Work in Zambia covered two farming areas, each for three seasons, the equivalent of six areas; and the costs fell within our estimated range for daily-visit techniques, averaging £7,100 per area.[15] However, this total cost included the pains of setting up the unit for collection. It also operated in a community settled in villages where the enumerators were able to cover seventeen-eighteen farmers a day, almost halving the enumerator manpower requiremtnts. Against this, the target sample size was 150 in each area, reducing the differentials in manpower between this case and our example.

In our comparison of models from extremes of the spectrum, there is considerable difference in costs per area, a difference compounded by the higher total cost and lower capacity of the daily-visit unit. With both units operating at capacity, costs per area covered of a program based on daily-visit data collection will be about four times that of a program based on single visit collection.

## Speed of Coverage

Finally, the speed of coverage for single-visit collection techniques is half that of detailed techniques. It is an aspect of the single-visit program which can be important for particular purposes where data is urgently required. The need to cover the full agricultural calendar with detailed collection may involve delay before the program can begin, increasing the eighteen-month span of the program itself. The greater flexibility in timing of single-visit data collection increases its advantages on this criterion. It is not a particular asset in an ongoing program; but for urgent work, be it cross-checking on broad census data or rapid data generation for investment appraisal, single visits give much faster results. A comparison illustrates this point.

In 1962, benchmark data, as well as information on the existing farming pattern and measurement of the resource use of the traditional practice, were required to evaluate the potential benefits from irrigation made possible by dam construction at Nyamba ya Mungu on the Pangani River in Tanzania. Surveys were carried out in three areas where irrigation was expected to be technically feasible. Each survey covered twenty-five farmers with a one-visit questionnaire. Although there were many faults, mainly that the preparation stages were telescoped drastically and the sample was too small, useful information was obtained in a period of four months with a team of three enumerators.

The comparison with detailed collection techniques can be very marked. For example, a comparative survey of Virginia and aromatic tobacco was requested and approved by the Western Region Research Committee in November, 1962, and a regular-visit technique was proposed for tobacco as a complex crop.[16] The season started in August for one crop and finished in August for the other, so work was delayed until the following August. The report was presented in June, 1965, tabulation having been started upon completion of collection in August, 1964. Thus the report appeared thirty-two months after initial request for the work. Because the difference in labor productivity of the crops was so big, caused by a massive peak requirement for picking and stringing aromatic, it would have been revealed by an examination of existing secondary sources, or by a one-visit survey at the end of the 1962-63 season, thereby giving results within twelve months. Aromatic tobacco was discontinued as a cash crop in 1967, two years after the report was issued, production in the area having reached 450,000 pounds in 1966. Had it been stopped because of lack of potential compared with Virginia at the end of the 1963 season, production would have been a mere 44,000 pounds and would have involved far fewer farmers.

Using Robertson and Stoner's format, estimates can be made of the cost of increasing precision.[17] Assuming a level of 10 percent standard error from seventy-five observations on the dominant crops in the system, and using the cost patterns set out for the two extreme types of survey organization, the costs of increases in precision are shown in Table 69. Sample size must be quadrupled to halve the standard error. Where small increases in sample size are required, the single-visit survey is more flexible. The unit can reduce coverage, spending longer in the field and on data tabulation, and increase precision. Above a certain level, the need to visit after harvest precludes an extended period in the field, and both methods are dependent upon increased resources at the supervisor and enumerator levels pro rata.

The table illustrates the dramatic rise in area costs under the assumptions made and emphasizes the case for careful consideration of the level of precision being sought. In practice, the assumption that a single economist can supervise and handle the administration of over thirty-five field staff as easily as five in the case of the single visit, and of over 200 field and office staff in the case of the detailed unit, is unrealistic. Provision for an executive officer, and then further overheads for the unit, would increase costs at an early stage in the sequence for detailed collection and would be necessary at the final stage in the single-visit program.

One final aspect is related not to the complexity of the farming system but to what might be called its stability. It is peculiar to a particular phase in the penetration of a farming system by a new enterprise diffusing through the community as an innovation. Collinson, in surveying aromatic tobacco growers in an area where the crop was newly introduced, has illustrated the phenomenon.[18] A frequent-visit survey was carried out, a sample of fifty farmers being selected from registered lists of aromatic tobacco growers before the start of the season. Of these fifty farmers, forty-three cooperated through the season but only thirty-three planted the crop and only thirty carried it through to harvest. Despite the precautions of a frame designed to cover the main attribute under investigation, observations on

TABLE 69

Cost of Increasing Precision Under Assumed Data Variance and Organizational Characteristics

| | | | Single Visit | | Detailed | |
|---|---|---|---|---|---|---|
| Level of Precision (% S.E.) | Observations Required | Sample Size | Total Cost per Area | Unit Cost per Farm | Total Cost per Area | Unit Cost per Farm |
| 10 | 75 | 100 | 1,987 | 19.9 | 8,530 | 85.3 |
| 7.5 | 135 | 180 | 2,944 | 16.4 | 13,858 | 77.0 |
| 5.0 | 300 | 400 | 5,577 | 13.9 | 28,510 | 71.3 |

Note: Potential coverage by single-visit technique is assumed reduced from three to two areas in increasing precision to a 7.5 percent standard error.

Source: Compiled by the author.

aromatic tobacco growing were depleted by 40 percent. A one-visit survey, with a sample of fifty-one farmers chosen after the season had begun, and rejecting those not growing tobacco, realized forty-six observations on the whole production cycle.

The two sequences described are extremes. Inevitably practical compromises will be made between the loss of accuracy and the cost of coverage. Detailed-visit techniques will have reduced visit frequencies and thus have increased coverage for a given set of enumerators. This will lower the costs of area coverage, with a decrease in precision caused by increased memory bias. Limited-visit surveys will be forced into two or three visits, reducing the coverage for the year. The survey program of any unit will be an amalgam of collection techniques tailored to different design conditions. Figure 6 shows such a program carried out between August, 1963, and February, 1965, involving six surveys and five different visit frequencies, three detailed and three limited techniques. The first condition to be met is to identify the data required to achieve the objectives of the investigation. Urgency may impose further conditions, but certain uses demand precision which can be achieved only by careful observation on the individual farm unit; this is particularly true where comparisons between farms—for example, of management or motivational differences—are required. The single-visit end of the spectrum will give cheaper and faster coverage and will more often meet the manpower limitations likely to exist in developing agriculture. On the other hand, particular local conditions preclude their successful application, and presurvey diagnosis of the complexity of the crop establishment pattern and of the output flows is necessary before limited-visit techniques should be considered. The baseline for performance must be the well-organized, properly sampled, and supervised frequent-visit technique, with the onus on limited-visit techniques to give comparable precision without excessive memory bias eroding the usefulness of the data.

## POSSIBLE COMPOSITE DESIGNS

Nevertheless, the scarcity of manpower and a profusion of alternative opportunities create pressures for limited-visit surveys once a clear use for data is defined and users are convinced of the potential of the approach. An attempt has been made to explore composite designs in order to economize on the use of field manpower. The approach taken was to cover easily enumerated data by a limited-visit technique to allow the selection of a subgroup for detailed investigation.

**FIGURE 6**

**A Composite Survey Program**

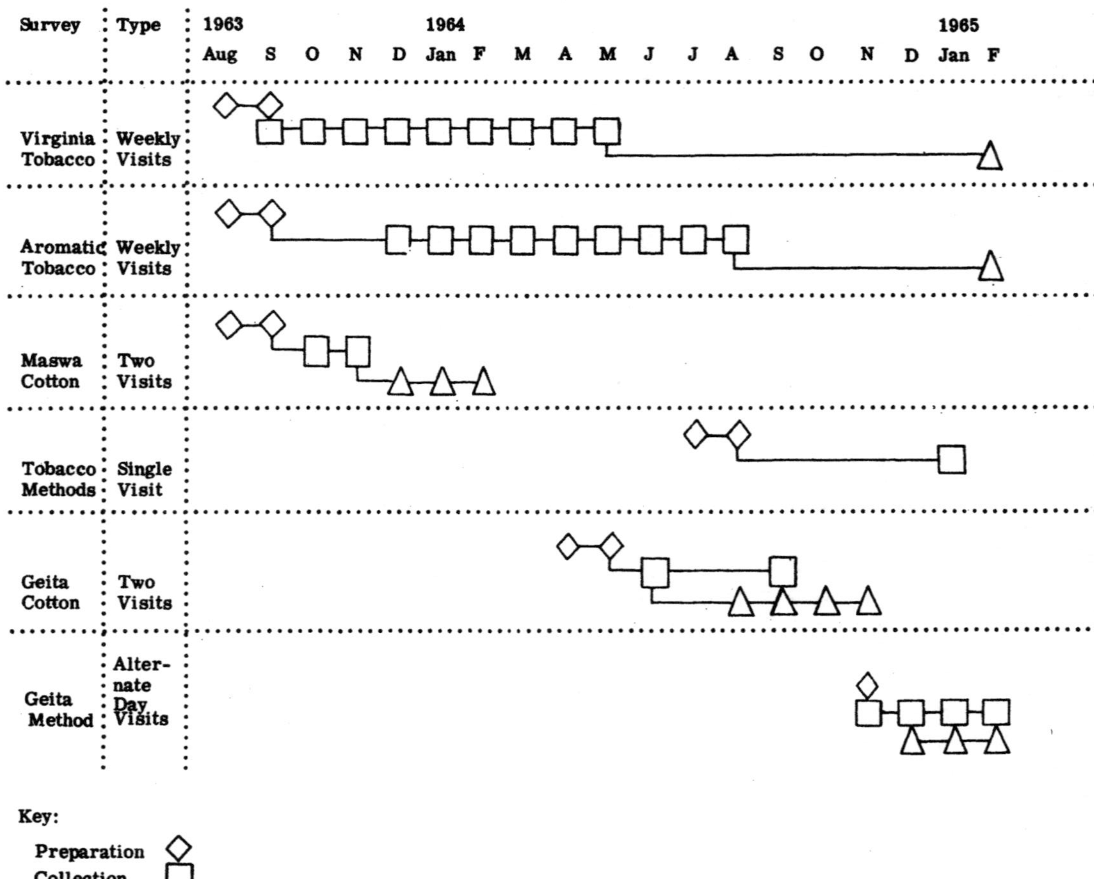

The problem of selection criteria is a repetition of that in selecting a representative farm: finding criteria to reflect the values of the population over a wide range of attributes. The problem is increased for this exercise because the criteria must be capable of enumeration by objective measurement in a limited-visit technique. Two criteria were used in testing: total available labor, enumerated from family size plus the incidence of hired labor, and acres cropped per unit of available labor, involving higher survey costs for acreage measurement. Both criteria focus on relatively easily measured aspects of the limiting factor in the system concerned. Two groups of five farmers closest to the average of the respective criteria were selected from data on forty-two farms used as a documented population. The means and coefficients of variation for the thirteen selection criteria in the discussion of representative farms were calculated for these two groups and compared with the values for the whole sample. The results are presented in Table 70.

With forty-five degrees of freedom, t tests on the means revealed no differences at the accepted significance levels. Out of twenty-five tests, t values reached 1.00 in nine cases and exceeded 1.68 in one case; $\bar{t}$ = 1.84 in the case of July labor as a percent of the total. Acres cropped per unit of available labor, as a factor reflecting efficiency of labor use on the critical operations in the system, did not contain scale factors well, thus giving large distortions in the sample means of cropped acreage and in the rate of work index, both components which complement the selection criterion in the system. Total labor available caused no dramatic distortions and is a particularly attractive component because it is very simple to collect on a question-and-answer basis.

Two practical problems have discouraged further work in this direction:

1. Unless farms in the selected subgroups are in close proximity, problems in enumeration by a single field officer would remain. It is unlikely they would be close.

2. There could be no guarantee of cooperation from the selected farmers, who will perhaps be few enough to feel isolated from the rest of the community. It might be a difficult point to put across in survey preparation where only one or two farmers are to be investigated more closely.

Both these factors might require compromise in the selection of groups, which would immediately distort the representativeness of

TABLE 70

Comparison of Selected Subsamples and the Whole Sample over Important Attributes

| | Whole Sample | | 5 Farmers near Mean Total Available Labor | | 5 Farmers near Mean Acres Cropped per Labor Unit | |
|---|---|---|---|---|---|---|
| | Mean | % C.V. | Mean | % C.V. | Mean | % C.V. |
| % of Cropped Area in Cotton | 53.8 | 39.4 | 58.0 | 20.9 | 51.6 | 27.3 |
| Total Labor Available (man-equiv.) | 4.11 | 38.9 | — | — | 4.96 | 22.2 |
| Total Labor Used (man-equiv.) | 1.32 | 38.6 | 1.02 | 23.3 | 1.54 | 29.0 |
| % Used in Cultivation | 48.2 | 31.9 | 41.0 | 19.9 | 49.6 | 14.9 |
| Rate of Work in Cultivation | 99.8 | 43.6 | 98.9 | 31.9 | 79.4 | 20.8 |
| Cropped Acreage (acres) | 10.9 | 39.0 | 9.1 | 30.7 | 14.1 | 27.0 |
| November Labor as % of Total | 13.3 | 31.0 | 15.4 | 34.9 | 13.8 | 38.4 |
| December Labor as % of Total | 13.9 | 30.0 | 13.5 | 31.3 | 14.8 | 16.9 |
| June Labor as % of Total | 12.3 | 38.0 | 12.8 | 22.5 | 13.5 | 19.3 |
| July Labor as % of Total | 11.5 | 32.0 | 10.2 | 28.4 | 8.2 | 54.8 |
| Total Cotton Production (lbs.) | 3,652 | 51.0 | 3,062 | 54.1 | 4,927 | 43.0 |
| Total Food Grains (lbs.) | 1,915 | 96.0 | 2,192 | 80.0 | 2,833 | 75.0 |
| Gross Return per Acre (lbs.) | 223 | 41.8 | 222 | 33.9 | 229 | 33.1 |

Source: Compiled by the author.

the subsample. Physical dispersion in particular has no obvious solution unless casual help can be enlisted in the area of each farm.

With the close reflection of the population realized by this comparison, such techniques justify further investigation. Even if it seemed too narrow a base for planning, it would afford a useful and relatively cheap check on the data realized from limited-visit surveys where the organizational and cooperation problems could be coverde.

## NOTES

1. D. C. Catt, "Surveying Peasant Farmers—Some Experiences," Journal of Agricultural Economics, XVII, 1 (1966); J. D. MacArthur, "The Economic Study of African Small Farms—Some Kenya Experiences," ibid., XIX, 2 (May, 1968).

2. J. D. MacArthur, "Labour Costs and Utilisation in Rice Production of the Mwea/Tebere Irrigation Scheme," East African Agricultural and Forestry Journal, XXXIII, 4 (1968); M. P. Collinson, "Aromatic Tobacco and Virginia Tobacco: A Comparative Survey of Two Tobaccos on Family Farms in the Tabora Region of Western Tanzania," East African Agricultural Economists Conference paper (1970).

3. Collinson, op. cit.

4. D. Pudsey, "A Pilot Study of Twelve Farms in Toro," (Uganda Dept. of Agriculture, 1966). (Mimeographed.)

5. K. E. Hunt, Agricultural Statistics for Developing Countries (Oxford: Oxford University Press, 1970).

6. M. P. Collinson, "Usmao Area," Agricultural Economic Survey no. 3 (Dar es Salaam: Tanzania Dept. of Agriculture, 1962). (Mimeographed.)

7. M. P. Collinson, "Usmao Area"; D. von Rotenhan, Land Use and Animal Husbandry in Sukumaland, "Africa Studies," XI (Munich: IFO, 1966).

8. A. Larsen, "A Choice of Sampling Population in Rural Sukumaland," East African Agricultural Economists Conference paper (1970).

9. J. E. Bessel, J. A. Roberts, and N. Vanzetti, "Survey Field Work," Agricultural Labour Productivity Investigation Report no. 1 (Universities of Nottingham and Zambia, 1968).

10. Collinson, "Usmao Area."

11. D. Pudsey, "An Economic Study of the Farming of the Wet Long Grass Area of Toro" (Kampala: Uganda Dept. of Agriculture, 1967). (Mimeographed.)

12. Collinson, "Aromatic Tobacco and Virginia Tobacco"; M. Upton, Agriculture in South-Western Nigeria, "Development Studies," 3 (University of Reading, 1967).

13. Bessel, Roberts, and Vanzetti, op. cit.

14. J. D. MacArthur, "The Economic Study of Small Farms—Some Kenya Experiences," Journal of Agricultural Economics, XIX, 2 (May, 1968).

15. J. E. Bessel, personal communication.

16. Collinson, "Aromatic Tobacco and Virginia Tobacco."

17. Robertson and Stoner, Shell Symposium on New Possibilities and Techniques for Land Use Surveys with Special Reference to Developing Countries.

18. Collinson, "Aromatic Tobacco and Virginia Tobacco."

# PART III

# PLANNING EXTENSION STRATEGY AND CONTENT

CHAPTER

# 15

# THE PLANNING TASK IN RELATION TO ADOPTION AND DIFFUSION RESEARCH

The planning task for farm economics in traditional African agriculture is to provide appropriate extension content, aimed at fuller satisfaction of farmers' priorities as well as the development of the economy. Only if farmer and government objectives can be mutually satisfied by selected improvements can development reasonably be expected. One other facet has to be contained by the planning process: the farmer should be able to perceive the suggested improvements as useful and to identify them with the fuller satisfaction of his own needs. Attractive presentation is important in stimulating the initial adoption of changes and complements the appropriateness of content, which encourages sustained acceptance.

Considerable efforts have been made in the field of adoption and diffusion research. The usefulness of the field in planning extension programs ranges wider than our immediate application, but an important part is to allow the estimation of the rate of adoption and diffusion of an innovation within a given farm population. The field has been divided into six areas after J. Moris, who centers his classification on the definition that farmer innovation is a behavioral change following the acceptance of new ideas or practices:[1]

1. The processes by which innovations diffuse from area to area and person to person.

2. The rate at which innovations diffuse over time, and the persons involved at various stages within the same population.

3. The characteristics of persons adopting at different stages of diffusion.

4. The phases of the individual adoption process: awareness, interest, evaluation, trial, and conviction.

5. The media influencing the phases of the adoption process.

6. The characteristics of the innovations which appeal to farmers, often dependent on their personal situation and perception.

C. M. Coughenor divides the field into three interconnected systems: the "client system," representing the farmers as the "market" for innovations; the "generating system," which is producing the technology and therefore dictating the type of change available; and the "link system" communicating the available improvements between the generators and the clients.[2] The last two systems are represented by agricultural research and extension. In an analysis of Coughenor's division, B. Brock criticizes sociologists for their preoccupation with the client system and reflects Coughenor's concern that adoption and diffusion research has been unbalanced, with insufficient attention to the generating and link systems.[3] An imbalance is in fact evident in Moris' classification. Brock says that the failure of change programs is usually rationalized in terms of the characteristics of the farmer community. While acknowledging that the field provides for the study of innovation characteristics, as an independent variable influencing the rate of diffusion, she believes it to be a line of work which has been neglected. She uses the work of P. Roy et al. in India as an example explicitly assuming that the changes offered are good for the farmer, and points out the danger in interchanging the expressions "adoption of recommended practices" and "willingness to change."[4] Brock asks whether many of the adoption scales drawn up of "receptiveness to new ideas" or "modernity" might not in fact be measuring "gullibility" or "willingness to follow directions." Her doubts echo the argument advanced here for the compound nature of the peasant farmer's objective function, and therefore for the need for compound planning criteria. Adoption and diffusion theory cannot yet identify the relationships between generating, linking, and client systems. Until it can do this, it will not have the capacity for ex ante estimates of the rates of adoption and diffusion for a given innovation within a specific farmer population. It has so far been limited to ex post studies tracing the rates of diffusion and relating these to innovation characteristics.

Our attempts to marry profitability and "acceptability," the latter measuring facets of farmers' nonmarket priorities as evaluation criteria, have sought to fill part of this gap. On adopting an innovation, the farmer will find that the change gives better satisfaction of his own needs, thus ensuring his sustained use of the new technique. With the distinction made between sustained acceptance and initial adoption, further criteria are required to evaluate the attractiveness of the change to the farmer in order to encourage him to try it out.

The characteristics of innovations which appeal to farmers are the six areas identified above as within the field of adoption and diffusion research. There is no standard ranking of characteristics; and the range studied, as well as the definition of each, varies between workers. E. M. Rogers identifies relative advantage, compatability, complexity, divisibility, and communicability as important.[5] Other researchers, notably Brander and Kearl, have used the term "congruity" to describe an affinity between present and proposed practices.[6] We have already identified the two criteria of profitability and acceptability; and in planning, "acceptability" is split in two—complexity and acceptability itself. Our concern here is other criteria which will give a measure of the attractiveness of the improvement to the farmer. It is important that all these can be scored to give a basis for comparison among alternative innovations on each of the criteria used. Two other criteria are adopted, taken from other researchers but defined specifically to allow scoring in the course of the planning sequence. These are divisibility and congruity. All five criteria are outlined and the basis for scoring each is described.

## SUSTAINED ACCEPTANCE

### Profitability

In the planning sequence, profitability remains the key criterion, the one which indicates the market benefits arising to both the economy and the farmer from each alternative element of extension content. It is measured by the increment in the net value of marketed output realized from the change. Clearly, an increase in marketed output may be qualified by other national or regional objectives, which might equally well be built into the planning model. For example, the provision of work for one laborer might be important in areas with a severe rural unemployment problem.

### Acceptability

Although acceptability has so far been used to cover all possible areas of disturbance to farmer utilities consequent on the introduction of a change in the system, the definition is narrowed for the purposes of scoring. Acceptability scores for each innovation are based on the "costs" of any change, measured by the loss of utilities derived from other products sacrificed by the resource allocation required to allow the change. Scores are also useful to compare alternative food producing activities, for selecting for replacement those which provide least farmer utility and therefore offer least resistance to change.

### Complexity

Other facets of what has, up to now, been summarized as acceptability are covered in the planning sequence by the independent criterion of complexity. This measures the amount of disturbance the changes create in the management routine of the farmer—for example, the degree of reorientation he must make in his usual sequences of field operations over the season. It is scored by counting the number of days shifted in each planning time period between crops, or between different planning time periods.

The complexity of the change, as well as its acceptability and profitability, will become apparent to the farmer only when he tries out the innovation. These three therefore are criteria to decide whether the farmer will sustain initial adoption.

## INITIAL ADOPTION

Initial adoption will be partly dependent on how well the advisory services communicate the idea of the profitability and acceptability of the practice being encouraged. Here the emphasis is on the role of the "link system," which is not our immediate concern. Accepting that extension methods are effective, two other criteria will influence the farmer's impression of the suitability of the innovation: congruity and divisibility.

### Congruity

As the word implies, congruity scores the innovation on how closely it compares with existing practice. The score is accumulated on the number of new facets the farmer is expected to absorb. Agronomic differences are predominant; pure stand instead of intercrop, a different time of planting, and the use of a purchased input contribute a unit each. Social ramifications are also scored: whether the change is consistent with the social role of those involved, whether it will penalize family members who benefited from the distribution of proceeds under the existing system. All these facets are potential barriers to initial adoption by the farmer.

## Divisibility

Divisibility refers to the scale at which an innovation can be introduced. It is a special criterion with two important aspects:

1. The possibility of the trial of an innovation, on a small scale, reduces a clash with the farmer's risk preferences. If the level of outlay required for trial is high in relation to his existing income, uncertainty of the outcome will inhibit his participation. If the innovation is divisible, it can be introduced on a scale which is consistent with the farmer's debt ceiling.

2. Divisibility brings a new plane into the evaluation. Clearly, where an innovation can be introduced on a small scale, it will mask the extent of the changes required in management routine and any potential clashes with food supply activities. It explains circumstances in which an innovation may be initially adopted but, once the scale of adoption increases, the problems created may lead to its rejection. The scale of adoption determines the level of impact which innovations will make on the system.

Divisibility is not scored except by the capital outlay requirement of the change.

Each of the five criteria outlined are important but are scored on their own values. The direct comparison of scores for innovations is possible only across a single criterion, and there is no denominator for aggregating scores of several criteria to give an index for the final selection of a best innovation. Innovations will have poor scores on some criteria and good scores on others. Selection will depend on the judgment of the planner and his subjective weighting of scores on each criterion for a particular planning decision.

With staff/farmer ratios greater than 1:1000 and accepting that diffusion over the community will follow if extension content is appropriate to the majority of farmers in the area, then once innovators are identified, continuity in extension is vital to development over a period of time. The reaction of a farmer to successful innovation and his subsequent action on further changes are aspects of the client system which have been largely ignored by adoption and diffusion research. Clearly, innovation and change are not once-and-for-all occasions but are by nature a dynamic process. Extension strategy for an area will be a sequence of innovations to be put across to farmers to increase marketed output.

This idea of continuity in innovation, and consequently continuity in diffusion, is a central concept of the planning sequence described in this section and is built around a relaxation in the attitude of an adopter to subsequent innovations. It is generally acknowledged that the initial breakthrough in innovation is the most difficult. Once established as an innovator and satisfied by the results of advice received, the farmer will be amenable to further advice because he has increased confidence in the ability of extension personnel. His inclination will be to increase the scale on which he uses the initial innovation and to adopt additional changes recommended to him. This will eventually precipitate a clash in the demand for farm resources, which jeopardizes satisfaction of his established nonmarket priorities and the reliability and pattern of food supplies. In addition to selecting the initial extension content, the planning sequence needs to anticipate and avert this type of clash by introducing complementary innovations in subsistence activities, which will increase the productivity of resources employed in food production while maintaining farmer utilities from preferred foods and insurance crops.

In this sequence the characteristics of scale and congruity are causal: increased scale or greater incongruity creates higher complexity and acceptability as well as profitability scores. Both increase the changes in management routine and thus the complexity score and, where large enough, will breach a resource constraint, jeopardizing the satisfaction of food supply priorities, and thus increase the unacceptability score. The key to effective extension strategy, and the objective of the planning task is to balance the innovations being offered against the relaxing farmer attitudes toward change, to give continuing satisfaction of farmer priorities and expanding market production. Figure 7 sets out the sequence schematically as a model. Successful results in each of seven time periods allow a continuous development of the sequence.

Each letter represents a different innovation and is circled in the time period in which it is introduced into the system. Each innovation is characterized by a degree of incongruity and the scale on which it is introduced, and may increase in scale between time periods. Any step in congruity or scale may produce repercussions in terms of complexity or acceptability. An example of the pattern of adoption by an innovator is described over the seven time periods of the model.

FIGURE 7

Model Adoption Sequence for an Innovator Farmer

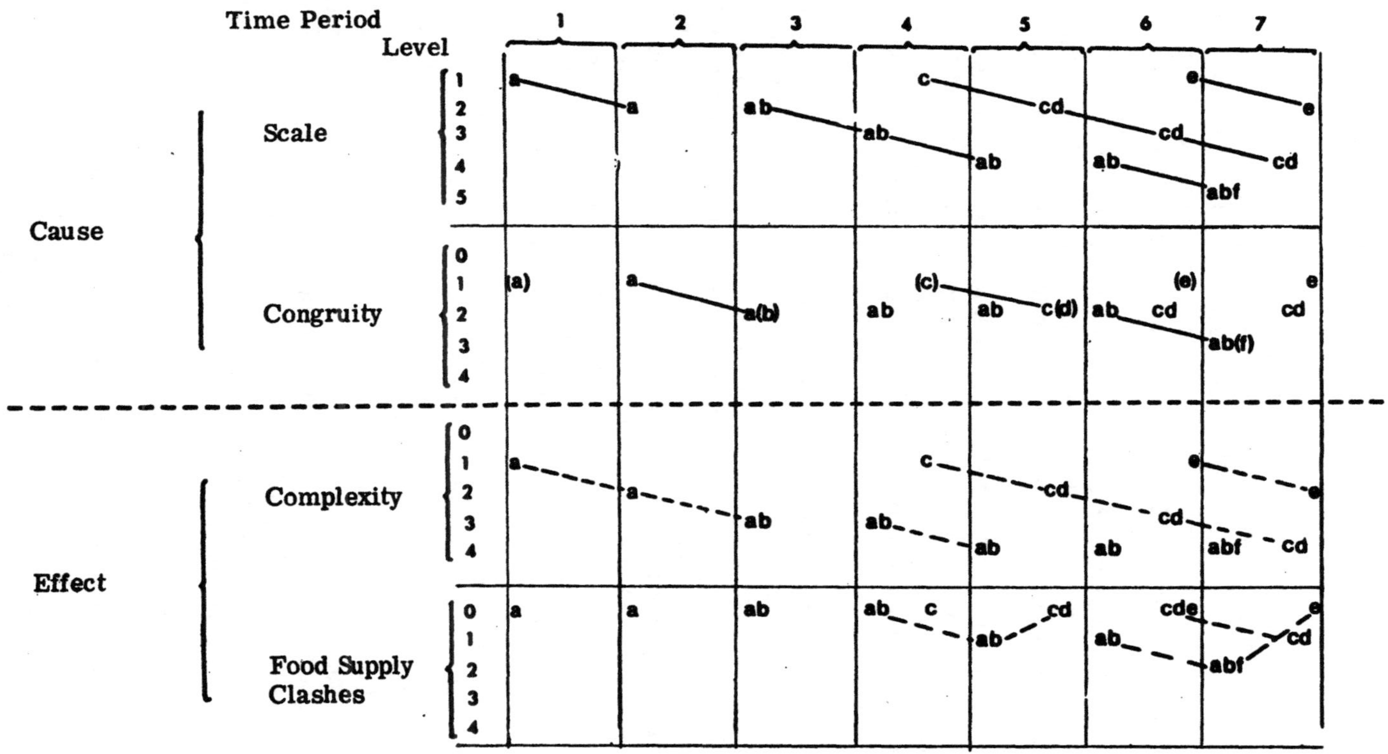

In time period 1, innovation a is adopted, having good congruity with existing practice at level 1, and on a small scale, also at level 1. It evokes a reaction within the system requiring a minor reorganization of management routine with a complexity count at level 1. The impact is not drastic enough to clash with food supply priorities.

Time period 2 shows an increase in the scale of adoption of innovation a, requiring some further reorganization of the management routine; the complexity score increases to level 2.

In time period 3 the extension service recommends a change in husbandry b associated with a, to give greater productivity; a stays on the same scale. The congruity level associated with ab increases to 2, and an adjustment in management routine increases complexity to level 3.

When time period 4 is reached, an increase in the scale of adoption of the joint innovation ab elicits no reaction within the system. However, anticipating a clash with food supply priorities when the scale of ab is further increased—a clash pinpointed in the planning sequence—innovation c is introduced on the advice of the extension service. A more efficient way of producing the main food which would be affected by the impending clash in demands for labor, it is introduced on a small scale at level 1 and is similar to the existing practice, so the congruity level is low at 1. The introduction of c creates no repercussions on complexity or acceptability.

In time period 5, a new innovation d is introduced on advice, associated with d, which improves the productivity of this new technique in producing the major food. At the same time the scale level of both ab and cd is increased: ab to level 4 and cd to level 2. This is a large change and is reflected in considerable adjustment of the management routine; complexity levels for both ab and cd are increased. In addition, the acceptability count of ab would have been lowered to level 1, indicating a clash with food supply priorities. However, the clash has been prevented by the more efficient food producing techniques cd.

For time period 6, further increases in the scale of ab will be possible only with cd at a higher level. Increasing the scale of cd will itself begin to clash with the supply of a second major food type. Innovation e is introduced to produce this second food more efficiently. At the same time cd is increased in scale toward the level required before a further increase of ab is possible. There is some further adjustment of the management routine and the complexity level for cd rises to 3.

The final scale increase in ab is made in time period 7 and innovation f is introduced, a further reduction in congruity which increases productivity. There are complementary increases of scale: in cd to meet the food supply requirements jeopardized by a higher level of ab; in e to avert the anticipated clash between an increase in scale of cd and the second major food type. Significant adjustments to the management routine increase the complexity levels of cd and e. The increased incongruity of ab with f added would have created a clash with food supply patterns, but it has been averted by the buildup of cd and e.

Over the period the farmer gains in confidence from results which improve the satisfaction of his own needs and yet safeguard his priority for a secure food supply. As he does so, he becomes amenable to changes further removed from existing practice and is stimulated to adoption on an increasing scale. The time scale of the model will vary with four factors: the economic potential of available technology, the gap to be covered in congruity between existing and improved practices, the quality of the planning sequence and the extension services, and the rate at which farmers' attitudes relax. The initial innovations will be limited in both incongruity and scale by, among other things, the farmer's existing debt ceiling. As his income rises his debt ceiling will also rise; and, as his confidence in changes increases, his propensity to invest will increase. The rate of both these changes will be an important determinant of the rate of development.

The model presented is the basis of the planning sequence which seeks to determine the full potential the farmer can achieve with his available resources and the rate at which he can be expected to reach the potential. The sequence falls into four parts.

1. Although data have been collected on the existing farming system, further information is required to provide planning coefficients for the improvements to be tested. The first part of the planning sequence is the compilation of an inventory of available technology. There are data problems associated with preparing these coefficients; these are discussed in Chapter 16.

2. The model relates the expected changes in the attitude of an adopting farmer to extension strategy over a period of time. Clearly, over the same time period other dynamic factors will alter the economic environment and may revise the desirable sequence of innovations. Factor and product prices or government policy objectives may alter. Where such changes are forseeable, they should be written

into the planning sequence. Where they are not foreseeable but occur, the planning sequence should be rerun to evaluate any changes implied for extension strategy. (Chapter 17 outlines both types of factors and how some of those which are foreseeable can be included in the planning model.)

3. To choose the practices which should feature in the system at the end of the planning period, the inventory of available technology must be related to the farmer's expected resource position. The solution which emerges is termed the "goal" system, toward which extension strategy is aimed during the period. (Planning the goal system is covered in Chapter 18.)

4. The crucial part of the planning sequence is the path from the existing to the goal system. It involves reconciling the increments in scale and incongruity, which the relaxing attitude and increasing risk preferences of the farmer will find acceptable in successive time periods. (Selecting this path is detailed in Chapter 19.)

## NOTES

1. J. Moris, "The Application of Adoption Theory to the Study of Agricultural Development in East Africa," East African Agricultural Economists Conference paper (1969).

2. C. M. Coughenor, "Towards a Theory of the Diffusion of Technology," in First Interamerican Research Symposium on the Role of Communications in Agricultural Development (1964).

3. B. Brock, "The Sociology of the Innovator," East African Agricultural Economists Conference paper (1969).

4. P. Roy, F. C. Fliegel, J. E. Kilvin, and L. K. Sen, Agricultural Innovation Among Indian Farmers (1968).

5. E. M. Rogers, "Categorising Adopters of Agricultural Practices," Rural Sociology, XXIII (1958).

6. Brander and Kearl, "Evaluation for Congruence as a Factor in the Adoption Rate of Innovations," Rural Sociology, XXIX (1964).

CHAPTER

# 16

# PREPARING AN INVENTORY OF AVAILABLE FARM IMPROVEMENTS

The investigational phase has provided data which will be used to simulate the factor and product relationships of the existing farming system. For planning we need the same type of coefficients for the improved technology available for extension to farmers, in order to measure its impact when interpolated into existing relationships. Compiling an inventory of available farm improvements requires further investigation covering the sources of improved technology.

For our planning sequence, based on Sukumaland farming, we shall limit the coverage to on-the-farm improvements which will intensify labor productivity and the return from land. The limitation is made to narrow the task and also because the area is on the fringe of a fertility clash, with increasing population densities forcing new areas into cultivation and limiting the fallow period. The inventory will cover improvements generated by agricultural experimentation within the area, which has been identified as the major source of improved farm technology in traditional agriculture.

The inventory needs to contain the same range of coefficients as were derived from the survey investigation for the existing farm system. In our example the labor coefficients and revised timing of changes in practice will be particularly important, since it will be these that create complexity and ultimately decide the acceptability, as well as the profitability, of alternative innovations. The inventory must also contain a detailed description of the practices and purchases involved, which will dictate the congruity and divisibility of the changes. Bringing together the range of possible farm improvements and the data essential to their evaluation is the groundwork of the planning sequence, which is of greater importance than the particular technique to be used in manipulating the data but is often overlooked. The technique used does affect the format of the data, however; and the inventory will be drawn up in a form suitable for use in a simplex tableau for linear programming.

Controversy has raged over the way agricultural researchers should design their experiments in order to allow economic interpretation of the results. A good deal of the literature has stressed the need for functional representation, and progress has been made in discussing experimental design to allow this.[1] But actual practice, particularly in developing areas where economists have had perhaps even less impact on agriculture research than in advanced agriculture, is still based on specified treatments. The alternatives required for a functional approach are more expensive and more difficult to administer and control, particularly in developing areas.[2] Nor has farm economics itself solved the practical problems of transfer from a functional approach to system models for decision making. G. Weinschenk has pointed out the difficulties of qualifying the relationship between input and output by a measure of managerial ability, and has concluded that the point estimates required are best taken directly from experimental results.[3] Very often the selection of a point from a function depends more heavily on information on the factors used to relate the experiments than on the function itself. Added to this, even the rough functional interpretations available are in terms of the land resource alone. As U. Renbourg, among others, has noted, data on complementary resource use often remain unrecorded.[4] This is the biggest problem in deriving planning coefficients on improved technology. Finding point estimates raises three sets of problems which are equally relevant to more refined techniques of functional representation:

1. Evaluating the criteria used in specifying treatment levels for experiments and interpreting these levels in terms of congruity with existing farm practice.

2. Measuring both the specified and unspecified nontreatment variables in terms of congruity with farmer practices and resource use required.

3. Qualifying the response levels by the reliability of the results from season to season.

These three problems are discussed more fully.

## SPECIFICATION OF TREATMENT LEVELS

In most experimental work, treatment levels and treatment interrelationships examined will be decided by bringing knowledge in the field of plant/environment relationships to bear on the local situation. The timing and combination of field practices will be based on theoretical considerations and on the results of experience in other areas. Where purchased inputs are included, the range of treatment levels may be dictated by considerations of supply; for example that a unit of 20 pounds of substance is the smallest quantity available. Characteristically, recommendations issued by research centers are a combination of a group of treatments found to complement each other, with interactions giving the highest yield per acre. However, each single treatment may represent a significant change in congruity, and both treatments and levels of treatment will influence complexity and profitability, all criteria important in the evaluation of potential. With congruity an important factor in initial adoption: we need to be able to specify the inputs and the output increments associated with each reduction in congruity, as well as to score each step on complexity and acceptability in the course of planning.

In the Sukumaland example, where labor supply limits the productive activity of the family, practices which demand increased labor, which focus or re-time labor use, are likely to give high complexity counts. Similarly, with a high proportion of interplanting in the system and the staggered planting of food crops a dominant practice, changes requiring pure stands and fixed planting times will represent important reductions in congruity. Clearly, however, this disaggregation to single steps of congruity is limited by the availability of data to relate to each step. It should be carried out as far as the experimental results will allow. Congruity should be a criterion in the decision as to which treatments should be included in an experimental program.

With treatments identified and ranked in order of congruity, the second and third groups of problems listed concern the levels of the input and output coefficients to be adopted for each reduction in congruity.

## MEASUREMENT OF NONTREATMENT VARIABLES

The agricultural researcher limits his interest to specified treatment variables around which his experiment is designed. In order to get clear results from his program, he is concerned to minimize all other sources which might vary the response to the specified treatments. One consequence is the blanket coverage of all treatments with other nontreatment variables. Some of these may be specified and may themselves represent significant reductions of congruity from existing farmer practice. For example, all plots may be sprayed to remove a pest as a potential source of variation. Spraying may well be a practice unknown to local farmers, who remain susceptible to pest attack. Achieving the results of the experiment involves persuading the farmers to spray—a considerable reduction in the congruity of any innovation package. Other nontreatment variables will be unspecified. Particularly important in this group is labor. It is usual practice to supply labor ad lib to experimental plots, to ensure that the results are not influenced by weed or tilth differences. Thus, while it is possible for a research station to flood a plot experiment with labor, it is grossly impractical on the farm. Davidson and Martin have pointed out the relative scarcity of labor and capital on the commercial farm.[5] Clearly, distortion of the economic balance between resources applied in experimentation may deviate the viability of the results. This is particularly true in the labor-limited traditional agricultural system. S. R. Wragg has noted:

> In the design of fundamental experiments it would be right and proper that the supply of resources not being considered as a part of the experimental treatment should be sufficient to ensure they did not limit production. Adopting this same principle for applied experiments. . . . is liable to produce results which are of little relevance in the real world in which they are meant to be applied.[6]

Two consequences arise for planning coefficients. Output data will be inflated due to management standards which are impossible to achieve on farms, and indeed are uneconomic. The data must be discounted for use in farm situations. Second, because labor use is not specified, it is unusual for records to be made of the quantities used. Even when records are available, the small sizes of experimental plots and the ad lib use of labor grossly exaggerate the input coefficients, making them unsuitable for planning purposes. Experimental records can be useful for identifying the congruity implications of the nontreatment variables and for checking the timing and operational sequence components of labor coefficients. The rates of work of labor, however, need considerable discounting in parallel with the output data, before use in planning can be justified. Before discussing methods for discounting experimentally derived input and output data, and alternative sources of planning coefficients, the third problem is outlined more fully.

## RELIABILITY IN RESULTS

The reliability of expected production levels has been noted as a criterion for the evaluation of innovations qualifying the profitability criterion. Particularly for food crops, high average levels of yield may be unacceptable if they vary widely from season to season. Because researchers are concerned to minimize all nontreatment sources of variation, design is often based on single-site plot experiments which avoid microclimatic variations. Per se this would be unimportant if the experiments were repeated over a number of seasons and an averaging of results over the period formed the basis of estimates to be used in planning. However, repetition of the same treatments until an adequate interseasonal population of results is obtained is both boring for the worker and perhaps wasteful of resources. Response rates tend to be based on researchers' "consensus" rather than statistical estimates; and a selected set of results, unlikely to be unfavorable, often backs up the recommendations made. Not only does this tend to inflate output levels, it also means that it is unusual to have any measure of the reliability of response levels from treatments for planning purposes. The variances between replicates of single-site experiments are much too sheltered to justify their use in planning.

In addition to these difficulties in deriving planning coefficients for improved farm practices, we have already noted that many of the attributes of the existing system measured by the survey are susceptible to interseasonal variation. Preparation of planning data will include the evaluation of the influence of the particular season on those susceptible variables important to the model.

Three approaches to planning coefficients by removing one or other of the distortions first outlined are discussed: an improvement in experimental technique to reflect farmer conditions more closely, the use of special surveys of innovators to see what results have been obtained under actual farm conditions, and the use of the trial farm unit to generate supplementary planning data.

## EXPERIMENTAL PROGRAMS AS A SOURCE OF DATA ON IMPROVEMENTS

Experimental programs can aid the construction of improved planning coefficients in two ways: by adopting designs which reflect farmer conditions more closely, and by serving as a source of time-series data when work has been continued in the same place over a long period.

J. D. Robinson has emphasized the need to get away from single-site experiments conveniently located on research stations and into extensive programs with sites on smallholder fields, although he notes that control is more difficult.[7] Substituting replication over a wide area for replication on a single site does create problems in holding down the level of experimental error, but it also allows an increase in the number of sites and a full reflection of microclimatic variations in the results. A. M. Scaife shows the coverage possible within such a program.[8] At the same time the labor which is applied by farmers to experiments sited on their land, although still a distortion because they inevitably give more than normal attention to the plots, is much more realistically determined than the ad lib application of the research stations. There tends to be less artificial inflation of output data, and results are therefore closer to the commercial possibilities. The variation measured between the same treatments over the replications gives a useful measure of the microclimatic influence. If the series is repeated over two or three years, a high proportion of the range of climatic effects can be recorded.

Recording of the timing and quantity of all treatment and non-treatment variables would present little extra work for the research center, though it may pose organizational difficulties with extensive programs. Labor used on experimental plots on smallholdings would be easily identified events for the farmer, and recording from memory would be a practical proposition. Visits for general supervision of these programs are necessarily fairly frequent.

So long as the specified variables remain constant, time series of basic trials on research stations also serve as a population giving measures of dispersion that are useful in calculating reliability. Again, the variation in results of such trials will be the effect of both micro and seasonal differences in climate. Analysis of variance on rainfall over the whole of the Sukumaland area shows variation to be greater between years than between locations. It can be assumed, therefore, that a series of experiments over several years, albeit on a single site, would reflect the probabilities affecting the single farmer. Stations often run plots continuously under different fertility regimes to observe long-term trends. The severity of the trend permitting, such series are useful for establishing the reliability of yield levels. Measures derived from this source serve to modify the averaged levels of output from a variety of crop practices, in the absence of a series on each particular practice. Individual practices may or may not improve reliability; this will depend on the sources of variation in the yields of particular crops. A useful series is shown in Table 71.

TABLE 71

Series of Results Illustrating the Effects of Tie-Ridging on Cotton and Bullrush Millet
(lbs./acre)

| | | 1947 | 1948 | 1949 | 1950 | 1951 | 1952 | 1953 | 1954 | 1955 | 1956 | 1957 | Mean | Coeff. Var. (%) |
|---|---|---|---|---|---|---|---|---|---|---|---|---|---|---|
| Cotton | Tied | 854 | 1,201 | 583 | 435 | 1,037 | 830 | 1,103 | 941 | 1,465 | 1,137 | 1,106 | 972 | 8.6 |
| | Untied | 585 | 1,082 | 320 | 387 | 972 | 843 | 748 | 777 | 1,150 | 1,076 | 1,118 | 821 | 10.2 |
| Bullrush Millet | Tied | 483 | 1,427 | 304 | 845 | 706 | 288 | 1,119 | 798 | 1,024 | 441 | 511 | 722 | 14.4 |
| | Untied | 460 | 1,090 | 90 | 908 | 620 | 342 | 1,018 | 719 | 804 | 486 | 359 | 627 | 14.3 |

Note: It is not known how consistent the specified variables are in this example.

Source: J. E. Peat and K. J. Brown, "Yield Responses of Raingrown Cotton in Lake Province, Tanganyika," Empire Journal of Experimental Agriculture, XXX (1962).

The series shows the results of trial plots and the effect of the practice of tie-ridging on average yield levels and yield reliability. Tie-ridging raises average levels in both cases and improves the reliability of the yield of cotton. Where reliability is of importance to the farmer, the improvement from new practices will weight evaluation favorably. The main question to be asked in planning is how the observed level of reliability should be allowed to modify the average output coefficients. In our application we have stressed reliability in food supply as a priority of the farmer. Assuming that from the investigation into the food economy we have established that the farmer has to fall back on cassava as a contingency starch staple in the dry season once in every six years, then, unless the period is exceptional, this gives five years out of six as the expected reliability level of his preferred starch staple (e.g., bullrush millet). How far must the average experimental yield of the millet be discounted in deciding the acreage requirement to meet these objectives? The normal curve gives the probability that 16 percent, or approximately one in six, of cases will fall below the level of -1 standard deviation. This meets the farmer's requirements and gives a basis for modifying the average experimental yield by -1 standard deviation, or -104 pounds for millet, to give an output coefficient which meets farmers' expectations of reliability, as a basis for the acreage of millet appropriate to subsistence requirements.

It may not always be appropriate to apply full-risk discounting to derive output coefficients. Where there is an expediency substitute, such as cassava in our example, an alternative is to link the innovations with the substitute. The substitute itself may offer the possibility of marked improvements in productivity, and developing the two crops as complements may be a better innovation than the single one heavily discounted.

Provisions of this type in experimental work can improve the usefulness of the results for planning. Recognition that extensive programs of trials are more appropriate to the needs of the farmers in developing areas is an important step toward data which are more relevant to the farm situation. Nevertheless, the practice of opting out of the economic environment by providing ad lib nontreatment variables to randomize and control unwanted influences, in order to highlight treatment comparisons, will inevitably leave a problem of inflated levels of both inputs and outputs.

## SURVEYS OF INNOVATIONS AS A SOURCE OF DATA ON IMPROVEMENT

S. R. Wragg has gone so far as to recommend that farm economists would be well advised to derive their planning data from carefully selected farm survey work rather than from experimental results.[9] Clearly, the output and input levels would be generated within the local economic circumstances. But as a source of data for planning coefficients, surveys of adopting farmers pose several problems, particularly in developing agriculture.

The main general problem is loss of control of the specified variables, both treatment and nontreatment. The coefficients derived are an amalgam from varying degrees of managerial efficiency in implementing the recommendations given. While this may give a truer picture of what the typical farmer can expect, in developing agriculture two particular problems raise more serious limitations to the usefulness of special surveys. The major point is that government research and advisory efforts are often the only stimulus to change on the farms. The pressing need is for ex ante evaluation to provide appropriate material for these efforts. An ex post survey cannot meet this need. Even if there were an entrepreneurial core in the community, survey investigation of these as a sub-population would be expensive. As since adopters by definition a low proportion of the total population, incorporating them into a sampling scheme would greatly complicate survey organization in the field. It would require a very wide deployment of enumerators to cover a useful sample of adopters. At the same time, they would rarely be innovating throughout an enterprise, which means that input and output data would have to be collected by plot, usually requiring a frequent-visit survey design. As an addition to the investigation of the existing traditional system, adopter surveys tend to swamp the already meager resources available to farm economics in traditional agriculture.

## TRIAL FARM UNITS AS A SOURCE OF DATA ON IMPROVEMENTS

The use of a farm unit as a tool in farm economics features extensively in the literature under a range of names from "experimental" farms to "example" or "demonstration" farms, and it has performed as large a range of functions. A. S. Barker defines the purpose of the example farm as evaluating the effect of new production methods on economic relationships.[10] Demonstration farms are more often seen as a tool of extension. Others, notably I. G. Reid in the discussion on Barker's paper, see units as a laboratory for the investigation of farmer problems, and stresses the need for representativeness.[11]

The same sort of confusion between fairly distinctive alternatives for investigation and extension has been apparent in work overseas. A. L. Jolly, a pioneer in the field in the 1950's, was convinced that the investigation of existing peasant farming systems had nothing to contribute to the improvement of traditional agriculture.[12] He believed that institutional factors inhibited small farm development, and he organized unit farms as a means of investigation of farming systems, unfettered by the handicaps imposed by the infrastructure of West Indian agriculture. E. Clayton supports Jolly's work:" The Unit farm is a practical method of investigating farm management possibilities in peasant farming."[13] G. P. Hirsch criticizes Jolly's method and conclusions severely but also dismisses existing agriculture as a basis for development. The emphasis in the discussion aroused by Jolly's work was on an investigational farm unit as a substitute for investigation in the farm population. M. P. Collinson has reported on experience with a trial farm in Tanzania.[14] This work in Sukumaland in the early 1960's came after an advance in techniques for whole-farm planning. Whereas it took Jolly seven years to produce one viable holding out of nine established, based on different crops, the profitability of the Tanzanian farm was never in question. Program planning was used to set up the farm, with the objective of studying the rate of development and the reasons for divergence between planned and actual performance.

As farm planning research on ordinary small farms progressed in Sukumaland, the data generated by the trial farm became a useful supplement to collected data. The performance of recommended practices under actual conditions on a small farm, albeit carefully managed, highlighted the inflated output levels of experimental work. The yield levels achieved offered a better basis for output coefficients for use in planning. The input data recorded on practices new to traditional agriculture could be integrated with the data on the main operations collected from small farm surveys. It became clear that this type of farm had a role in adapting experimental data for planning purposes, as well as in providing data required on nontreatment variables usually not observed in the course of experiments. At the same time specified nontreatment variables can be closely controlled and their impact on economic relationships in the system quantified.

In advanced agriculture high cost is the main criticism of this type of unit. It arises from the typically capital-intensive large-scale nature of farm organization and the large number of type groups, creating the need for extensive replication of trial farms. Neither point is valid in traditional agriculture, where capital investment is typically very low and the limited production opportunities and methods give a homogeneity over wide areas which are readily identifiable. The reality of the factor relationships in a working farm situation will solve the major shortcomings of experiments as a source of planning data. Nevertheless, there are clear criticisms on statistical grounds. A single unit is susceptible to all the problems discussed in the selection method for deriving a representative farm. Also, it will be characterized by a single site and thus have the disadvantages of single-site experiments vis-à-vis microclimatic sources of variation in planning variables. Meeting these two criticisms involves an interseasonal sample of input and output data to give estimates of the distribution of attributes influenced by the climate, and tying the farm into the population of farms in the area, to relate measures to the local average of the attributes concerned.

The trial farm unit reported on by Collinson was run for three years, a period inadequate to sample the interseasonal variation. In addition to this, a major objective of the farm was to study the potential rate of development from internally generated surpluses, an objective which was inconsistent with the need to carry the same methods and practices over a long enough period to observe the interseasonal variation. However, the unit gave results which showed up the discrepancies between experimental yield levels and those obtainable under farm conditions. Table 72 shows the yields obtained on the major crops. The final row in the table shows the output coefficients used in the original plan, derived by discussion with agronomists on the station, who subjectively discounted the available experimental results.

Relating the trial farm unit to the local population of farms in order to derive input coefficients for planning is straightforward. Operations unaffected by climatic or other identifiable sources of variation, such as soil types, can be related; and differences noted will be due to managerial or motivational sources of variation. These give an adjustment factor for the unit. As observations accumulate over a number of seasons, the distributions of more variable attributes can be related to cross-section data, subject to the same sources of variation, to verify the weighting. In Table 73 data from the unit for the ridging operation on hill sand soils are compared with cross-section data for the same attribute. (Ridging on these sandy soils is little influenced by climatic factors.)

TABLE 72

Output Coefficients: Realized and Expected
(lbs./acre)

| | Maize | | Cotton | | | |
|---|---|---|---|---|---|---|
| | No Fertilizer | With Fert. | With Fert. | With Fert. & Insect. | Groundnuts | Rice |
| 1962-63 | 574 | – | 820 | – | 688 | 1,342 |
| 1963-64 | – | 986 | 805 | – | 692 | 993 |
| 1964-65 | – | 1,861 | – | 943 | 354 | 1,254 |
| Average | 574 | 1,423 | 812 | 943 | 578 | 1,196 |
| Planned | 1,200 | 1,500 | 800 | 1,200 | 600 | 4,000 |

Source: Compiled by the author.

TABLE 73

Rates of Work Compared as a Basis for Weighting Planning Coefficients

| Data Source | Year | Acres Cropped per Available Labor Unit | Ridging (man-days/acre) Cotton | Maize |
|---|---|---|---|---|
| Trial | 1962-63 | 3.18 | 10.8 | 10.3 |
| Farm | 1963-64 | 3.47 | 8.7 | 11.4 |
| | Mean | 3.32 | 9.8 | 10.8 |
| Survey | 1963-64 | 2.92 | 10.6 | 14.1 |

Source: Compiled by the author.

The relationship between the work rates and acres cropped per available labor unit as a measure of labor efficiency is consistent. In this case weightings of +20 percent would be justified on trial farm labor data used for planning coefficients. This weighting would be confirmed or modified as the farm built up interseasonal data to allow a more reliable averaging of its own rates. These same interseasonal distributions can be used to locate seasonal peculiarities in cross-section data.

The farm should reflect the conditions of local farming, particularly the objectives of local farmers, their crop opportunities, and methods. A simulation of the system of relationships is important. From several points of view installing a farm family on the holding is important; the temptation to vary the labor force to meet seasonal contingencies is more easily avoided and the type and extent of family contingencies can be recorded. However, full control of crops and crop acreages is necessary, and the farmer's choice of food and cash crops should be limited to those of the local population under current research. The management regime should be wholly imposed on the farm: selection of an existing farm would compromise the final control of decisions, so a unit should be established for the task. As a corollary to this, the farmer must be insulated from the effects of the practice being tried. He should be paid a wage regardless of the viability of the unit and a bonus dependent on the implementation of the planned program and based on his efforts rather than the farm results. If the efforts are correctly managed, the results will follow; and management is the responsibility of the economist in charge. Repetition is important to establishing populations of attributes over time, although it is somewhat inconsistent with the need for the continual incorporation of research findings into the system. The crop program can be planned to the farmer's labor capacity. Within this crop, acreages can validly be subdivided to give continuity to a series of records while allowing new practice to be introduced. Too fine a subdivision, which would allow the "scale" effect to distort the rates of work, should be avoided. Subdivisions compromise any optimizing of farm productivity, but the units are useful as demonstration holdings. Other ancillary benefits include that of bringing the technical research staff face to face with the questions of adapting results and experimental designs to the economic circumstances of the farmer.

Of the three alternatives covered, the survey of adopters is the least useful because of the essentially ex ante requirement in planning which it cannot, by definition, meet. Both better design of experimental programs and trial farm units offer the means to deflate input and output coefficients and to generate reliability measures to qualify the averages usually specified. Of the two the trial farm unit, designed for the purpose, is a more complete technique for deriving relevant planning coefficients.

The review of available experimental work to build the inventory of possible improvements can be done only at the source. As well as treatment levels, the original experimental record will usually give the levels and timing of specified nontreatment variables, and often the timing of ad lib labor operations. The inventory of improvements for Sukumaland has been limited to results available prior to the 1966 season and available for testing on the trial farm unit over the period 1962-65. Aspects of the agronomy of cotton, maize, groundnuts, and cassava has been covered by experimental work at the Western Research Center, Ukirigru. In addition, because intercropping is dominant in local subsistence agriculture, use is made of studies of intercropping made at Mwanhala, a substation of Ukirigru some 150 miles to the south. This work, done in 1957-58, was reported by A. C. Evans in 1961.[15] Other published work presenting results on some of the aspects covered are by Evans and Shreedran, K. J. Brown, and Scaife, as well as publications of the Cotton Research Corporation.[16]

Available results were examined in the light of data on the local farming system collected over the period 1961-65. Treatments and specified nontreatment variables were scored for congruity with established practices, to identify steps in the process of improvement.

## Varieties

Varietal changes have two potential sources of incongruity. The first is taste differences from the established variety. If improved varieties are available, the new ones should be scored on important taste criteria during the inventory of possible innovations. D. Jowett has demonstrated the use of rank-correlation techniques for scoring taste preferences.[17] The scores should be used to rank changes involving the new varieties on a scale of congruity. Clearly, if the new varieties are within the range of existing taste preferences, congruity, in this respect at least, is complete. The other aspect is seed replacement. If the farmer must buy seed each year to maintain a hybrid, it will usually be a departure from present practice and so reduced congruity.

The pool of improvements being considered here does not contain any new food crop varieties. Maize, groundnut, and cassava varieties were recommended by the center, but seed was not generally available and no tests on taste had been done. Regular issues of new cotton seed were made by an established procedure which farmers had long used. Ginning is done centrally and cotton is sold on the seed; farmers must collect new seed for the following season's plantings.

## Purity of Stand and Plant Population

Changing from mixed to pure cropping is a large step for a traditional farmer and has implications for each of the intercrops currently dominating subsistence production. This factor is often overlooked by extension programs, which tend to concentrate on improving the performance of a single constituent. Foods grown in a mixture tend to be complementary in their role in the household—perhaps in taste, perhaps as insurance. Change may precipitate important direct clashes in food supply priorities, as well as the significant reduction in congruity implied. Such is the case here, where maize, groundnuts, and cassava are grown wholly as intercrops in the existing system, and yet from the result available all are recommended in pure stands. Improved intercrops of maize and groundnuts from Mwanhala have been included for comparison. Changes in plant population by changed spacing also reduces congruity and, in extreme cases, an increased population may have important repercussions on the labor requirements for the planting operation. Groundnuts raise this problem in Sukumaland. Existing practice is to intercrop maize with a legume population of about 3,500 stands per acre, while the recommended density for pure groundnuts is 79,380 stands per acre.

## Time of Planting

Time of planting alternatives have dominated research efforts in the area aimed at exploiting an "early planting effect," which is believed to be associated with a flush of nutrients at the new rains but is not fully understood. Because the labor requirements for cultivation and planting create the peak which limits the size of farms, time of planting is also a key economic variable. Changed times are not necessarily incongruous, though they may have a considerable direct impact on both complexity and acceptability. Congruity is reduced by a practice which includes a fixed time of planting. Flexibility in planting time, to meet seasonal contingencies of both weather and labor supply, is an important attribute in the existing system.

## New Practices

A range of increasingly incongruous new techniques is set out below. Their importance is dependent in part on timing, which is geared to time of planting.

1. Extra weeding. Normal practice by farmers is two weedings in cotton, one in maize mixtures. Specified nontreatment conditions of the experimental work assume an earlier first weeding, thus creating the need for a third in cotton and a second in maize. Pure maize does not shade out late weed growth as effectively as does an intercrop canopy. In all cases the extra weeding will fall in a slack period of the season, but the earlier first weeding may heighten existing peak labor requirements.

2. Tie-ridging is a recommended technique for blocking off the ridges, holding heavy rainfall <u>in situ</u> to allow its use to be prolonged. It also prevents excess accumulation at low points, which would break the ridges and carry away top soil. The maximum effect, ties must be made before the trough in rainfall probability in January, and they thus increase the requirements at the existing peak period.

3. The use of fertilizer. As well as implications for increased labor at busy times of the year, the use of fertilizer has two elements of incongruity: it involves a new technique for application and also requires the farmer to purchase an input, both of which are departures from existing practice and are barriers to adoption.

4. The use of insecticide parallels the use of fertilizer. It is more demanding managerially, requiring six applications and mixing of the insecticide. It also involves long-term investment in a pump, with implications for farmer time preferences. But since it is applied during the slack period of the season, it does not increase peak labor requirements.

The permutation of alternatives is extensive, even within the four crops being considered. It has been accepted that point estimates must be made within the context of data available from experimental results. In practice, selection has been limiting to improvements considered for test on the trial farm unit at Ukiriguru in 1962-65. Exceptions are variations of planting times, for which output levels are interpolated, and the one case of intercropped maize and groundnuts from work at Mwanhala.

Initially a total of fifty-two innovation alternatives are considered, mainly on cotton, maize and groundnuts. Two additional ones each are considered on cassava and sweet potatoes. The sweet potato changes are taken from the existing system itself (always a possible source of improvement), where the sweet potato requirement for food is satisfied partly from pure stands and partly from intercrops. Pure stands, requiring modified time of planting, are considered as alternative source to the intercrops.

Cotton

The use of fertilizer and insecticides, each at a single level, is considered separately and together at four different times of planting, with and without the practice of tie-ridging, and are compared with "controls" without purchased inputs. This gives thirty-two alternative combinations, differing in congruity but all with the same spacing and weeding regime, except for the very late-planted alternatives, which do not need the final weeding.

Maize

The joint use of fertilizer and insecticide is considered over five times of planting, of which two are variables, spreading the planting over periods up to three months long. Alternatives without purchased inputs are included, while spacing and weeding regimes remain the same.

Groundnuts

A constant spacing and weeding regime is considered over five times of planting, of which one is variable, with planting spread over a six-week period. One intercropped activity is considered as an alternative to pure stands.

Cassava and Sweet Potatoes

A constant spacing and weeding regime is considered at two different planting times for each crop.

Table 74 gives an array of available experimental results on different times of planting for cotton. It demonstrates the type of grouping which has been used to derive point estimates for the different activities selected.

The output coefficients from this type of array have then been modified by results on the trial farm unit. This has been done directly, where a comparison has been possible, and by the relationships established from direct comparison for practices not tested on the farm. It is at this stage of the planning sequence, while modifying the experimental results, that outputs should be discounted for farmers' expectations of reliability. Most of the data available from the trial farm are averages from only two seasons and are inadequate to measure the dispersion of results. For convenience, it is assumed that expectations have been built into the coefficients used. The trial farm unit has also provided input data on work rates and timing for practices not covered by farm survey but required in planning. These include the application of fertilizer, both in the seedbed and as a top dressing, the application of insecticides, tie-ridging, the harvesting of pure stands not grown in the existing system, and harvesting at higher yield levels when operational overheads are reduced. A factor of 1.2 has been used to upgrade trial farm rates to average local levels.

The planning coefficients for fifty-two selected innovations are set out in Tables 75 and 76, and show reducing congruity from left to right. Table 75 sets out thirty-two innovations for the cotton crop; and innovation A has the greatest congruity with existing practice, differing only in the higher plant population and a requirement for a third weeding.

TABLE 74

Sample Array of Results of Time of Planting Trials on Cotton
(lbs./acre)

| | Time of planting | | | | | | | | | | | | |
|---|---|---|---|---|---|---|---|---|---|---|---|---|---|
| | | November | | | | December | | | January | | | February | |
| Season | 0 | 10 | 20 | 30 | 10 | 20 | 30 | 10 | 20 | 30 | 10 | 20 | 30 |
| 1953/54 | | | | 936 | | 1,035 | | 967 | | 396 | | | |
| 1955/56 | | | 880 | | 553 | | 520 | | 336 | | | | |
| 1957/58 | | 911 | | | 821 | | | 684 | | | 613 | | |
| 1958/59 | | | 1,255 | | 1,216 | | 1,077 | | 635 | | | | |
| 1959/60 | | | 1,044 | | 924 | | 663 | | 397 | | | | |
| Sub Grouping | | | A | | | B | | | C | | | | D |
| Center Date | | | Nov. 23 | | | Dec. 19 | | | Jan. 10 | | | | Jan. 30 |
| Mean Yield | | | 1,005 | | | 910 | | | 782 | | | | 475 |

Source: A. C. Evans and S. Sheedran, "Intercropping Studies II," East African Agricultural and Forestry Journal, XXVI (1960-61).

TABLE 75

Per-Acre Input and Output Data for Thirty-Two Possible Cotton Innovations

| | Time of Planting | | | | Plus Fertilizer | | | | Plus Insecticide | | | | Plus Fertilizer & Insect. | | | |
|---|---|---|---|---|---|---|---|---|---|---|---|---|---|---|---|---|
| | A | B | C | D | E | F | G* | H | I | J | K | L | M | N | O* | P |
| Date of Planting | 12/7 1/22 | 1/1 | 12/7 | 1/22 | 12/7 1/22 | 1/1 | 12/7 | 1/22 | 12/7 1/22 | 1/1 | 12/7 | 1/22 | 12/7 1/22 | 1/1 | 12/7 | 1/22 |
| Fertilizer Use (shs.) | 0 | 0 | 0 | 0 | 45 | 45 | 45 | 45 | 0 | 0 | 0 | 0 | 45 | 45 | 45 | 45 |
| Insecticide Use (shs.) | 0 | 0 | 0 | 0 | 0 | 0 | 0 | 0 | 25 | 25 | 25 | 25 | 25 | 25 | 25 | 25 |
| Pump Cost (shs.) | 0 | 0 | 0 | 0 | 0 | 0 | 0 | 0 | 150 | 150 | 150 | 150 | 150 | 150 | 150 | 150 |
| Labor: Seedbed Fert. (man-days) | 0 | 0 | 0 | 0 | .7 | .7 | .7 | .7 | 0 | .7 | 0 | 0 | .7 | .7 | .7 | .7 |
| Months Applied | 0 | 0 | 0 | 0 | NDJ | Nov. | Dec. | Jan. | 0 | 0 | 0 | 0 | NDJ | Nov. | Dec. | Jan. |
| Topdress Fert. | 0 | 0 | 0 | 0 | 2.1 | 2.1 | 2.1 | 2.1 | 0 | 0 | 0 | 0 | 2.1 | 2.1 | 2.1 | 2.1 |
| Months Applied | 0 | 0 | 0 | 0 | JF | JF | Jan | Feb | 0 | 0 | 0 | 0 | JF | JF | Jan. | Feb. |
| Insecticide Use | 0 | 0 | 0 | 0 | 0 | 0 | 0 | 0 | 3 | 3 | 3 | 3 | 3 | 3 | 3 | 3 |
| Months Applied | 0 | 0 | 0 | 0 | 0 | 0 | 0 | 0 | FMAM | MAM | FMAM | MAM | FMAM | MAM | FMAM | MAM |
| Harvesting: Total Man-Days | 24 | 25 | 26 | 22 | 26 | 27 | 28 | 23 | 26 | 27 | 27 | 25 | 28 | 29 | 31 | 25 |
| Months Used | MJJA | MJJ | MJJ | MJJA | MJJA | MJJ | MJJ | MJJA | MJJA | MJJ | MJJ | MJJA | MJJA | MJJ | MJJ | MJJA |
| January Tie- | DJ | | | | DJ | | | | DJ | | | | DJ | | | |
| Ridging | 5 | 5 | 5 | 5 | 5 | 5 | 5 | 5 | 5 | 5 | 5 | 5 | 5 | 5 | 5 | 5 |
| Output Experimental (lbs.) | – | 650 | 800 | 400 | – | 900 | 1,100 | 500 | – | 850 | 1,100 | 550 | – | 1,200 | 1,500 | 750 |
| Trial Farm | 520 | 650 | 650 | 350 | 665 | 750 | 850 | 400 | 665 | 750 | 800 | 450 | 850 | 900 | 1,100 | 550 |
| Tie-Ridged(+ 18%) | 615 | 650 | 765 | 410 | 785 | 885 | 1,000 | 475 | 785 | 885 | 945 | 530 | 1,000 | 1,060 | 1,300 | 650 |

*Innovations allowing direct comparison between experimental and trial farm results.

Notes: The fertilizer regime adopted was twenty units of phosphate in the seedbed and twenty units of nitrogen topdressed four weeks after germination.

The insecticide regime adopted was six sprays, each of one pound a.i. 75 percent DDT wettable powder, at ten-day intervals starting ten weeks after germination. A knapsack sprayer was used.

Rates of picking have been assumed to increase from 20 pounds per man-day at present farm yields of 475 pounds per acre to 35 pounds per man-day at 1,100 pounds per acre. Estimates were made by regressing available rates of work on yield levels out of the same population.

With tie-ridging, an increase of harvest labor of one man-day is assumed. The total man-day requirement for harvesting is based on trial farm yields.

The output from innovations with a variable planting time was built up by assuming an even distribution over the period and applying the appropriate yield levels to quarter-acres planted at four times during the period.

Nontimely postharvest operations are excluded throughout the planning sequence.

TABLE 76

Per-Acre Input and Output Data for Possible Food Crop Innovations

| | Maize: Time of Planting | | | Maize: Plus Fertilizer and Insect. | | | | | | | Groundnuts: Time of Planting | | | | Maize & Groundnuts | | Cassava | | Sweet Potato | |
|---|---|---|---|---|---|---|---|---|---|---|---|---|---|---|---|---|---|---|---|---|
| | MAA | MAB | MAC | MAD* | MAE | MAF | MAG | MAH | MAI* | MAJ | GNA | GNB* | GNC | GND | GNE | MGN | CASS A* | CASS B* | SPOT A | SPOT B |
| Date of Planting | 11/30<br>2/28 | 11/30<br>1/15 | 12/15 | 1/15 | 2/7 | 11/30<br>2/28 | 11/30<br>1/15 | 12/15 | 1/15 | 2/7 | 12/15<br>1/25 | 12/30 | 12/7 | 1/15 | 2/7 | 12/15 | 11/25 | 3/15 | 11/15 | 1/15 |
| Fertilizer & DDT Dust. Cost (shs.) | 0 | 0 | 0 | 0 | 0 | 55 | 55 | 55 | 55 | 55 | 0 | 0 | 0 | 0 | 0 | 0 | 0 | 0 | 0 | 0 |
| + Labour: Fert. (man-days) | 0 | 0 | 0 | 0 | 0 | 4 | 4 | 4 | 4 | 4 | – | – | – | – | – | – | – | – | – | – |
| Months Applied | 0 | 0 | 0 | 0 | 0 | DJFM | DJF | Jan | Feb | FM | – | – | – | – | – | – | – | – | – | – |
| Dust (man-days) | 0 | 0 | 0 | 0 | 0 | 5.6 | 5.6 | 5.6 | 5.6 | 5.6 | – | – | – | – | – | – | – | – | – | – |
| Months Applied | 0 | 0 | 0 | 0 | 0 | JFMA | JFM | JF | FM | MA | – | – | – | – | – | – | – | – | – | – |
| Total Harvest Labour (man-days) | 4.0 | 4.7 | 6.7 | 4.7 | 2.7 | 9.3 | 10.0 | 12.0 | 9.3 | 5.3 | 17.0 | 18.0 | 20.0 | 16.0 | 14.0 | 15.0 | – | – | – | – |
| Months Used | MJ | MJ | May | MJ | Jun | MJ | MJ | May | MJ | Jun | AM | AM | AM | AM | May | AM | – | – | – | – |
| Output (lbs.) Experimental | – | – | 1,100 | 780 | 450 | – | – | 2,000 | 1,800 | 800 | – | 800 | 1,200 | 550 | 240 | M 1,000<br>730 | 20,000 | 7,000 | – | – |
| Trial Farm, Discounted | 600 | 700 | 1,000 | 700 | 400 | 1,400 | 1,500 | 1,800 | 1,400 | 800 | 430 | 550 | 750 | 320 | 110 | M 700<br>GN 320 | 9,000 | 3,000 | 4,000 | 4.000 |
| Green Maize Mar. | 300 | 350 | 500 | 0 | 0 | 700 | 750 | 900 | 0 | 0 | – | – | – | – | – | 0 | – | – | – | – |
| Apr. | 300 | 350 | 1,000 | 700 | 200 | 700 | 750 | 1,800 | 1,400 | 400 | – | – | – | – | – | 525 | – | – | – | – |
| May | 300 | 350 | 0 | 350 | 400 | 700 | 750 | 0 | 700 | 800 | – | – | – | – | – | 525 | – | – | – | – |

*Innovations allowing direct comparison between experimental and trial farm results.

Notes: Each maize enterprise can produce dry maize and/or green maize. The months of potential green maize supply vary with the planting time.

Harvesting labor has been estimated from trial farm data. In the existing system maize, legumes, and cassava are intercropped; maize, cassava, and sweet potatoes are eaten from the field. No labor input data for pure stands was available.

Seed costs for groundnuts have been deducted from the discounted yields. At fifty pounds per acre they are a significant requirement. Seeds for the other crops are not significant in terms of total output and have been ignored.

DDT dust is used in areas of the crop suffering from stalk borer attack. Full treatment is two applications each of 5 pounds of 5 percent powder, shaken into the plant funnels, costing 20 shillings per acre. An allownace of 10 shillings has been included

The fertilizer regime is forty units of nitrogen per acre, applied as a topdressing three weeks after germination.

Table 76 is divided between the food crops being considered as possible innovations. Again, within the columns for each crop, the alternatives are ordered by reducing congruity and, except for maize, the differences are less significant.

The two tables make up an example of an inventory of innovations, to be used in illustrating the planning sequence for deriving extension content for a type of farming area. The possible permutations have already been considerably simplified by the selection of times of planting and of single levels of the limited number of treatments included.

## NOTES

1. S. R. Wragg, "Co-operative Research in Agriculture and the Provision of Input/Output Coefficients," Journal of Agricultural Economics, XXI (1970).

2. J. D. Robinson, "Extensive Trials with Perennial Crops on Smallholders' Fields in Africa," Experimental Agriculture, VI, 4 (1970).

3. G. Weinschenk, "Quantitative Research in Agricultural Economics," in Proceedings of the 1964 Conference of International Agricultural Economists (London: Oxford University Press, 1966).

4. U. Renbourg, "Discussion of Weinschenk," op. cit.

5. H. A. Davidson and B. R. Martin, "Relationships Between Field and Experimental Yields," Australian Journal of Agricultural Economics, IX (1965).

6. Wragg, op. cit.

7. Robinson, op. cit.

8. A. M. Scaife, "Maize Fertilizer Experiments in Western Tanzania," Journal of Agricultural Science, LXX (1968).

9. Wragg, op. cit.

10. A. S. Barker, "The Example Farm in Management Investigation," Journal of Agricultural Economics, XI, 4 (1956).

11. I. G. Reid, discussion of Barker's paper, Journal of Agricultural Economics, XI, 4 (1956).

12. A. L. Jolly, Report on Unit Farms to the Caribbean Secretariat (Trinidad: Imperial College of Tropical Agriculture, 1953). (Mimeographed.)

13. E. Clayton, "Note on Research Methodology in Peasant Agriculture," Farm Economist, VIII, 6 (1957).

14. M. P. Collinson, "Experience with a Trial Management Farm in Tanzania, 1962-65," East African Journal of Rural Development, II, 2 (1969).

15. A. C. Evans, "Time of Planting Studies in Tanzania," East African Agricultural and Forestry Journal, XXVI (1964-65).

16. A. C. Evans and S. Sheedran, "Intercropping Studies II," East African Agricultural and Forestry Journal, XXVI (1960-61); K. J. Brown and J. E. Peat, "Yield Responses of Raingrown Cotton in Lake Province, Tanganyika," Empire Journal of Experimental Science (1962). Scaife, op. cit.; Cotton Research Corporation, Progress Reports from Experimental Stations: Western Research Centre, Tanzania (annually).

17. D. Jowett, "Use of Rank Correlation Techniques to Determine Food Preferences," Experimental Agriculture, II, 3 (1966).

CHAPTER

# 17

# THE PLANNING SEQUENCE: CONSTRAINTS AND COEFFICIENTS IN SIMULATION

The planning sequence has two stages. First, a system is derived which optimizes the satisfaction of farmer objectives within the technology, opportunities, and resources available to the farmer. This stage defines extension strategy, the aim being to move from the existing system to this optimum as a goal. The second stage identifies steps toward the goal within the risk preferences and inherited attitudes of the farmer. This stage defines extension content at points in time over the adoption period and, in the course of an essentially subjective evaluation of the balance between the incentives offered by innovations and the barriers formed by their incongruity, complexity, and acceptability scores, in order to decide the content and scale of the steps, the length of the adoption period will emerge for the area concerned.

The model of adoption set out in Chapter 14 covers the period over which an innovator is expected to absorb the available technology. In the limited sense that it reaches over a period of years, the model is dynamic. The planning sequence, as it stands, is static. In practice the optimum solution will be altered by new technology being developed, by changes in market conditions, and by the effects of government policy. As far as possible, such changes should be anticipated and built in as conditions of the planning sequence. Thus a second dynamic aspect is superimposed on the adoption model, incorporating exogenous and endogenous system variables likely to alter over the period and possibly changing both the extension strategy and the appropriate content of extension programs during the adoption period.

The two stages of the planning sequence are described in Chapters 18 and 19. This chapter examines the types of conditions in the economy/(the exogenous variables) and within the system itself (the endogeneous variables), which might alter the constraints of the planning model by varying over the adoption period. In the course of the chapter, the simulated model of the existing systems to be used as an example in the planning sequence is set out. The simulation has been based on survey data to show the balance of the use of labor resources in different crop activities with the consumption of foods and sale of marketed produce. From this system as a base, a modified LP core program requiring a 32K store was used to test the impact of the range of innovations already described. The agenda of this program allows the addition or overwriting of rows and columns as well as the overwriting of any single or series of coefficients in the matrix. Once designed, the main series of programs required for the planning sequence generated about 200 program solutions requiring about half an hour of computer time. Programming using a computer allows the testing and scoring of a wider range of changes, but the selection of extension content requires a subjective evaluation over the five planning criteria, a process which cannot be computerized without a common denominator of scale values for all the criteria.

## ENDOGENOUS CONSTRAINTS ON THE MODEL

The farm survey provides data on the levels of labor supply over the season, on the food consumption pattern over the season, and on present capital outlay—the three main endogenous constraints on the present farm system. It also provides the input/output coefficients relating labor required to crop activities at particular times of the season. In discussing the constraints on the Sukumaland system as an example, various hypothetical features are included to demonstrate the incorporation of some of the aspects stressed in the investigation section.

### Labor Supply Constraints

In building a model of the existing system, the season must be divided into periods. This may be based on regular time intervals—weeks, fortnights, or months—or on periods which vary with the necessary timeliness of particular tasks. Careful examination of the spread of operational timing over the sample of surveyed farms and the timing requirements of new practices provide the basis for a decision on the periods which are subsequently assumed homogeneous for the supply and demand for labor. The division which is made should not allow unrealistic substitution of essentially different labor streams, but neither should it create artificial rigidities in timing.

The interval boundaries are a matter of convenience, and the interpretation of planning results should include an appraisal of artificial limitations on flexibilities created by the assumptions made. For our example monthly intervals are used, justified by the large crop acreages, the labor-intensive operations, and the practice of staggered planting.

In Sukumaland crop labor needs are most intensive during the period November to February. The operations carried out in November, December, and January are physically very arduous, and in these months total labor supplied by the family falls and the incidence of hired labor is at its height. Inquiries on the food supply pattern show this as a possible dietary constraint of family capacity, with nutritional status improving in February as the new season's crops become available. Actual levels of labor supplied over this critical period are accepted as constraints on capacity from October to February. Survey data suggest no period at which social custom will limit supply, and there is no regular or seasonal off-farm work. The level of labor supplied by the family in peak periods is accepted as a limit of potential supply for other months. The single exception here is the month of October, when the rains are uncertain until the end of the month and time available for seasonal work is limited to twenty man-days.

Casual labor is hired for the cultivation and weeding operations during the November - February peak; and while the level of hire is modest, it represents an important phenomenon from three points of view.

1. Hired casual labor is the only cost item in the existing farming system, farmers paying 3.50 shillings per man-day hired in cash and kind. The purchased inputs required for most innovations will be competing for the limited cash outlay on labor, unless credit is available.

2. Hired casual labor is a scale-increasing input. In the light of the incipient fertility clash, emphasis on intensification is necessary. This increases the importance of credit for purchased inputs and indicates against credit for casual labor hire in this area. Already the initial selection of possible innovations has precluded consideration of the purchase of ox equipment and the hire of tractors on grounds of accelerating fertility losses.

3. Only 62 percent of the farmers in the survey hired labor, which raises the question of the average level of hired use to be incorporated into the model.

Since it is likely that the 62 percent will cover the potential innovators and casual labor is available for hire with no identifiable constraints on the level or timing of its availability, labor buying is accepted as a possibility for farmers of the area and the time of hire will be flexible, depending on requirements. The quantity hired will be limited by the capital outlay constraint and the profitability of alternative capital uses. For the simulation of the existing system the level of use will be the average of those farmers hiring labor.

The survey season is judged to be climatically normal and the input/output coefficients recorded are accepted as typical. Table 77 sets out the resource relationships and crop activities of the simulated system. It includes the capital outlay of 150 shillings, the cost of the forty-three man-days of casual hired labor, itself a significant constraint on changes to the system.

## Food Supply Pattern

The survey provides data on the types and quantity of food being supplied from each crop activity identified as sufficiently important to warrant independent representation in the planning model. As acceptable levels of supply, these are incorporated into the simulated model. Again, for the example, the survey season has been accepted as normal and the amounts available have been used as food requirement conditions for the system and output coefficients for the activities. Preferences for green maize as a flow over the three months March through May, and for fresh sweet potatoes for the eight months February through September, have been included in the simulation. The production of these preferred foods shows the importance of timing the present crop activities into sequence. Old cassava, as an insurance against maize or sweet potato failure, is included as a separate activity.

Table 78 sets out the way in which requirements are supplied from the range of crop activities represented in the model and the timing of supplies for green maize and sweet potatoes. The table also shows the value of cash output from the cotton enterprise. It completes the factor/product balance for the model of the existing system.

Over the adoption period the levels of these constraints may alter, and it is important to outline the possible alterations which require consideration when setting up the planning sequence for a particular area. On the labor supply side, intentional changes, perhaps

TABLE 77

Crop Activities, Acreages, and Monthly Labor Use in the Sukuma Sample

| Crop | | | Early Maize, Legumes, Sw. Pot. | Early Maize, Cassava, Legumes, Sw. Pot. | Early Sweet Pot. | Late Maize, Cassava, Legumes, Sw. Pot. | Rice | Late Maize, Legumes, Sw. Pot. | Late Sweet Potato | Old Cassava | Cotton | Total |
|---|---|---|---|---|---|---|---|---|---|---|---|---|
| Planting Time (center date) | | | Nov. 1 | Nov. 20 | Dec. 15 | Jan. 15 | Feb. 10 | Feb. 15 | Mar. 10 | — | Dec. 15 | |
| Acreage | | | .558 | 1.46 | .05 | 1.46 | .30 | .58 | .05 | (1.00) | 4.03 | 8.51 |
| Labor | Available | | Used (Man-days) | | | | | | | | | Total Use |
| | Family | Hired | | | | | | | | | | |
| October | 20 | — | 7 | — | — | — | — | — | — | 7 | — | 14 |
| November | 55 | 2 | 7 | 37 | — | — | 1 | — | — | — | 12 | 57 |
| December | 55 | 19 | 6 | 7 | 2 | 7 | 0 | — | — | — | 52 | 74 |
| January | 55 | 20 | — | 7 | — | 29 | 3 | — | — | — | 36 | 75 |
| February | 60 | 2 | — | — | — | 15 | 13 | 14 | — | — | 20 | 62 |
| March | 60 | — | — | — | — | — | 5 | 6 | 2 | 8 | 12 | 33 |
| April | 60 | — | — | — | — | — | 4 | — | — | — | — | 5 |
| May | 60 | — | 2 | 6 | — | — | 4 | — | — | — | 32 | 44 |
| June | 60 | — | — | — | — | 6 | 4 | — | — | — | 40 | 50 |
| July | 60 | — | — | — | — | — | 3 | 2 | — | — | 28 | 33 |
| August | 60 | — | — | 9 | — | — | — | — | — | — | 12 | 21 |
| September | 60 | — | — | — | — | 6 | — | — | — | — | — | 6 |
| Family Total | — | — | 22 | 66 | 2 | 63 | 38 | 22 | 2 | 15 | 201 | 474 |
| Hired Total | — | 43 | — | — | — | — | — | — | — | — | 43 | |
| Capital | 150 | — | — | — | — | — | — | — | — | — | 150 | 150 |

TABLE 78

Food Supply Pattern and Market Production
in the Sukuma Sample

| | | EML | ECML | EP | LCML | R | LML | LP | OC | COT |
|---|---|---|---|---|---|---|---|---|---|---|
| Food Needs | (lbs.) | Food Supplies | (lbs.) | | | | | | | |
| Dry Maize | 1,300 | 116 | 511 | – | 511 | – | 174 | – | – | – |
| Green Maize (Mar.) | 175 | 175 | – | – | – | – | – | – | – | – |
| Green Maize (Apr.) | 175 | 29 | 146 | – | – | – | – | – | – | – |
| Green Maize (May) | 175 | – | – | – | 146 | – | 29 | – | – | – |
| Fresh Cassava (Sept.-Nov.) | 750 | – | 760 | – | – | – | – | – | – | – |
| Fresh Cassava (Dec.-Feb.) | 750 | – | – | – | 760 | – | – | – | – | – |
| Fresh Sw. Pot. (Feb.-Mar.) | 200 | 93 | 117 | – | – | – | – | – | – | – |
| Fresh Sw. Pot. (Apr.-May) | 300 | – | 117 | 200 | – | – | – | – | – | – |
| Fresh Sw. Pot (June-July) | 300 | – | – | – | 219 | – | 81 | – | – | – |
| Fresh Sw. Pot (Aug.-Sept.) | 200 | – | – | – | – | – | – | 200 | – | – |
| Legumes | 200 | 29 | 73 | – | 73 | – | 29 | – | – | – |
| Rice | 300 | – | – | – | – | 300 | – | – | – | – |
| Cassava Reserve | 4,000 | – | 1,577 | – | 993 | – | – | – | 1,431 | – |
| Cash | | – | – | – | – | – | – | – | – | 912 |

Notes: Some items are slightly oversupplied due to acreage rounding. Dry and green maize are alternative end uses.

incorporated in the initial extension content as a deliberate objective, may increase labor capacity over the critical period. At the same time the relaxation of farmers' attitudes to change, assumed in the adoption model, may break down traditions which inhibit adoption; sex-linked labor specializations are a case in point. Indeed, the benefits arising from this type of change can be quantified with the use of the model. When innovations alter the timing of the labor bottleneck, this in itself may be sufficient to stimulate the family to greater efforts during the new peak, putting in more time in the fields or increasing their rates of work on newly critical operations. None of these possibilities is anticipated for the example, and the labor supply constraints are assumed to stay uniform over the adoption period.

There are many possibilities of change in the food supply requirements constraining the system. Farmer preferences may alter with new tastes, or natural pest and disease factors may increase the costs of maintaining supplies. The most significant possibility is a switch to food purchase.

At this point there is a short digression into the problem of evaluating the rationale of the farmer in growing rather than buying his foods. Where cash-crop enterprises have higher productivity, the correct course of action for the farmer will be to grow these and buy foods with the proceeds. But the problem of pricing home-produced foods makes the evaluation difficult. If average local purchase prices are used, they fail to measure the uncertainties associated with local retail outlets as a source of supply and with the level of income from the cash crop alternatives because of both yield and price variations. Accurate pricing of the subsistence activities must consider the reliability of outside sources of supply and this is proportionately more important for the preferred foods.

It is a truism to see the farmer as a maximizer of his own balance of utilities. He maximizes satisfactions subject to his level of knowledge and capacity as a manager. While a significant proportion of his satisfactions is not related to the market, a further proportion clearly is, since he produces some goods for sale. Measuring his resource endowment in terms of market potential can give an improved base for valuing his nonmarket activities, and this is attempted here. Ignoring price uncertainties, the return to cotton growing provides a market measure for Sukumaland, the example area. The approach used optimizes the farmers' resource use in available production opportunities, production methods, and current management levels, and relaxes the food supply conditions. Food supply activities are free to enter the solution, at retail price levels up to the required food levels, and at sale price levels above this. Hired labor can be brought in at any time, but the quantity is limited by the existing capital constraint of 150 shillings and the going wage rate of 3.50 shillings per man-day. The net value produced by optimally solving this system is 1,914 shillings, compared with a net cash value of 762 shillings produced by the actual, observed system. The difference of 1,152 shillings represents a valuation in market terms of the alternatives forgone by the subsistence production of the existing system, and thus a valuation of the composite utilities of the food supply pattern produced as a priority by the farmer.

Subdivision of this value begins to erode the "composite" nature of the consumption pattern. There is some justification for identifying production activities for subsistence, since the farmer has the choice of planting each mixture every season. There is the same justification for allocating the market valuation between them on the basis of their relative requirements of critical labor resources. In the existing system, labor supply is critical, to the extent of requiring hired resources, in the period November to February; but with labor hire of two, nineteen, twenty, and two man-days, respectively, the four months vary in importance. To allocate the market valuation, total labor use was expressed as a percentage of family labor availability for that month; and this percentage used to weight labor absorbed by each of the subsistence activities. Activities not requiring any labor in these months (late sweet potatoes and old cassava) were valued at the average local purchase price and the value was deducted from the market valuation before allocation among the other six activities. Total value of production of the two subsistence activities affected was 53 shillings, leaving a value, for allocation on the basis of critical labor use, of 1,099 shillings.

A market valuation of 1,099 shillings, distributed over a total of 154.6 weighted man-days, gives an average of 7.11 shillings per day. Table 80 shows this value allocated to the subsistence activities and adjusted to a per-acre basis and compares this market valuation with that based on food purchase prices.

Further subdivision of the valuation between the constituent foods of each subsistence activity cannot be justified, because the foods are true joint products. Weightings would be required for the three possible roles of each food in providing quantity, insurance, and taste. Such weighting could only be arbitrary.

The method provides a base for comparing the opportunity cost of subsistence activities and the technical innovations within the con-

TABLE 79

Labor Use in the Existing System,
Weighted for the Critical Months

| Crops<br>Months | Weight | EML | ECML | EP | R | LCML | IML |
|---|---|---|---|---|---|---|---|
| | | | | Man-days | | | |
| Nov. | 102 | 7.1 | 37.2 | 0 | .9 | 0 | 0 |
| Dec. | 124 | 7.2 | 9.1 | 1.9 | 0 | 9.1 | 0 |
| Jan. | 126 | 0 | 9.2 | 0 | 3.8 | 36.8 | 0 |
| Feb. | 102 | 0 | 0 | 0 | 12.6 | 14.9 | 14.8 |
| Total | 154.6 | 14.3 | 55.5 | 1.9 | 17.3 | 50.8 | 14.8 |

Source: Compiled by the author.

ditions of the existing system. Present subsistence activities, valued on this basis and contributing to the solution value, are allowed to compete with new cash-crop technology and food-buying alternatives up to the level of family food requirements. The solution chooses new cash-crop techniques and food buying, or fewer cash crops and less food production, or some combination of the two.

This analysis suggests one indicator for the planning sequence: that crops already grown in pure stands offer the best vehicles for initial innovations, while the composite nature of the mixture of constituents in the intercropped activities demands several complementary innovations. A degree of relaxation of the farmer's attitudes will be required before he can be expected to tolerate the complexities of changes which cover several crops. It is assumed for our example that the farmer places high values on his subsistence activities, which he could not cover by producing cotton and buying foods. It is further assumed that there are no changes in the food supply pattern over the adoptive period. Both these assumptions mean that the satisfaction of the existing food supply pattern by subsistence production is a condition of planning over the whole adoption period, and that the reorientation of farm resources toward the market will be partly limited by the increases in productivity which can be achieved on the subsistence activities.

TABLE 80

Comparison of Alternative Valuations
for Subsistence Activities

| | Early Maize Legumes Sweet Pot. | Early Maize Cassava Legumes | Early Sweet Pot. | Rice | Late Maize Cassava Legumes Sweet Pot. | Late Maize Legumes Sweet Pot. |
|---|---|---|---|---|---|---|
| Market Valuation (shs.) | 101.67 | 394.61 | 13.49 | 123.00 | 261.19 | 105.23 |
| Shs. per Acre | 175.29 | 270.28 | 269.80 | 410.00 | 247.39 | 181.43 |
| Valuation on Food Purchase | 86.42 | 281.78 | 10.00 | 100.00 | 262.80 | 71.34 |
| Shs. per Acre | 149.00 | 193.00 | 200.00 | 300.00 | 180.00 | 123.00 |

Note: Per-acre valuations for late sweet potatoes and old cassava, at average purchase prices, are 200 and 45 shillings, respectively.

Source: Compiled by the author.

### Capital Outlay Constraint

As with labor supply constraints, the observed level of capital outlay is adopted as an initial constraint on the planning model; that is, 150 shillings used in the existing system for the hire of casual labor. It represents about 16 percent of a gross cash income from the system of 912 shillings.

It is implicit in the adoption model that farmers' incomes will rise as a result of innovation. One-sixth of a rising level of income will give an increasing capital outlay; and, in addition, as income rises, a larger proportion will be made available for reinvestment as the marginal propensity to consume falls. However, given the present near-subsistence levels of consumption and the high demand for school and medical services, no allowance has been made in the example for an increase in the proportion of income allocated to reinvestment. Initial program solutions give optimal solution values of about 2,500 shillings gross cash income for the system. From this it is estimated that capital outlay will rise from 150 shillings to 400 shillings over the adoption period.

## EXOGENOUS CONSTRAINTS ON THE MODEL

As we have indicated, the planning sequence as it stands is static, but because the adoption model covers a period of time, a capacity to cope with change must be built in. The increasing level of capital outlay is one important variable implicit in the adoption model itself; others are changing technology, market conditions, and government policy objectives. Of these, new technology likely to become available over the adoption period is an unknown; and, like any other unforeseen contingency, it must be dealt with by altering the constraints and coefficients of the model and rerunning the sequence. In itself this presents no problems, except that significant changes in the wide variety of conditions surrounding the model must be picked up. On the other hand, some degree of anticipation of changing market conditions and government policy objectives is possible, for the planning of extension programs will be done within the context of a national development plan, which will itself anticipate market changes, and of the institutional developments resulting from government policies.

Certainly there is a degree of circularity in this macro/micro link. In economies where the growth of agricultural production is a base for national development, the evaluation of farming potential must itself play an important part in policy formulation. At the beginning of the study we identified two objectives for farm economics—guiding farmer resource use and contributing to more effective policy formulation—and stressed our preoccupation with its role in extension. In this context, the conditions to be incorporated in the micro planning sequence must be accepted as predetermined by national objectives and plans. In a macro role to guide policy formulation, shadow pricing of both products and factors would replace actual and expected prices in the model.

The three important groups of exogenous variables entering the planning model are product prices, factor prices, and institutions in the infrastructure, including marketing, transport, retailing, and credit facilities. All three groups will be weighted heavily by government policy objectives and will usually be documented within national development plans. The levels and changes in levels of exogenous variables for the model, over the adoption period, must be decided in the light of present and anticipated plans because they affect the production environment of the area concerned.

The producer prices of export crops will ultimately be dependent on the world market, which is exogenous not only to the farming system in question but also to the economy, and as such is out of the control of government. Within these external market conditions, however, the government may manipulate producer prices and marketing and processing margins to meet policy objectives.

### Current Levels of Exogenous Variables in the Example

In the example area of Sukumaland, organized markets exist for cotton and maize and, intraregionally, for rice. Cotton and particularly maize prices are fixed nationally. Within the cotton prices dictated by world trade, marketing and processing margins are fixed on a cost-plus basis, giving as high a return as possible to the producer, subject to a stabilization formula to even out short-term variations. Maize prices are based on a policy of national self-sufficiency with a nationwide buying and selling organization. The price in particular areas is a national price modified only by local transport differentials. Recently there has been a move to locate marketed production in the areas of highest potential. The whole policy of self-sufficiency has required an internal producer price above import parity to bring forth the amounts needed, though rationalization of production to areas of highest potential should allow a move toward parity level.

An import-substitution policy dictated by foreign exchange considerations influences the prices of manufactured inputs. The effects of this policy are modified by economic growth objectives which require subsidies on those inputs required for innovation. The case of cotton fertilizer in Sukumaland is an example of both import concessions—no tariffs are levied against agricultural inputs—and direct subsidy—the price to the producer is on the order of 15 percent below cost. In this case the subsidy is met within the industry, with the body of growers subsidizing the innovators. In Sukumaland the availability of credit for agricultural inputs, with existing programs for cotton fertilizer and insecticides, are important institutional conditions to be incorporated in the model. At the same time, the lack of reliable retail outlets in the rural areas militates against a move to food purchase on the part of the farmer. Even with maize, the one food staple with a national collection and distribution network, purchase is discouraged by producer/retail margins greater than 100 percent.

Table 81 sets out the current levels of exogenous variables and institutional conditions which form the context for planning at the start of the adoption period.

## Expected Changes in the Levels of Exogenous Variables over the Adoption Period

Market and policy changes will be the two types of variation affecting the level of exogenous variables over the period. Market changes will be limited to projections of world supply and demand for export produce. For specialized crops sold very locally, or for the few cases where export crops from the area dominate world trade, the repercussions of increased production from the program should be reflected in falling price expectations, unless commensurate growth in demand is anticipated. No such changes in cotton prices have been included for the example.

The main source of changes is policy, both for internal prices and for institutional development. While changes of either type might be dramatic, given political stability, basic policy objectives are usually long-term. Development plans are an obvious source of information on policy and programs which might require new conditions to be built into the sequence. Two plan provisions are included for the Sukumaland example. First, existing credit programs are extended to include maize fertilizer and insecticides during the period. Second, a local fertilizer factory is scheduled to begin output, and part of the establishment costs are to be passed on to the farmer in the form of higher prices. The cotton fertilizer subsidy level will remain, but the change in manufacturer's price will be absorbed by the farmer. This price increase will have occurred by the time an extension program is ready for implementation.

TABLE 81

Current Levels of Exogenous Variables for Sukumaland

| | Cotton | Dry Maize | Rice | Ground-Nuts | Dry Cassava | Fresh Sw. Pot. | Other Legumes |
|---|---|---|---|---|---|---|---|
| Producer Price (cts./lb.) | 45.3 | 15 | 20 | 45 | 1 | 1 | 25 |
| Retail Price (cts./lb.) | | 25 | 30 | 90 | 3 | 5 | 40 |

| Inputs (shs.) | Fertilizer Cotton (per acre) | Fertilizer Maize (per acre) | Insecticide Cotton (per acre) | Insecticide Maize (per acre) | Pumps (each) | Labor Cultivation (per man-day) | Labor Harvest (per man-day) |
|---|---|---|---|---|---|---|---|
| Price | 40.00 | 45.00 | 25.00 | 10.00 | 150.00 | 3.50 | 4.50 |
| Credit Available | * | * | * | * | * | | |

Note: Cotton price is a weighting of two grades in expected proportions: Grade A, 90 percent at 48 cts. per lb.; Grade C, 10 percent at 21 cts. per lb.

Source: Compiled by the author.

The final step in building these dynamic elements into the planning sequence is to fit expected changes in both exogenous and endogenous variables into the adoption period. Four expected changes have been identified:

1. A rise in the price of cotton fertilizer to the grower.

2. An extension of the credit program presently covering cotton inputs to fertilizer and insecticide for use on maize.

3. A willingness on the part of the farmer to consider improvements on intercropped foods.

4. An increasing level of capital outlay as income rises from improved productivity due to innovation.

These are examples of four different types of change liable to occur over the adoption period, any of which may alter the optimum solution, the ranking of innovations, and consequently the appropriate extension content. Of these four, two are outside both the system and the model, and the timing of their incidence can be predetermined. It is assumed that the price rise for cotton fertilizer will be in force at the start of the program implementation. And, for this example, in order to be able to assess the impact of credit facilities on program potential, that credit for all inputs to be evaluated will be made available in the second season of program implementation. This question of credit is an important one for this example, since it offers an alternative to the use of hired labor for an increase in the scale of the system. Credit reduces the risk the farmer attaches to higher capital outlay. At some point, however, the further use of credit facilities is inhibited by the farmer's debt ceiling. The example assumes that this does not occur at the levels of credit involved in the improvements chosen in the planning sequence and, further, that the farmer's own capital will continue to be spent on hired labor.

The third and fourth changes, on the other hand, are indeterminate, both depending on the rate of adoption over the period. Capital outlay is constantly changing over the period, under the earlier assumption, at a rate of 16 percent of income received for any season. It would be more satisfactorily treated as fully dynamic by the use of interperiod transfers in a programming model. Since the other types of changes cannot be treated so easily by such a technique, and since our purpose is to demonstrate a planning sequence rather than to manipulate a theoretical model, capital outlay will be treated at two levels only. The existing level of 150 shillings will prevail until the

TABLE 82

Four Sets of Conditions Created by Changes in Variables over the Adoption Period

| | A | B | C | D |
|---|---|---|---|---|
| Credit Availability | No | Yes | Yes | Yes |
| Intercrop Changes | No | No | Yes | Yes |
| Capital Outlay Level (shs.) | 150 | 150 | 150 | 400 |

Source: Compiled by the author.

final year of the adoption period, when the set of conditions for that year will include an outlay of 400 shillings.

Each change will vary the set of conditions within which the planning sequence must operate over the adoption period. We can identify four sets labeled A-D, which are given schematically in Table 82. Other conditions remain constant.

Of the four sets, A covers the present farm system under existing market and infrastructural conditions, including the higher price for cotton fertilizer. D covers conditions expected at the end of the adoption period and therefore, importantly, the conditions under which the optimum or goal system will be derived. B and C cover two intermediate positions: B the introduction of credit facilities, the timing of which is predetermined as year 2 of the adoption period, laid down within the national development plan; and C is where the farmer's attitudes have relaxed sufficiently for him to consider changes in his intercropped subsistence activities. The timing of C is indeterminate *ex ante* but will emerge in the course of planning the steps over the adoption period.

The changes covered are very limited and illustrate the way variables can be usefully incorporated in the planning sequence even though a static planning technique is being used. Solution under conditions at Point D will provide the goal for extension strategy in the form of the key innovations which enter the optimum systems and also the scale at which they can be expected to be established by the end of the adoption period. The comparison of alternative innovations over the criteria of profitability, congruity, complexity, and acceptability under each set of conditions will provide the basis for the selection of appropriate and progressive extension content for the area.

CHAPTER

# 18

# THE PLANNING SEQUENCE: SETTING THE OBJECTIVE FOR EXTENSION STRATEGY

The first of the two stages of the planning sequence proper identifies a system capable of both satisfying farmer objectives and maximizing the increase in marketed production, within the conditions expected at the end of the adoption period. It forms the goal toward which extension strategy over the period is directed. The conditions include those variable ones incorporated as set D. Perhaps the most important of the unchanging variables is the range of available technology, including the practices followed in the existing system, and those drawn up for consideration in the inventory of possible improvements.

Prior to solving the program, the innovations are set up into a matrix in the same format as the activities of the existing system. Labor profiles are built up from the components of operational sequence, timing aid work rates (derived from survey investigation), or from experimental and trial farm records where new operations, or revised sequences and crop calenders, are implicit in the new practices. Table 83 shows both input and output coefficients for the cotton innovations, identified by their key letters. Table 84 is a similar table for the possible food crop innovations.

Maize and cassava are joint product activities producing dry or fresh green maize and fresh or reserve cassava, respectively. Each output possibility has been entered in the matrix independently and acreage values added to give the crop acreage in any solution.

Before solving, the matrix can be reduced significantly by applying simple criteria derived from an analysis of the existing system. It seems certain that future as well as present limitations on the scale of the system will be labor use in the period December-

TABLE 83

Thirty-two Innovation Combinations for Cotton: Per-Acre Input/Output Data Required for a Program Matrix

| | A | B | C | D | E | F | G | H | I | J | K | L | M | N | O | P |
|---|---|---|---|---|---|---|---|---|---|---|---|---|---|---|---|---|
| Planting Time Center Date | 12/7 1/22 | 12/30 | 12/7 | 1/22 | 12/7 1/22 | 12/30 | 12/7 | 1/22 | 12/7 1/22 | 12/30 | 12/7 | 1/22 | 12/7 1/22 | 12/30 | 12/7 | 1/22 |
| Labor Use (man-days) | | | | | | | | | | | | | | | | |
| October | 3 | – | 3 | – | 3 | – | 3 | – | 3 | – | 3 | – | 3 | – | 3 | – |
| November | 7 | 3 | 13 | – | 7 | 3 | 14 | – | 7 | 3 | 13 | – | 7 | 3 | 14 | – |
| December | 7(10) | 13 | 8 | 10 | 8(11) | 14 | 8 | 11 | 7(10) | 13 | 8 | 10 | 8(11) | 14 | 8 | 11 |
| January | 7(9) | 8(13) | 5(10) | 9(14) | 8(10) | 8(13) | 7(12) | 9(14) | 7(9) | 8(13) | 5(10) | 9(14) | 8(10) | 8(13) | 7(12) | 9(14) |
| February | 5 | 5 | 4 | 6 | 6 | 7 | 4 | 8 | 5 | 5 | 4 | 6 | 6 | 7 | 4 | 8 |
| March | 4 | 4 | 2 | 5 | 4 | 4 | 2 | 5 | 5 | 5 | 3 | 6 | 5 | 5 | 3 | 6 |
| April | 2 | 2 | – | 3 | 2 | 2 | – | 3 | 3 | 4 | 1 | 4 | 3 | 4 | 1 | 4 |
| May | 8(9) | 9 | 12 | 6(7) | 9(10) | 10 | 13 | 6(7) | 10 | 11 | 13 | 7(8) | 11 | 12 | 14 | 7(8) |
| June | 8 | 8(9) | 7(8) | 9 | 9 | 9(10) | 8(9) | 10 | 9(10) | 9(10) | 8(9) | 11 | 10(11) | 10(11) | 10(11) | 11 |
| July | 8 | 8 | 7 | 7 | 8 | 8 | 7 | 7 | 8 | 8 | 7 | 8 | 8 | 8 | 8 | 8 |
| August | 3 | 3 | 1 | 2 | 3 | 3 | 4 | 2 | 3 | 3 | 3 | 2 | 3 | 4 | 4 | 3 |
| September | – | – | – | – | – | – | – | – | – | – | – | – | – | – | – | – |
| Total | 62 | 63 | 62 | 57 | 67 | 68 | 70 | 61 | 67 | 69 | 68 | 63 | 72 | 75 | 76 | 67 |
| Cash costs (shs.) | 0 | 0 | 0 | 0 | 45 | 45 | 45 | 45 | 35 | 35 | 35 | 35 | 80 | 80 | 80 | 80 |
| Value of No Ties net out- | 236 | 249 | 294 | 159 | 284 | 295 | 340 | 136 | 295 | 315 | 338 | 175 | 373 | 340 | 433 | 177 |
| put (shs.) Tied | 278 | 294 | 347 | 188 | 335 | 348 | 401 | 169 | 345 | 372 | 399 | 207 | 440 | 401 | 511 | 209 |

Notes: The pump has been depreciated over three years at five acres a year and added to recurrent cost for insecticide inputs at 10 shillings per acre.

As previously noted, postponable operations such as cotton grading and groundnut shelling are omitted. Solutions have been checked to see that enough slack occurs in total family labor use to cover these within proper periods.

The coefficients in parentheses show the extra labor required for tie-ridging, realizing an improved return per acre, as indicated in the final row of the table.

Twenty Innovation Possibilities for Food Crops: Per-Acre Input/Output Data Required for a Program Matrix

| | MAA | MAB | MAC | MAD | MAE | MAF | MAG | MAH | MAI | MAJ | GNA | GNB | GNC | GND | GNE | MGN | CASSA | CASSB | SPOTA | SPOTB |
|---|---|---|---|---|---|---|---|---|---|---|---|---|---|---|---|---|---|---|---|---|
| Center Planting Date | 11/30 1/28 | 11/30 1/15 | 12/15 | 1/15 | 2/17 | 11/30 2/28 | 11/30 1/15 | 12/15 | 1/15 | 2/7 | 12/15 1/25 | 12/30 | 12/7 | 1/15 | 2/7 | 12/15 | 11/25 | 3/15 | 11/15 | 1/15 |
| Labor (man-days) | | | | | | | | | | | | | | | | | | | | |
| October | 5 | 7 | – | – | – | 7 | 8 | – | – | – | – | – | – | – | – | – | 7 | – | – | – |
| November | 6 | 7 | 8 | – | – | 7 | 9 | 8 | – | – | 12 | 7 | 26 | – | – | 15 | 19 | – | 30 | – |
| December | 6 | 7 | 17 | 8 | – | 7 | 9 | 17 | 8 | – | 13 | 26 | 11 | 20 | – | 15 | 4 | – | – | – |
| January | 6 | 8 | 6 | 17 | 8 | 7 | 9 | 11 | 17 | 8 | 13 | 8 | 4 | 16 | 26 | 10 | 4 | – | – | 30 |
| February | 6 | 4 | 2 | 6 | 17 | 7 | 7 | 7 | 11 | 17 | 3 | – | – | 5 | 15 | – | 4 | 14 | – | – |
| March | 3 | – | – | 2 | 6 | 5 | 2 | – | 7 | 10 | – | – | – | – | – | – | 3 | 16 | – | – |
| April | 1 | – | – | – | 2 | 2 | – | – | 0 | 7 | 8 | 9 | 15 | 6 | – | 10 | – | 8 | – | – |
| May | 1 | 3 | 7 | 3 | – | 3 | 5 | 12 | 5 | 0 | 9 | 9 | 5 | 10 | 14 | 10 | – | 3 | – | – |
| June | 3 | 2 | – | 2 | 3 | 4 | 5 | – | 4 | 5 | – | – | – | – | – | – | – | – | – | – |
| Total | 37 | 38 | 40 | 38 | 36 | 49 | 54 | 55 | 52 | 47 | 58 | 59 | 61 | 57 | 66 | 60 | 41 | 41 | 30 | 30 |
| Food Supply (lbs.) | | | | | | | | | | | | | | | | | | | | |
| Dry Maize | 600 | 700 | 1,000 | 700 | 400 | 1,400 | 1,500 | 1,800 | 1,400 | 800 | – | – | – | – | – | 700 | – | – | – | – |
| Green Maize (Mar.) | 300 | 350 | 500 | – | – | 700 | 750 | 900 | – | – | – | – | – | – | – | – | – | – | – | – |
| Green Maize (April.) | 300 | 350 | 1,000 | 700 | 200 | 700 | 750 | 1,800 | 1,400 | 400 | – | – | – | – | – | 525 | – | – | – | – |
| Green Maize (May) | 300 | 350 | – | 350 | 400 | 700 | 750 | – | 700 | 800 | – | – | – | – | – | 525 | – | – | – | – |
| Fr. Cass. (Sept.-Nov.) | – | – | – | – | – | – | – | – | – | – | – | – | – | – | – | – | 4,500 | 1,500 | – | – |
| Fr. Cass (Dec.-Feb.) | – | – | – | – | – | – | – | – | – | – | – | – | – | – | – | – | 4,500 | 1,500 | – | – |
| Sw. Pot. (Feb.-Mar.) | – | – | – | – | – | – | – | – | – | – | – | – | – | – | – | – | – | – | 4,000 | – |
| Sw. Pot. (June-July) | – | – | – | – | – | – | – | – | – | – | – | – | – | – | – | – | – | – | – | 4,000 |
| Legumes | – | – | – | – | – | – | – | – | – | – | 430 | 550 | 750 | 320 | 110 | 320 | 9,000 | 3,000 | – | – |
| Cass. Ins. | – | – | – | – | – | – | – | – | – | – | – | – | – | – | – | – | 9,000 | 3,000 | – | – |
| Cash Costs | 0 | 0 | 0 | 0 | 0 | 55 | 55 | 55 | 55 | 55 | 0 | 0 | 0 | 0 | 0 | 0 | 0 | 0 | 0 | 0 |
| Net Cash Value | 90 | 105 | 150 | 105 | 60 | 110 | 170 | 215 | 155 | 65 | 194 | 248 | 337 | 144 | 50 | 249 | – | – | – | – |

Notes: Cash returns are shown for maize and groundnuts as possible competitors for cotton. The return for maize is based on the sale of dry maize only.

As with all planning data, postponable operations not influencing peak periods are omitted. Similarly, no harvest labor requirements are included for those food crops gathered directly from the field: sweet potatoes and cassava.

Source: Compiled by the author.

January. Those innovations showing a low return to labor required over this period can be excluded from programming. COT D, COT H, COT L, and COT P, with all the tie-ridging alternatives, are excluded as because the return they offer to December-January labor is lower than that of the existing cotton enterprise. Tie-ridging intensifies the productivity of land, but the return to the extra labor required is relatively low. If accelerating fertility losses were considered a major problem, the evaluation of tie-ridging would be more complex.

All activities from the existing system and the inventory of improvements are included in a linear program bounded by the constraints outlined with the variable conditions at the levels of Point D, the final year of the adoption period. Existing and potential food enterprises compete to satisfy farmers' food requirements, in the same supply pattern as the existing system; and cotton activities compete to maximize the productivity of residual family labor resources and the use of 400 shillings' capital for labor buying. Credit is available to meet the costs of purchased inputs for all innovations on both food and cash crops.

Table 85 sets out the solution values for the optimum system. Only three minor activities of the existing system are retained: rice, which had no alternative and was forced into the solution as a required food, and the two existing early and late plantings of sweet potatoes in a pure stand. The important feature of the solution on the subsistence side is that none of the intercropped activities is retained. All the maize, legume, and cassava requirements, together with the balance of the sweet potatoes, are met by innovations. The cotton enterprise entering the solution, COTO, is a higly intensive, fixed-time-of-planting activity.

Table 86 compares the value of the marketed output from cotton in the goal solution and in the existing system.

Within the conditions expected at the end of the adoption period, the net value of cash income can be raised from 762 to 2,059 shillings, an increase of 170 percent. Achieving this potential requires radical changes in the methods of food production and highly intensive management of the cotton crop. The extent of reorganization of food production techniques implies a protracted adoption period, though, on the other hand, the threefold increase in cash returns bodes well for incentives to change.

Although this solution gives the optimum result under the conditions imposed on the system, it is optimal only in terms of

TABLE 85

Optimal Resource Allocations Under Planning Conditions for End of Adoption Period

| | EP | R | LP | MAF | MAG | GNC | CASSB | CASSA | SPOTEM | SPOTJJ | Total Foods | COTO | Total All | Labor Family | Labor Hired |
|---|---|---|---|---|---|---|---|---|---|---|---|---|---|---|---|
| Acreage | .075 | .300 | .050 | .929 | .233 | .267 | 1.091 | .248 | .050 | .075 | 3.308 | 5.826 | 9.134 | — | — |
| Labor Use | | | | | | | | | | | | | | | |
| October | — | — | — | 6.50 | 1.87 | — | — | 1.74 | — | — | 10.11 | 17.48 | 27.59 | 20.00 | 7.59 |
| November | — | .90 | — | 6.50 | 2.10 | 6.94 | — | 4.71 | 1.50 | — | 22.65 | 81.56 | 104.21 | 55.00 | 49.21 |
| December | 2.25 | — | — | 6.50 | 2.10 | 2.94 | — | .99 | — | — | 14.78 | 46.61 | 61.39 | 55.00 | 6.39 |
| January | — | 3.00 | — | 6.50 | 2.10 | 1.07 | — | .99 | — | 2.25 | 15.91 | 40.78 | 56.69 | 55.00 | 1.69 |
| February | — | 12.30 | — | 6.50 | 1.63 | — | 15.27 | .99 | — | — | 36.69 | 23.30 | 59.99 | 59.99 | — |
| March | — | 5.10 | 1.50 | 4.65 | .47 | — | 17.46 | .74 | — | — | 29.92 | 17.48 | 47.40 | 47.40 | — |
| April | — | 5.10 | — | 1.86 | — | 4.01 | 8.73 | — | — | — | 19.70 | 5.83 | 25.53 | 25.53 | — |
| May | — | 3.90 | — | 2.79 | — | 1.34 | 3.27 | — | — | — | 11.30 | 81.56 | 92.86 | 60.00 | 32.86 |
| June | — | 3.60 | — | 3.72 | — | — | — | — | — | — | 7.32 | 58.26 | 65.58 | 60.00 | 5.58 |
| July | — | 3.30 | — | — | — | — | — | — | — | — | 3.30 | 46.61 | 49.91 | 49.91 | — |
| August | — | — | — | — | — | — | — | — | — | — | — | 23.30 | 23.30 | 23.30 | — |
| September | — | — | — | — | — | — | — | — | — | — | — | — | — | — | — |
| Total | 2.25 | 37.20 | 1.50 | 45.52 | 10.27 | 16.30 | 44.73 | 10.16 | 1.50 | 2.25 | 171.68 | 442.77 | 614.45 | 511.13 | 103.32 |
| Capital | — | — | — | — | — | — | — | — | — | — | — | — | — | — | 400.00 |

Source: Compiled by the author.

TABLE 86

Present and Potential Values of Marketed Production

| System | | Present | Potential |
|---|---|---|---|
| Gross Value of Cotton Production | (shs.) | 912.27 | 2,988.93 |
| Cash Costs | (shs.) | | |
| Cotton Crop Inputs | | — | 466.08 |
| Food Crop Inputs | | — | 63.91 |
| Hired Labor | | 150.00 | 400.00 |
| Net Value of Cotton Production (shs.) | | 762.27 | 2,058.94 |

Source: Compiled by the author.

profitability. As we have frequently emphasized, a range of planning criteria is required to cover the diverse objectives of the traditional farmer. The consequence of relatively poor scores for the solution activities on congruity, complexity, and acceptability will be a prolonged period for adoption. The longer the time scale, the higher the costs of extension and the worse the benefit cash flow, both direct and from diffusion effects. Although the other criteria will be most useful in selecting the content of each improvement package, they are also important in qualifying the selection of a goal system on a profitability criterion.

Several researchers have demonstrated that various solutions lie close to the optimum.[1] The case has been made that alternatives should be available to allow the farmer to choose a subjectively preferred enterprise combination, in an attempt to allow flexibility to meet peculiarities of the objective function of individual farmers. Monte Carlo techniques have been promoted on these grounds.[2] Similarly, we cannot afford to have a goal solution which may be only marginally more profitable than alternatives which have greater congruity with existing practice and are less complex and more acceptable to farmers. It would prolong the adoption period.

The program was allowed to generate solutions for each possible cash crop activity in turn, solutions which were scored for all four criteria. Before setting out the results, the details of measurement for each criterion for our example are described.

## PROFITABILITY

The objective value of the basic solution for each set of conditions A-D shows the performance of the present cotton enterprise under the particular conditions. The difference between this and the objective value of the solution incorporating a particular innovation is a measure of its profitability. Improvements in the efficiency of the system, by a new method of food production, can similarly be scored by the change in objective value.

## CONGRUITY

For the example, sequence congruity is scored very simply. Weighting by observed prejudices or ingrained traditions would increase the counts for particular innovations. The congruity score remains the same under all conditions. Each of the following items is scored one: a fixed time of planting where it is variable in the existing system and vice versa, a change in the fixed time of planting, an extra weeding, a significant change in plant population, the purchase of fertilizers or insecticides, the application of fertilizers or insecticides, and the introduction of a pure stand for an intercrop.

The use of purchased inputs involves a double penalty—the need to go out and purchase the requisite, and learning to apply it in the field—both of which are scored. As a criterion related to initial adoption rather than sustained acceptance, congruity is particularly important at the beginning of the adoption period. By the end it is assumed that farmers will have confidence in the advice of the extension services.

## COMPLEXITY

The term "complexity" has a restricted meaning, referring to the number of changes in the timing of labor inputs between monthly periods as a consequence of the new activities in the solution. As a basis for scoring it could be refined, since clearly changes in the type of work are as important as changes in quality. Weighting for this would improve the measure. As it stands, it is fairly simple and forms one basis for measuring the impact of a change, or group of changes, on the present management routine of the farmer.

Scoring is demonstrated in Table 87. The labor profiles for subsistence crops (totaled) and cash crops in the existing system are related to those derived from the acreage values of the solution. Differences in the monthly totals are summed to give the complexity score. For the subsistence profile an element of weighting is introduced, with differences in the period November-February doubled to reflect the increased disturbance created by changes over the critical part of the season. The example given is for the innovation COTO under conditions at point C.

Summing the differences, including the weighted differences of the subsistence profile, gives a complexity score of 295 for COTO under conditions at point C.

## ACCEPTABILITY

This final criterion scores the degree to which an innovation impinges on the existing resource allocation to subsistence activities and is a measure of the extent of the clash which will arise unless improved practices are introduced. Acceptability scores are useful to show which of the existing food enterprises is the least productive, in terms of farmers' satisfactions per unit of labor absorbed, and so will offer the least obstacle to change. The calculation is based on the fact that, at a certain level of expansion, the labor demands of an innovation can be met only from labor used for food production in the existing system. The score is based on the quantity of food sacrificed by the change, with one score point per pound; but in order to cover the insurance role of some foods, and preferences for others, quantities are weighted. The weightings are entirely subjective, from a description of the whole of the goods in the system: maize as a basic grain staple is weighted by 2; green maize, fresh sweet potatoes, and legumes as preferred staples by 3; fresh cassava as a substitute for sweet potato by 1; and reserve cassava by 0.5. These values have been used to multiply the quantities produced by those subsistence activities which are replaced in the optimal solution, in order to give a food value score for each one. Table 88 shows the food value scores for four enterprises which are replaced. It carries the process a step further by relating to these scores the labor used in order to obtain the productivity of labor in these terms for each activity in each month. Only the four critical months November-February are covered, to simplify minimizing the food values scores to be sacrificed for each innovation. Old cassava is omitted because it does not use labor in this period.

TABLE 87

Calculating a Complexity Score

| | | O | N | D | J | F | M | A | M | J | J | A | S |
|---|---|---|---|---|---|---|---|---|---|---|---|---|---|
| Subsistence | Old | 14 | 45 | 22 | 39 | 41 | 21 | 0 | 12 | 10 | 7 | 9 | 6 |
| | New | 7 | 19 | 17 | 23 | 42 | 27 | 19 | 9 | 11 | 3 | 0 | 0 |
| | Differences | 7 | 26 | 5 | 16 | 1 | 6 | 19 | 3 | 1 | 4 | 9 | 6 |
| | N-F weighted | X2 | 52 | 10 | 32 | 2 | | | | | | | |
| Cotton | Old | 0 | 12 | 52 | 36 | 20 | 12 | 0 | 32 | 40 | 28 | 12 | 0 |
| | New | 14 | 63 | 36 | 32 | 18 | 14 | 5 | 31 | 5 | 8 | 6 | 0 |
| | Differences | 14 | 51 | 16 | 4 | 2 | 2 | 5 | 1 | 35 | 20 | 6 | 0 |

Source: Compiled by the author.

TABLE 88

Basic Data for the Calculation of Acceptability Scores

| | EML | | ECML | | LCML | | LML | |
|---|---|---|---|---|---|---|---|---|
| Acreage | | .58 | | 1.46 | | 1.46 | | .58 |
| Weighted score | | 1206 | | 3927 | | 3592 | | 766 |
| Month | Man-Day Use | Food Value Score (per man-day) | Man-Day Use | Food Value Score (per man-day) | Man-Day Use | Food Value Score (per man-day) | Man-Day Use | Food Value Score (per man-day) |
| November | 7 | 172 | 37 | 106 | – | – | – | – |
| December | 6 | 201 | 7 | 561 | 7 | 561 | – | – |
| January | – | – | 7 | 561 | 29 | 124 | – | – |
| February | – | – | – | – | 15 | 249 | 14 | 55 |

Source: Compiled by the author.

A sample calculation is made, again for COTO under conditions at point C. The profile required for the new cotton activity is compared with that for cotton in the existing system and the differences listed, with the appropriate signs. The data are shown in Table 89.

A particular aspect of our example is the casual hire of labor over this period. Hire is flexible and can readily be shifted between time periods. The proposed activity, COTO, requires an additional fifty-one man-days in November but releases a total of twenty-two man-days from the other three months. All this released labor is within the total hired for the respective months and can be transferred to November, leaving a net additional requirement for COTO of twenty-nine man-days. (Clearly family labor cannot be transferred between time periods, and the degree of flexibility is governed by the existing level of hired labor.) The remaining deficit of twenty-nine man-days in November must be met from labor being used for subsistence production in the existing system.

Table 88 shows two existing activities as possible sources of November labour; EML and ECML. Of these, ECML is the cheaper source at a food value score of 106 per man-day required. Twenty-nine man-days gives a total of 3,074 as the acceptability score for COTO under conditions at point C. The highest possible score of 9,490 implies the sacrifice of all food activities for which alternative production techniques are available. (Under the assumptions that

TABLE 89

Comparison of Existing and Proposed
Cotton Labor Profiles

| | Nov. | Dec. | Jan. | Feb. |
|---|---|---|---|---|
| Present Cotton | 12 | 52 | 36 | 20 |
| Proposed COTO | 63 | 36 | 32 | 18 |
| Differences | +51 | -16 | -4 | -2 |
| Hired Labor Use, Present System | 2 | 19 | 20 | 2 |

Source: Compiled by the author.

the food supply pattern must be met from the farmers' resources, existing production activities with no alternatives are not available for substitution.)

The method of calculation for all four criteria having been outlined, Table 90 sets out the scores for all the cash-crop innovations under conditions at point D at the end of the adoption period. COTO, as a key innovation in the optimal solution, can be compared with the alternatives.

The comparison shows COTM and COTO well ahead of other possibilities on profitability. Both have poor congruity scores, but this is unimportant at the end of the adoption period. COTO has the better complexity score and COTM the better acceptability score. Reverting to a more detailed examination of the two as enterprises, other points can be made. COTM has a staggered planting regime with two advantages: greater reliability in achieving the assumed yield level and a flatter labor profile, allowing more efficient use of hired labor. The second of these may be of considerable significance. The complexity score is higher on COTM, indicating that it is further away from current management routine. In practice there may be difficulty in hiring a number of laborers for a short period; and the highly peaked demands of COTO, although more directly related to present routines, assume that this is possible at a much higher scale. The flatter requirements of COTM allow a longer hire for few laborers for the same increase in scale. Until new technology is found, further development of the system will depend on scale increases, and this consideration may be important.

The hire of semipermanent seasonal labor through the critical five-month period October-February is much more profitable with COTM than with COTO. A single laborer would provide twenty-five man-days each month. Table 91 compares the potential of the alternatives against this fixed supply of labor.

As a corollory to this, COTM would carry wage rates 50 percent higher than those of COTO as the improved method of cotton growing which has been adopted. These points become particularly important with government seeking absorptive capacity for urban unemployed in the rural sector.

The final point in the comparison relates to acceptability, on which COTM has a significantly better score than COTO. The food supply activities remain very stable in the solutions throughout the comparison of alternative cotton innovations. This is again a

TABLE 90

Criteria Scores for Cash-Crop Innovations Under Conditions at the End of the Adoption Period

| | Congruity Score | Profitability (shs.) | Complexity Score | Acceptability Score |
|---|---|---|---|---|
| Base Solution | At Point D | 1,183 | – | – |
| Optimal Solution | | 2,059 | | |
| COT A | 3 | 198 | 402 | 7,164 |
| B | 2 | 227 | 403 | 8,890 |
| C | 3 | 326 | 336 | 8,725 |
| E | 5 | 385 | 386 | 6,220 |
| F | 4 | 368 | 396 | 9,490 |
| G | 5 | 458 | 341 | 8,725 |
| I | 5 | 497 | 409 | 5,806 |
| J | 4 | 565 | 432 | 8,725 |
| K | 5 | 545 | 359 | 8,725 |
| M | 7 | 851 | 401 | 5,965 |
| N | 6 | 519 | 407 | 9,275 |
| O | 7 | 876 | 373 | 8,725 |

Source: Compiled by the author.

TABLE 91

Productivity of Seasonal Labor Hire with Alternative Key Innovations COTM and COTO

| | Acres | Oct. | Nov. | Dec. | Jan. | Feb. | Gross Margin (shs.) |
|---|---|---|---|---|---|---|---|
| COTM | 3.13 | 9.4 | 21.9 | 25.0 | 25.0 | 18.8 | 1,167 |
| COTO | 1.79 | 5.4 | 25.0 | 14.3 | 12.5 | 7.2 | 775 |

Source: Compiled by the author.

phenomenon of the flexible labor buying assumed, which counterbalances the particular requirements of individual innovations. There is one most efficient set of food production activities, that which minimizes the demand for labor over the whole critical period of October to February. Comparison of the existing and solution food-labor profiles over this period, giving the same food supply pattern, reveals the improvement in labor productivity resulting from the new techniques.

The plans give a saving of some 35 percent over the existing labor profile, representing labor resources released for reorientation to market production. An interesting further point is the saving in land, caused by restricting our consideration to intensifying innovations and by the fact that the cultivation and weeding operations are the ones limiting the scale of the system. It confirms that limiting consideration to intensifying innovations has delayed the rate of fertility loss from ever-increasing acreage requirements. For our immediate purpose, however, the similarity in the food activity patterns in the COTM and COTO solutions implies that both will be equally difficult to implement. However, the lower acceptability score for COTM shows that it can be expanded to a higher level without clashing with existing food activities for labor resources. This, together with a flatter labor profile which is more conducive to labor hire, marks it out as of greater potential for rapid development than COTO. COTM is selected as the key innovation and, with its complementary food activities, makes up the goal system. Extension strategy for the area will be aimed at exploiting the full potential of this innovation by progression of content, in a sequence of extension programs for cooperating farmers.

TABLE 92

Comparison of Food-Crop Labor Requirements in the Existing and Optimal Systems of COTO and COTM

| Source | Acres | Oct. | Nov. | Dec. | Jan. | Feb. | May | June | Total |
|---|---|---|---|---|---|---|---|---|---|
| Existing System | 5.42 | 14 | 45 | 22 | 39 | 41 | 12 | 10 | 183 |
| COTO Solution | 3.31 | 10 | 23 | 15 | 16 | 37 | 11 | 8 | 120 |
| COTM Solution | 2.57 | 13 | 30 | 17 | 18 | 24 | 8 | 8 | 118 |

Source: Compiled by the author.

A final analysis examines the contribution to the potential value of net marketed output of 2,059 shillings attributable to each of the changes introduced in the four sets of conditions A-D. This analysis is a bridge between the goal and the selection of a sequence of extension packages forming the path from the existing systems. It seeks to confirm that cash-crop improvements are the major source of potential and should be the focus for extension strategy, with changes in food production techniques aimed at sustaining the acceptance of more intensive cash-crop management. It is quite possible that the major source of potential could be a change of technique in food production, particularly where productivity of a highly labor-intensive activity could be dramatically increased, releasing large quantities of scarce seasonal labor for market production.

At the same time, the analysis will also confirm COTM as the particular key innovation for the planning sequence, although we have identified COTM as most suitable under conditions at point D. It may be that particular changes in conditions which are determinate (implying that they can be influenced) have penalized certain simpler innovations while adding only marginally to potential. Two examples are particularly relevant. Credit facilities are to be introduced in year 2, and assessing their contribution to total potential confirms this as an appropriate time for their introduction. But if they make little contribution, the decision might be altered. Similarly, we have seen that large reorganization of subsistence activities implies a difficult extension task. It is important to confirm that the potential increase in marketed output is worthwhile, for if it is not, extension might be better directed toward the cash-crop side of the farm, ignoring the more complex changes needed for higher productivity on the food crops. In either case, a removal of credit facilities or a decision to ignore changes in food growing techniques might bring an alternative to COTM to the fore.

Table 93 shows the objective values of optimal solutions under the sets of conditions A-D and the cash crop activities which feature in the solutions.

The analysis confirms that COTM and COTO appear in optimal solutions early in the sequence of conditions over the adoption period. However, the proportions attributed in this table are partially determined by the relative location of the conditions; and the best comparison of potential is by a series of -1 solutions, where each condition is removed from the program in turn. Table 94 gives the results for this -1 analysis.

The analysis gives a measure of net potential of each condition, but some qualification is required before a decision is taken on either the need for credit facilities or the desirability of including changes in methods of food production in extension strategy.

The potential from improved cotton management is very high. However, it is assessed with a capital outlay, on the part of the farmer, of 400 shillings (under final conditions at B). At this level of outlay the return to casual hired labor is considerably decreased because it is required for a longer period of the season, and intensification, by the use of fertilizer and insecticides on cotton, can compete. With a capital outlay limited to 150 shillings at the start of the adoption period, intensifying practices cannot compete with the high productivity of casual labor. Credit is therefore necessary early in the adoption sequence to realize the potential from improved cotton management.

Another factor underlines the importance of improving food production techniques. As long as the intercropped methods persist, both selective retail developments in key staples and interarea specialization in food production will be frustrated. Since the development of an internal exchange system represents an important step in increasing production opportunities, added weight is given to the inclusion of changes in food production techniques in the extension strategy. At the same time its relatively late position in the sequence

TABLE 93

Changes in Potential as Conditions Alter over the Adoption Period

| | Existing System | A Better Cotton | B + Credit | C + Improved Food Methods | D + Extra Capital |
|---|---|---|---|---|---|
| Optional Solution Objective Values (shs.) | 762 | 997 | 1,354 | 1,825 | 2,059 |
| Increments in Potential (shs.) | — | 235 | 357 | 471 | 234 |
| As Percent of Full Potential | — | 18 | 27 | 36 | 18 |
| Cotton Activities | COT | COTI 1.16 | COTI .68 | COTM 4.04 | COTO 5.83 |
| and Acreages | 4.03 | COTG 2.51 | COTM .91 | COTO 1.23 | |
| | | | COTO 2.23 | | |

Source: Compiled by the author.

TABLE 94

Measure of the Net Potential of Each Variable Condition

| | Improved Cotton Management | Credit | Improved Food Techniques | Extra Capital |
|---|---|---|---|---|
| Potential from Other 3 Conditions (shs.) | 1,183 | 1,670 | 1,729 | 1,825 |
| Net Potential (shs.) | 876 | 389 | 330 | 234 |

Source: Compiled by the author.

of the adoption period also seems justified, since initial success with cotton and credit should encourage the relaxation of attitudes that is conducive to the incorporation of food crop innovations.

We turn now to planning the progression from the existing system to full exploitation of COTM, as the selected key innovation under expected conditions at the end of the adoption period.

## NOTES

1. U. Renbourg, Studies in the Planning Environment of the Agricultural Firm (Uppsala, Sweden: University of Uppsala, 1962).

2. S. C. Thompson, Monte Carlo Programming Techniques (University of Reading, Dept. of Agriculture, 1970).

# CHAPTER 19

# THE PLANNING SEQUENCE: SELECTING EXTENSION CONTENT OVER THE ADOPTION PERIOD

This second stage in the planning sequence proper combines, over the adoption period, the relaxation anticipated in farmers' attitudes to change and a progression of increasingly alien, but also increasingly profitable, innovations. The procedures used in selecting the progression of extension content are not a formal technique. Formality is precluded by the lack of an empirical basis for the measurement of changing farmer attitudes, as well as by the lack of a denominator for the common evaluation of the criteria. A dynamic capital outlay constraint would serve to measure the rate of change of farmer attitudes and control the progression, but to be realistic it would have to cover two facets which are difficult to quantify:

1. The relationship between increasing income levels and the costs farmers are prepared to incur, that is, their changing marginal propensity to consume.

2. The effect of credit facilities in raising the level of costs farmers are prepared to incur, i.e., the relationship between cash outlay and debt ceilings.

Both these could be dependent on the efficiency of the marginal capital to be employed in the circumstances of the particular area being planned.

In the sequence, the rate of change of farmers' attitudes is estimated, and so the progression of content is dependent on judgment. This essentially subjective process is aided by the criteria designated as important for planning and by the methods derived for an objective scoring of each innovation on these criteria. In implementation in the field, the extension services will be essentially advisory. The farmer himself will decide the pace of development, which will vary between farmers, a point which will be discussed further in drawing together the conclusions for extension organization in Chapter 20.

The planning criteria will change in relative importance over the adoption period. Profitability will be consistently important as the one criterion measuring the positive incentives to change. Of the negative criteria representing potential barriers to change, congruity and divisibility will be of greatest importance early in the period. Both are overt characteristics of innovations influencing the farmer on initial adoption. As attitudes toward change improve later in the adoption period, and as the scale and incongruity of both increase, it will be complexity and acceptability which, with profitability, dominate the farmers' decisions on whether to sustain the adoption.

There are four steps in selecting the progression of extension content. In the first step a progression in congruity is developed toward the key innovation at a rate which meets the expected rate of relaxation of farmer attitudes. It continues up to the congruity level of the key innovations in the optimal solution under the conditions at the end of the adoption period.

The second step evaluates the scale on which innovations are to be adopted by assessing the likely reaction of farmers to the results of improved management. It relates the scale increases to the timing of the congruity progression derived in the first step. Step 2 also analyzes present and planned labor profiles to find the points in the progression at which a clash occurs between the labor requirements of the planned change and those of existing subsistence production. It measures the extent of the clash and the labor which must be released for the change to be realized.

The third step further analyzes the present and planned labor profiles in order to give the alternative food value scores from the sacrifice of different subsistence activities as marketed production expands. It also evaluates the alternative improved production techniques, which would maintain the existing food supply pattern but at the same time release sufficient labor to meet the needs of the cash-crop changes. Alternatives are evaluated on their attractiveness for initial adoption by the farmer, i.e., their congruity, and on their efficiency in generating spare labor for the cash-crop innovations concerned, which is a measure of their profitability.

The fourth step superimposes a food-innovation sequence on the progression of cash-crop innovations in a way which allows an expansion toward full exploitation of the key innovation, at a rate expected to be taken by the adopting farmer.

At the first and third steps permutations of alternatives are possible, requiring subjective selection. In step 1 several key innovation progressions are possible, and the balancing of the criteria to derive each package is a matter of judgment. In step 3 a three-way compromise is required which will reach back into step 2. The acceptability score for the key innovation package where a clash occurs will minimize the food value points sacrificed. To replace those foods requires innovations, and there may be alternatives with varying congruity scores and varying efficiency in releasing the labor required for the key innovation. The choice of an improved food technique will require a balance of all three facets, which will, again, be a matter of judgement.

The remainder of the chapter follows through the four steps in the Sukumaland system. Table 95 sets out the scores on the criteria of profitability, congruity, and complexity for the intermediate sets of conditions over the adoption period. Acceptability scores are given only for conditions under set C, since it is anticipated that adoption will only be on a modest scale early in the period. It is peculiar to the example that the condition of casual hired labor allows a flexibility in accommodating reallocations at peak periods by shifting the timing of labor hire. Clashes with resource allocations to food production will occur only as the demand for casual labor becomes greater than the current level, later in the adoption period.

The condition of hired casual labor particular to the example has two important repercussions on this second stage of the planning sequence. The first is in the selection of cotton innovations to lead up to COTM as the key innovation of the goal solution. With the exception of innovations involving the use of fertilizer, which cannot compete on profitability or congruity in the early adoption period, the total labor requirements of other cotton innovations over the critical cultivation months are the same and only the timing differs; this can be met by changes in the timing of casual labor hire. Thus, until the scale increases above the present level, or until fertilizer is brought in, innovations can be compared on the criteria of congruity and profitability. Second, even when labor reallocations from existing food activities are required, this flexibility gives advantages. Requirements can be met by casual hired labor released by a food crop innovation in any month, not necessarily that month in which labor is needed. This gives greater scope in the choice among alternative improved food techniques and a greater opportunity to minimize the food value points sacrificed.

TABLE 95

Criteria Scores Under Condition Sets A-C

| | | A Existing Foods No Credit | | B Existing Foods Credit | | C Alternative Foods, Credit | | |
|---|---|---|---|---|---|---|---|---|
| Basic Solution Value | | 762 | | 762 | | 1,035 | | |
| Optimal Solution Value | | 997 | | 1,354 | | 1,825 | | |
| Difference | | 235 | | 592 | | 790 | | |
| Cotton Innovation | Congruity Score | Profit | Complex. | Profit | Complex. | Profit | Complex. | Accep. |
| COT A | 3 | 117 | 115 | 117 | 115 | 280 | 326 | 3,802 |
| B | 2 | 125 | 64 | 125 | 61 | 207 | 299 | 8,725 |
| C | 3 | 197 | 137 | 197 | 137 | 264 | 264 | 4,211 |
| E | 5 | 23 | 84 | 211 | 109 | 414 | 322 | 3,751 |
| F | 4 | 0 | 31 | 202 | 61 | 840 | 288 | 8,890 |
| G | 5 | 41 | 104 | 233 | 128 | 353 | 264 | 3,346 |
| I | 5 | 121 | 82 | 345 | 130 | 504 | 331 | 3,405 |
| J | 4 | 121 | 96 | 403 | 78 | 463 | 303 | 8,165 |
| K | 5 | 118 | 121 | 345 | 137 | 420 | 282 | 4,692 |
| M | 7 | 72 | 89 | 561 | 124 | 786 | 322 | 3,074 |
| N | 6 | 2 | 38 | 374 | 72 | 464 | 310 | 8,163 |
| O | 7 | 79 | 113 | 493 | 161 | 702 | 295 | 3,074 |

## STEP 1: SELECTION OF A SEQUENCE OF COTTON INNOVATIONS

COTM has been selected as the key innovation, and extension strategy is aimed toward its full exploitation under the conditions expected at the end of the adoption period. A sequence from the existing cotton enterprise to COTM requires a compromise between building up the incongruity level and the scale of innovation, and at the same time maintaining incentives for sustained acceptance by good profitability, complexity, and acceptability scores. Table 96 sets out a selected path. It is noteworthy that although the sequence is related to the four sets of conditions, the time scale is not specified. Since credit facilities are predetermined for year 2, conditions at A will hold for only one year. But C, and therefore the timing of D and the period under B, will be dependent on the scale of each change to be assessed in step 2.

COTM has seven elements of incongruity which must be gradually built into the cotton enterprise: a high plant population, an extra weeding, a varied time of planting, and the purchase and application of both fertilizer and insecticides.

Under conditions A, at the beginning of the adoption period, intensification by the use of purchased inputs must complete with casual labor for the 150 shillings capital outlay limit of the farmer. Purchased inputs are less profitable and of lower congruity, and are unlikely to attract farmers to adoption. Indeed, until credit is made available, there is little basis for extension content which will stimulate interest among farmers in the changes with good congruity scores. COTC, with a score of 3, has the highest profitability of all possible changes but also has a bad complexity score created by its varied time of planting. Also, COTC is essentially "scale dependent," for its high profitability comes from the increased area which can be cultivated by a spreading of the labor demand because of its varied time of planting. But at the start of the adoption sequence the scale of innovation will be small. COTB, with the best congruity, requiring an extra weeding and proper spacing, gives the next highest profitability and a very low complexity score. It is selected as the innovation to be promoted under conditions at A, in the absence of credit facilities. It will be "passive" extension content, requiring the advisers in the field to take the initiative in promotion, a feature which brings difficulties in identifying innovators.

Once credit facilities are available, for the second season of the sequence, COTJ emerges as an innovation of outstanding potential

## TABLE 96

### Criteria Scores on a Selection Sequence of Cotton Innovations

| | | A | | B | | C | | | D | | |
|---|---|---|---|---|---|---|---|---|---|---|---|
| | Congruity | P | C | P | C | P | C | A | P | C | A |
| COT B | 2 | 125 | 64 | 125 | 61 | 207 | 299 | 8,725 | 227 | 403 | 8,890 |
| J | 4 | 121 | 96 | 403 | 78 | 463 | 303 | 8,165 | 547 | 359 | 8,725 |
| I | 5 | 121 | 82 | 345 | 130 | 504 | 331 | 3,405 | 565 | 438 | 8,725 |
| M | 7 | 72 | 89 | 561 | 124 | 786 | 322 | 3,074 | 851 | 401 | 5,965 |

P = Profitability.
C = Complexity.
A = Acceptability.

Source: Compiled by the author.

with a congruity score of 4 (increased by the purchase and application of insecticides) and has better profitability than any innovation package with a congruity count lower than 7. Furthermore, it has a good score for complexity, involving changes in management routine at off-peak periods. COTJ thus seems a good fulcrum for the sequence and has a special appeal for initial adoption by farmers. The use of insecticide has two additional attractive aspects:

1. It solves a problem which is clearly visible to the farmer.

2. It is the only purchased input which, because it is carried out at flowering, can be applied or saved according to the appearance of insect pests and the potential of the stand of cotton raised.

However, once the scale of adoption increases to a level at which new food techniques are required in order to release labor resources, problems arise. With its fixed time of planting and a consequent high peak requirement for labor, COTJ has a very poor acceptability score. Clashes will therefore occur relatively early in the adoption period. Two more elements of incongruity remain to be built into the sequence: available time of planting and the use of fertilizer. A varied time of planting will delay the clash with food labor allocation and allow further expansion. COTI incorporates this into the sequence and reflects improved profitability and a good acceptability score under conditions at C. The remaining element of incongruity, the use of fertilizer, brings us to COTM, the key innovation.

## STEP 2: ASSESSING THE SCALE OF COTTON INNOVATIONS AND LOCATING POTENTIAL CLASHES WITH SUBSISTENCE PRODUCTION

The first part of the second step is to match the increasing scale and incongruity of innovations with the expected reactions of farmers, given successful results over the adoption period. In the course of analyzing the labor profiles to locate potential clashes with existing food resource allocations, points C and D will be related to a time scale giving the expected length of the adoption period.

### Year 1

In the absence of credit facilities, extension efforts are directed to improving spacing, retiming existing weedings to give earlier cleaning, and adding a third weeding. These practices should be particularly encouraged on early-planted cotton, to give the biggest visual impact in the field.

### Year 2

With the introduction of credit facilities, extension content will continue to be improved spacing and weeding, but additionally farmers will be encouraged to take up insecticides. Those farmers coming forward to try insecticides will be identified as potential innovators who present a logical focus for extension effort. It is assumed that insecticides and pumps are both made available as hard credit, with the cost of the pump to be repaid over four years. Insecticides are offered in acre packs, and it is assumed that interested farmers will try them out on one acre. The objective value realized from the system, is 845 shillings, an increase of 83 shillings, or 11 percent, on the existing system. The main impact of the change will be made by drawing the farmers' attention to the increase in the number of flowers surviving to give a mature fruit.

### Year 3

Extension effort is toward an increase in scale of the innovation. The farmer takes on credit enough insecticide for three acres, while 1.03 acres continue to be cultivated under traditional practice. The objective value is 1,029 shillings, an increase of 267 shillings, or 35 percent, on the existing system. At this scale of innovation it is worthwhile checking the labor profile over the critical months October-February, May, and June. The analysis is set out in Table 97.

Although rounding to whole numbers distorts the arithmetic to some extent, there is no intrusion into resource allocation to food production, and the pattern of hired labor requirement remains the same as in the existing system.

### Year 4

Insecticide, proper spacing, and retimed weeding have consolidated themselves as innovations. The farmer is prepared to follow these improved methods on all his cotton acreage. Extension effort is on a congruity change, bringing in one acre of COTK, an earlier-planted crop, as a start to easing the spread of planting times toward

TABLE 97

Analysis to Check for Possible Intrusion by Innovation In Year 3 into Labor Allocation for Subsistence Crops

| | Oct. | Nov. | Dec. | Jan. | Feb. | May | June |
|---|---|---|---|---|---|---|---|
| Availability (man-days) | 20 | 55 | 55 | 55 | 60 | 60 | 60 |
| Existing Food Needs | 14 | 45 | 22 | 39 | 41 | 12 | 10 |
| Residual Available | 6 | 10 | 33 | 16 | 19 | 48 | 50 |
| COTJ 3.00 Acres | 0 | 9 | 39 | 24 | 15 | 33 | 27 |
| COT 1.03 Acres | 0 | 3 | 13 | 9 | 5 | 8 | 10 |
| Total Cotton | 0 | 12 | 52 | 33 | 20 | 41 | 37 |
| Hired Labor Required | 0 | 2 | 19 | 17 | 1 | 0 | 0 |

Source: Compiled by the author.

COTI, which has a varied time of planting. The farmer is agreeable, and a further increase in objective value is achieved at 1,133 shillings, representing a rise of 371 shillings, or 49 percent, on the net income from the existing system. The earlier planting distorts the labor profile, and Table 98 analyzes the extent of the changes required.

There is no intrusion into food crop resource allocation, though a changed pattern of casual labor requirement will need a retiming of labor hire by the farmer. This will form important extension content during the season. The practical implication is that the single laborer needed should be taken on in mid-November rather than the end of that month.

Table 95 compares this level of objective value with the potential under conditions at B, without the dramatic increase in congruity score associated with COTM. Full potential has been sacrificed to move a step forward on the congruity ladder and introduce an acre of earlier-planted cotton. The implication is, however, that further profitable changes will require increased scale or greater incongruity, and both are likely to breach food resource allocations.

TABLE 98

Analysis to Check the Degree of Intrusion by Innovations in Year 4 into Labor Allocation to Subsistence Crops

| | Oct. | Nov. | Dec. | Jan. | Feb. | May | June |
|---|---|---|---|---|---|---|---|
| Residual Family Labor Available | 6 | 10 | 33 | 16 | 19 | 48 | 50 |
| COTK 1.00 Acres | 3 | 13 | 8 | 5 | 4 | 13 | 8 |
| COTJ 3.00 Acres | 0 | 9 | 39 | 24 | 15 | 33 | 27 |
| Total Cotton | 3 | 22 | 47 | 29 | 19 | 46 | 35 |
| Hired Labor Required | 0 | 12 | 14 | 13 | 0 | 0 | 0 |

Source: Compiled by the author.

Year 5

Substituting a later-planted acre of cotton as the second part of the move from COTJ to COTI would reduce profitability. A positive step forward is to increase scale by adding a late-planted acre, to give three acres of COTI (a composite of an acre each of COTK and COTJ already in the system and the additional late-planted acre) and keep two acres of COTJ, accepting the foreseeable clash with food resource allocation. The acceptability score of 3,405 for COTI under conditions at C indicates that the impact will be smaller than alternatives with similar congruity levels. The five acres of cotton gives an objective value of 1,356, shillings, or a 78 percent increase on the existing system. Table 99 shows the inroads made into labor used for food production.

The hired labor requirement is twenty-two man-days above the purchasing power of the capital funds constraining the solution (if the four days in May, being at a slightly higher rate, are specifically included in the surplus). Under the existing conditions and with capital outlay limited to 150 shillings, the labor must be found from more efficient production of foods. In comparing goals we have already shown a potential saving of sixty-five man-days, though the figure

TABLE 99

Analysis to Show the Level of Intrusion into Food-Producing Resources in Year 5

| | Oct. | Nov. | Dec. | Jan. | Feb. | May | June |
|---|---|---|---|---|---|---|---|
| Residual Family Labor Available | 6 | 10 | 33 | 16 | 19 | 48 | 50 |
| COTI 3.00 Acres | 9 | 21 | 21 | 21 | 15 | 30 | 27 |
| COTJ 2.00 Acres | 0 | 6 | 26 | 16 | 10 | 22 | 18 |
| Total Cotton | 9 | 27 | 47 | 37 | 25 | 52 | 45 |
| Hired Labor Required | 3 | 17 | 14 | 21 | 6 | 4 | 0 |

Source: Compiled by the author.

varies with the minor variations in the food solution against each cash-crop innovation.

Sufficient changes in food crop production must have been introduced by year 5 to release twenty-two man-days of this sixty-five man-day potential. In practice, flexibility will be greater; but by imposing a set of conditions A-D, the alternatives have been reduced. Income has already risen to a gross of 1,423 shillings in year 4; and a capital outlay assumption based on 16 percent of this would increase available funds from 150 shillings to 228 shillings, allowing the hire of twenty-two more man-days of labor, enough for our needs in year 5.

However, keeping to our conditions, the implication of this clash is that improved food techniques must be worked up to the scale necessary to release the required labor by year 5. The improved technique introduced may require small quantities of purchased inputs which would reduce the objective values slightly.

## Year 6

In year 6 the final congruity step, the purchase and application of fertilizers, sees COTM as the key innovation established on part of the cotton acreage. Although the total cotton acreage is maintained at 5.00 acres, the use of fertilizer on three of these will intensify labor needs over the October-February period by nine man-days, and the increased production will require three extra man-days in May and June. The surplus to be found from existing subsistence activities will increase to thirty-four man-days, or over 50 percent of the slack which we know can be provided by improved food techniques. The net cash income realized is 1,599 shillings, 110 percent higher than the existing system.

## Year 7

In a sense year 7 is the final year of the sequence, since the improved subsistence pattern is finalized. The farmer extends his cotton acreage to the maximum possible under this set of conditions and uses fertilizer on the full 5.46 acres. This gives the optimal solution for COTM under these conditions, with a net cash income of 1,886 shillings, 148 percent higher than the existing system. Gross value of production is 2,473 shillings, 16 percent of which is 396 shillings, allowing the assumption of 400 shillings for the new debt ceiling of the goal solution which is duly implemented in the eighth season. Table 100 sets out the reconciliation to show the labor required from food activities to allow the small expansion of acreage and the intensification of labor use by the extended use of fertilizer.

Total hired labor required is 107 man-days, of which thirty-nine can be provided from the 150 shillings allowed as working capital; sixty-eight must be found by the release of labor through greater productivity in food production. (Sixty-eight is three days higher than the sixty-five days released in the solution for COTM under conditions at point D. The minor variations in food activities entering the different solutions are reflected here.)

## Year 8

The goal solution for COTM requires an increase in funds from the farmer for the hire of additional casual labor. His debt ceiling is assumed to rise to 400 shillings under conditions at D. This allows an expansion to 6.70 acres of COTM and a final net cash income of 2,098 shillings, 175 percent higher than the existing system but declining to 2,033 shillings once the cost of the purchased inputs for food crops is deducted.

TABLE 100

Analysis to Show Level of Intrusion into Food Producing Resources in Year 7

| | Oct. | Nov. | Dec. | Jan. | Feb. | May | June |
|---|---|---|---|---|---|---|---|
| Residual Family Labor Available | 6 | 10 | 33 | 16 | 19 | 48 | 50 |
| COTM 5.46 Acres | 16 | 38 | 44 | 44 | 33 | 60 | 54 |
| Hired Labor Required | 10 | 28 | 11 | 28 | 14 | 12 | 4 |

Source: Compiled by the author.

The clashes with resource allocations to subsistence activities begin in year 5, with an anticipated increase in the scale of the cotton acreage. Intrusion will intensify until, by year 7, all those existing subsistence activities with substitutes will have been replaced. Prior to year 5 retiming of the hire of casual labor will meet the changes likely to arise in the labor profile, although improved food-growing techniques will be introduced in year 3 and scaled up to give the release of labor required for cotton innovations by year 5.

## STEP 3: SELECTION OF IMPROVED FOOD GROWING TECHNIQUES

Step 3 is the most complex in planning the progression of extension content. It selects the improved food growing techniques which best meet the increased labor needs of the progression of cotton innovations. In making the selection, a compromise is required between the food value points sacrificed by dropping existing food activities and the congruity and efficiency of the new techniques which replace them. Each activity dropped releases a number of days of labor for a given food value score. The new techniques themselves absorb labor, and it is the net amount released while the food supply pattern is maintained which measures the efficiency of the substitution. Balancing this with the congruity of the new food techniques introduced is a subjective process. However, because optimizing solutions have indicated the cash crop enterprise as the key source of potential, the improved food crop techniques are built around the selected progressing of cotton innovations. This weights the balance in compromise toward efficiency and congruity as the dominant facets in selecting the innovations as food crops and the present activities to be run down. The basic data required for this step are assembled in Tables 101-103. Table 101 sets out in row a the monthly labor use for each substitutable activity and in row b the efficiency scores in food value points for each monthly labor coefficient.

The table shows, for example, that if an innovation requires labor from subsistence activities in December, EML is the cheapest, and therefore most efficient, source, up to a total of six man-days with a score of 201 food value points per day released; above this, LCML is the source of a further seven man-days, followed by ECML, the most expensive source, also with seven man-days. This, however, is a gross release of labor. The techniques being brought in to maintain the patterns of food supply may reabsorb some December labor, and final efficiency will depend on the net amount left surplus for reallocation to the cotton innovations.

Table 102 summarizes the present and possible sources of the ten food types for which substitutions will be made by the seventh year of the adoption period.

The apparent complexity of the intercrops breaks down when analyzed by food type, and the ten types can be produced from five innovations. The four intercrops produce the same foods; and only the timing of green maize, sweet potato, and fresh cassava output distinguishes them. Single maize, cassava, and groundnut innovations can produce the foods from several activities in the existing system. As long as output coefficients have been properly discounted for reliability, covering the insurance function of intercropping, and the fertility maintenance role of intercrops is not being ignored, the substitution is much simpler than would appear from an examination of the combinations and mixtures occurring in the system.

Of these innovations, groundnut and cassava alternatives are of the same congruity levels, leaving efficiency as the only base for choice between them. Sweet potato activities are specific, with no alternatives, and only the range of possibilities for maize requires a compromise between selection criteria. Table 103 sets out the scores for maize, showing the substitution possibilities for each innovation, its congruity score, and its efficiency in terms of dry maize produced per man-day of critical labor absorbed. The innovations

TABLE 101

Labor Use and Food Value Scores of Replaceable Subsistence Activities

| | Acres | Food value points | | Oct. | Nov. | Dec. | Jan. | Feb. | May | June | Total |
|---|---|---|---|---|---|---|---|---|---|---|---|
| OC | .95 | 713 | a | 7 | — | — | — | — | — | — | — |
| | | | b | 102 | — | — | — | — | — | — | 102 |
| EML | .58 | 1206 | a | 7 | 7 | 6 | — | — | 4 | — | 24 |
| | | | b | 174 | 174 | 201 | — | — | 301 | — | 50 |
| ECML | 1.46 | 3927 | a | — | 37 | 7 | 7 | — | 6 | — | 57 |
| | | | b | — | 106 | 561 | 561 | — | 654 | — | 69 |
| LCML | 1.46 | 3592 | a | — | — | 7 | 29 | 15 | — | 6 | 57 |
| | | | b | — | — | 511 | 124 | 239 | — | 599 | 63 |
| LML | .58 | 766 | a | — | — | — | — | 14 | — | — | 14 |
| | | | b | — | — | — | — | 55 | — | — | 55 |

Source: Compiled by the author.

TABLE 102

Present and Possible Future Activities as Sources of Required Foods

| Food | | Existing Activities EML | ECML | LCML | LML | OC | Innovations | | |
|---|---|---|---|---|---|---|---|---|---|
| Dry Maize | | * | * | * | * | * | All Maize Innovations | | |
| Green Maize | March | * | — | — | — | — | MAC | | MAA |
| | April | * | * | — | — | — | MAH | MAD<br>MAE | MAB |
| | May | — | — | * | * | — | | MAI<br>MAJ | MAF<br>MAG |
| Sweet Potato | Feb.-Mar. | * | * | — | — | — | SPOTA | | |
| | June-July | — | — | * | * | — | SPOTB | | |
| Legumes | | * | * | * | * | — | All Groundnut innovations | | |
| Fresh Cassava | Sept.-Nov. | — | — | — | — | — | CASSA | | |
| | Dec.-Feb. | — | — | * | — | — | | | |
| Cassava Reserve | | — | * | * | — | * | CASSB | | |

Note: Asterisk indicates which of the foods in left-hand column are provided by traditional activities' Columns.

Source: Compiled by the author.

TABLE 103

Criteria Scores for Maize Innovations

| Maize Innovation | Congruity Score | Pounds per Man-Day Required | Green Maize Supplied | | | Possible Substitutes for |
|---|---|---|---|---|---|---|
| | | | Mar. | Apr. | May | |
| MAA | 2 | 20 | * | * | * | all |
| MAB | 3 | 19 | * | * | * | all |
| MAC | 4 | 25 | * | * | — | EML-ECML |
| MAD | 4 | 21 | — | * | * | LML-LCML |
| MAE | 4 | 16 | — | * | * | LML-LCML |
| MAF | 6 | 37 | * | * | * | all |
| MAG | 6 | 32 | * | * | * | all |
| MAH | 7 | 33 | * | * | — | EML-ECML |
| MAI | 7 | 34 | — | * | * | LML-LCML |
| MAJ | 7 | 32 | — | * | * | LML-LCML |

Source: Compiled by the author.

are subgrouped, divided by a significant jump in congruity level caused by the introduction of purchased inputs.

In our example the complexity of this stage is reduced with the flexibility given by the use of casual hired labor. Up to the limit of casual labor used, the cheapest source can supply labor for any month by a shift in the timing of casual hire. Year 5 is a key year in the adoption period; the first clash can be expected between an increase in labor requirements on a cotton innovation and existing labor use on food crops.

It is planned first as the fulcrum for the progression of improvements on subsistence activities. Procedure for the example follows three steps:

1. Selection of food crop innovations for year 5 on the basis of efficiency and, for maize, congruity.

2. An evaluation of the scale on which existing activities must be sacrificed before the selected combination will release the twenty-two man-days of labor required by the cotton innovations in year 5.

3. Calculation of the acceptability score and comparison with alternative sacrifices.

GNC and CASSA are selected for the improved production of legumes and cassava, since both are the most efficient from their respective sets. Both new sweet potato activities are specific to particular foods required and must be included. In addition, an expansion of the existing sweet potato activity, EP, will be needed to compensate for a loss of production from intercrops in April-May. As we have already noted, the only significant choice set is for maize. Here MAF is chosen as the most efficient producer of maize and the one which features in the optimal solution; it has a high congruity level, but it is assumed that this can be built up in years 3 and 4.

We have previously noted the stability of the food solutions over the conditions at C and D and over the range of cotton innovations. From the optimal solutions we know that improved food techniques will release about sixty-five man-days of labor over the critical months for reorientation to the market.

Table 104 confirms this for the selected innovations.

The full substitution of selected innovations for year 5 would save sixty-seven man-days during the critical months. If 33 percent

Labor Requirements of Existing and Selected
Substitute Activities for the Foods Supplied

| | Pounds | Innovation Selected | Acreage Needed | Labor Required O | N | D | J | F | M | J | Total |
|---|---|---|---|---|---|---|---|---|---|---|---|
| Dry Maize | 1300 | | | | | | | | | | |
| Green Maize (Mar.) | 175 | | | | | | | | | | |
| " (Apr.) | 175 | MAF | 1.18 | 8.3 | 8.3 | 8.3 | 8.3 | 8.3 | 3.5 | 4.7 | 49.7 |
| " (May) | 175 | | | | | | | | | | |
| Sw. Pot. (Feb.-Mar.) | 200 | SPOT A | .05 | — | 1.5 | — | — | — | — | — | 1.5 |
| " (Apr.-May) | 100 | (EP) | (+).03 | — | — | 1.0 | — | — | — | — | 1.0 |
| " (June-July) | 300 | SPOT B | .07 | — | — | — | 2.0 | — | — | — | 2.0 |
| Legumes | 200 | GNC | .27 | — | 7.0 | 2.9 | 1.1 | — | 1.3 | — | 12.3 |
| Fr. Cass. (Sept.-Nov.) | 750 | | | | | | | | | | |
| Fr. Cass. (Dec.-Feb.) | 750 | CASSA | .61 | 4.1 | 11.6 | 2.4 | 2.5 | 2.4 | — | — | 23.1 |
| Cass. Reserve | 4,000 | | | | | | | | | | |
| Total Labor Needed, Critical Period (rounded) | | | 2.21 (ac.) | 12 | 28 | 15 | 14 | 11 | 5 | 5 | 90 |
| Requirements of Existing Substitutable Activities (rounded) | | | 5.03 (ac.) | 14 | 44 | 20 | 36 | 29 | 8 | 6 | 157 |
| Saving by Full Substitution | | | 2.82 (ac.) | 2 | 16 | 5 | 22 | 18 | 3 | 1 | 67 |

Source: Compiled by the author.

of the foods supplied by the substitutable activities were provided by improved techniques, the surplus of twenty-two man-days required for year 5 would be realized. Table 105 demonstrates the results of a 33 percent substitution on the profiles for the whole system.

This system has a hired labor need which can be covered by the 150 shillings capital sum available to the model. The acceptability score will, similarly, be 33 percent of the total food value points of those enterprises for which substitution was possible, i.e., 3,401 points. An across-the-board substitution of this type is possible for the two reasons noted. First, each innovation can supply the foods formerly produced by several intercropped activities. Specialized innovations are very limited, being only on the sweet potato crop and absorbing minimal amounts of labor. Second, the flexibility given by retiming casual hired labor means that the labor profile for the food innovations will alter very little if substitutions are selective to minimize acceptability scores. Table 106 sets out the data to allow the comparison of selective changes with the across-the-board solution of Table 105.

The extent of selective changes is limited by the flexibility in retiming the hire of casual labor, shown by the final row in the table, or, more precisely, by the need to use family labor in each month of the critical period. Selections which leave family labor unused reduce the efficiency of the solution because they require innovation on a larger scale; however, where the gains in acceptability are very high, some loss of efficiency may be worthwhile. This does not occur here, but we are concerned to demonstrate the possibility of more selective sacrifices to reduce the "cost" of innovations by lowering the acceptability scores.

The lowest-valued traditional activities are LML and EML, both with an average of fifty-five points per man-day used. LML offers no opportunity for selective substitution, since all its labor is required in February, which shows no hired requirement. EML, on the other hand, requires labor in months for which hired labor is used in the across-the-board solution. The extent of selective substitution with EML is limited by the two man-days hired in October. A further six man-days are generated from this low-value activity, to feed back into high-value enterprises. OC, which gave up two man-days in the 33 percent substitution across the board, can be restored to its existing scale, saving 204 food value points, and the remaining four days will save 276 points by reinstating part of the ECML areage. The sum will be

$$3{,}401 - (2 \times 102) + (4 \times 69) + (6 \times 55) = 3{,}251,$$

TABLE 105

Labor Profile for Year 5, Showing Savings in Labor
Use on Food Production Absorbed by Expanded Cash Crop
(man-days)

| | O | N | D | J | F | M | A | M | J | J | A | S | Total |
|---|---|---|---|---|---|---|---|---|---|---|---|---|---|
| All Remaining Traditional Food Activities | 9 | 30 | 15 | 27 | 31 | 15 | 5 | 9 | 8 | 5 | 10 | — | 164 |
| New Food Activities | 4 | 9 | 5 | 5 | 3 | 3 | 2 | 2 | 1 | — | — | — | 33 |
| Cotton Activities | 9 | 27 | 47 | 37 | 25 | 25 | 17 | 52 | 45 | 40 | 15 | — | 339 |
| Total Labor Use | 22 | 66 | 67 | 69 | 59 | 43 | 24 | 63 | 54 | 45 | 25 | — | 536 |
| Hired Labor Needs | 2 | 11 | 12 | 14 | — | — | — | 3 | — | — | — | — | 42 |

Note: The rice and early and late sweet potato activities, which are not altered, are included with the residual of the intercrops in the traditional profile.

Source: Compiled by the author.

TABLE 106

Data for Selective Sacrifices to Improve the Acceptability Score

| | | Labor Use, Critical per. | | | | | | | | | | |
|---|---|---|---|---|---|---|---|---|---|---|---|---|
| | Acreage | O | N | D | J | F | M | J | Total Use in Period | Saved by 33% sub. | Average Value (pts. day) | Further Possible Subst. |
| EML | .58 | 7 | 7 | 6 | – | – | 2 | – | 22 | 7 | 55 | 15 |
| ECML | 1.46 | – | 37 | 7 | 7 | – | 6 | – | 57 | 19 | 69 | 38 |
| (EP) | .05 | – | – | 2 | – | – | – | – | (2) | – | – | – |
| (R) | .30 | – | 1 | – | 3 | 12 | 4 | 4 | (24) | – | – | – |
| LCML | 1.46 | – | – | 7 | 29 | 15 | – | 6 | 57 | 19 | 63 | 38 |
| LML | .58 | – | – | – | – | 14 | – | – | 14 | 5 | 55 | 9 |
| (LP) | .05 | – | – | – | – | – | – | – | (0) | – | – | – |
| OC | .95 | 7 | – | – | – | – | – | – | 7 | 2 | 102 | 5 |
| Total | 5.38 | 14 | 45 | 22 | 39 | 41 | 12 | 10 | 183 | 49 | – | – |
| Of Which Substitutable Labor | – | 14 | 44 | 20 | 36 | 29 | 8 | 6 | – | – | – | 157 |
| Hired Labor Needed with 33% Sub. | | 2 | 11 | 12 | 14 | – | 3 | – | 42 | | | |

Note: The traditional activities to be retained are in parentheses.

Source: Compiled by the author.

a saving of 4 percent of the score for substitution across the board. Further selections will be increasingly marginal, since the activities available are very close in food value scores per day used. Each has a month in which no hired labor is used, implying a loss of efficiency due to idle family labor.

This selective substitution alters the pattern of hired labor requirements, partly by the shift of labor into the months of October for OC, and November, December, January, and May for ECML. It also alters the pattern of food production, which must be checked to see if the supply pattern can be maintained with the substitution without impairing the efficiency in releasing labor. The change maintains dry maize and groundnut production, the labor being equally productive for these foods in ECML as in EML. Cassava production is raised above requirements and the area required for CASSA as an innovation can be reduced. A small increase in sweet potato and green maize production is required. The extended scale of innovation on MAF and SPOTA exactly balances the release of labor resources from the reduced area of CASSA. On balance, then, acceptability can be increased by the selective rundown of a low-score intercrop, EML, without impairing the efficiency of the substitution. Table 107 describes the system in year 5.

Definition of the system gives us three fully defined points in the progression over the adoption period, years 1, 5 and 7, by which the substitution of improved food producing techniques is complete. Although a further clash, requiring the release of an additional twelve days of labor to the cotton innovations, will occur in year 6, the rate of progression, requiring the release of the total potential in year 7, will cover this. The scale of changes in food crop innovations for year 6 will be selected with the tolerance of the farmer in mind and checked to see if sufficient labor is released. It will be a part of the final step in planning the progression of extension content.

## STEP 4: BUILDING THE FOOD CROP INNOVATION SEQUENCE

The final stage is relatively straightforward. Several benchmarks have already been established over the development period, and this fourth stage is a progression from one to the next, indicating changes in either congruity levels or scales for each year. The system has been fully defined at years 1, 5, and 7; and year 5, with significant changes in both cash and food crop activities, serves as a fulcrum. The progression is taken in two parts: from year 1 to year 5 and from year 5 to year 8.

TABLE 107

The Final System for Year 5

| | EML | ECML | EP | R | LCML | LML | LP | OC | MAF | GNC | CASSA | SPOTA | SPOTB | COTJ COTI | TOTAL |
|---|---|---|---|---|---|---|---|---|---|---|---|---|---|---|---|
| Acreage | .19 | 1.10 | .06 | .30 | .97 | .37 | .05 | .95 | .42 | .10 | .13 | .02 | .03 | 2.00 3.00 | 9.69 |
| Labor (man-days) | | | | | | | | | | | | | | | |
| Oct. | 2 | – | | | | | | 7 | 3 | | 1 | | | 9 | 22 |
| Nov. | 2 | 28 | | 1 | | | | – | 4 | 3 | 2 | 1 | | 27 | 68 |
| Dec. | 2 | 5 | 2 | – | 5 | | | – | 3 | 1 | – | – | | 47 | 65 |
| Jan. | – | 6 | – | 3 | 19 | | | – | 3 | – | 1 | – | 1 | 37 | 70 |
| Feb. | – | – | – | 12 | 10 | 9 | | – | 4 | – | – | – | – | 25 | 60 |
| May | 1 | 4 | – | 4 | – | – | | – | 2 | 1 | – | – | – | 52 | 63 |
| June | – | – | – | 4 | 4 | – | | – | 1 | – | – | – | – | 45 | 54 |
| Foods (pounds) | | | | | | | | | | | | | | | |
| Dry Maize | 38 | 385 | – | – | 340 | 111 | – | – | 426 | – | – | – | – | – | 1,300 |
| Green Maize (Mar.) | 57 | 0 | – | – | – | – | – | – | 119 | – | — | – | – | – | 176 |
| " " (Apr.) | 10 | 110 | – | – | – | – | – | – | 55 | – | – | – | – | – | 175 |
| " " (May) | – | – | – | – | 97 | 19 | – | – | 60 | – | – | – | – | – | 176 |
| Sweet Pot. (Feb.-Mar.) | 30 | 88 | – | – | – | – | – | – | – | – | – | 80 | – | – | 198 |
| " " (Apr.-May) | – | 88 | 240 | – | – | – | – | – | – | – | – | – | – | – | 328 |
| " " (June-July) | – | – | – | – | 146 | 52 | – | – | – | – | – | – | 120 | – | 318 |
| " " (Aug.-Sept.) | – | – | – | – | – | – | 200 | – | – | – | – | – | – | – | 200 |
| Legumes | 10 | 55 | – | – | 49 | 19 | – | – | – | 75 | – | – | – | – | 208 |
| Rice | – | – | – | 300 | – | – | – | – | – | – | – | – | – | – | 300 |
| Fr. Cassava (Sept.-Nov.) | – | 572 | – | – | – | – | – | – | – | – | 178 | – | – | – | 750 |
| " " (Dec.-Feb.) | – | – | – | – | 504 | – | – | – | – | – | 246 | – | – | – | 750 |
| Cassava Reserve | – | 1,188 | – | – | 669 | – | – | 1,425 | – | – | 718 | – | – | – | 4,000 |

Note: There are some rounding errors in the total foods supplied.

Source: Compiled by the author.

## Development from Year 1 to Year 5

A start on food crop innovations is assumed to be acceptable to the farmer in year 3, giving two seasons to work up to the congruity and scale planned for year 5. There are no other conditions to be met for this period of the sequence, though of course the labor required for the innovations will need reconciliation with the system profile. Because the new methods are more efficient than those being replace, it is assumed that sufficient labor will be released as long as the small plots of the new crops can be substituted for traditional crops. In the model this presents no problem, and the best way to ensure it in practice is to fit the new plots into the farmer's established cultivation routine. When he is preparing land for his early maize mixture, a small piece of the prepared plot should be given over to observation of an innovation already agreed on. This causes the least disturbance of his routine and ensures that the innovation is tried under the same soil and climatic conditions as the existing practice it is designed to replace.

The scale of the plots required in year 5 is relatively small. There are no complications of congruity with CASSA, GNC, and the sweet potato innovations. These can be introduced as small observation plots during years 3 and 4.

Either improved cassava or groundnuts are introduced on a .05 acre plot in year 3, with the other and the sweet potato changes in year 4. Neither crop requires a purchased input, and giving the farmer a choice may enhance cooperation. Maize is more complex; it is included on a larger scale than the others and as using purchased inputs in the plan for year 5. Although MAF is the most efficient maize innovation, it uses both fertilizer and insecticide; and the late-planted plot gives only a modest yield. MAC has a congruity level closer to existing practice and gives a yield 2.5 times the local maize. A small .05 acre plot, planted with the farmers' early maize, should be promoted as a demonstration in year 3. To give a dramatic impact to the introduction of purchased inputs, the size of the plot is increased to .15 acre, still planted early, in year 4. This corresponds to MAH, with the highest yield level of 1,800 pounds per acre, or 4.5 times the local maize. The plot size can be maintained for year 5, with later plantings on similar plots to reach the .42 acres of MAF needed.

## Development from Year 5 to Year 8

Under the assumptions made, COTM will be established on a scale derived from the programmed optimal under conditions at C in year 7, since we have seen this to be almost identical with the pattern of the optimal solution under goal conditions at D. In the progression over the latter half of the period there will only be one intermediate step between years 5 and 7, for both of which food solutions are available. The intermediate step in year 6 will be almost wholly a scale increase. Although a new cassava innovation (CASSB) is required by year 7, it is a minor modification of CASSA, already introduced. The main factor influencing the extent of progress in year 6 is the need to release an additional twelve man-days of labor for the increased cotton requirement in that season. With a further twelve days required for the cotton crop, a substitution of improved technology for a further 12/67, or 18 percent of food supplies across the board, will be the minimum level of change needed. The scoring of acceptability and comparison of alternative selective changes can be repeated as for year 5. However, the easiest way to plan for year 6 is to ensure a smooth progression through from year 5 to year 7, checking that the level of changes chosen will release the required labor. The scale of food innovations is pitched between the known levels of year 5 and year 7, maintaining individual acreage in the proportions which will be required in the final solution, to ensure a balance in the food sacrifices and labor releases from the existing activities to be dropped. Table 108 shows the intermediate levels planned for.

The only change in the type of innovation is a splitting of the cassava production between early- and late-planted crops. The late-planted CASSB saves more labor after a certain level of cash and food crop innovation and features in the final solution.

TABLE 108

Innovation Acreages for Year 6, Interpolated Between Acreages in Year 5 and Final Acreages in Year 7

| Year | MAF | CASSA | | SPOTA | (+EP) | SPOTB | GNC |
|---|---|---|---|---|---|---|---|
| 5 | .42 | .13 | | .02 | +.01 | .03 | .10 |
| | | CASSA | CASSB | | | | |
| 6 | .80 | .25 | .20 | .04 | +.02 | .04 | .20 |
| 7 | 1.18 | .50 | .35 | .05 | +.03 | .07 | .27 |

Source: Compiled by the author.

TABLE 109

Acreages in the Changing System over the Development Period

| Condition | A | | B | | | C | | D |
|---|---|---|---|---|---|---|---|---|
| Year | 1 | 2 | 3 | 4 | 5 | 6 | 7 | 8 |
| EP | .05 | .05 | .05 | .05 | .06 | .07 | .08 | .08 |
| R | .30 | .30 | .30 | .30 | .30 | .30 | .30 | .30 |
| LP | .05 | .05 | .05 | .05 | .05 | .05 | .05 | .05 |
| EML | .58 | .58 | .58 | .58 | .19 | – | – | – |
| ECML | 1.46 | 1.46 | 1.46 | 1.46 | 1.10 | .50 | – | – |
| LCML | 1.46 | 1.46 | 1.46 | 1.46 | .97 | .90 | – | – |
| LML | .58 | .58 | .58 | .58 | .37 | – | – | – |
| OC | .95 | .95 | .95 | .95 | .95 | .50 | – | – |
| MAC | – | – | (.05) | – | – | – | – | – |
| MAH | – | – | – | (.15) | – | – | – | – |
| MAF | – | – | – | – | .42 | .80 | 1.18 | 1.18 |
| GNC | – | – | (.05) | (.10) | .10 | .20 | .27 | .27 |
| CASSA | – | – | – | (.05) | .13 | .25 | .50 | .50 |
| CASSB | – | – | – | – | – | .20 | .35 | .35 |
| SPOTA | – | – | – | – | .02 | .04 | .05 | .05 |
| SPOTB | – | – | – | – | .03 | .04 | .07 | .07 |
| Total Foods | 5.43 | 5.43 | 5.43 | 5.43 | 4.69 | 3.40 | 2.85 | 2.85 |
| Congruity Score | | | | | | | | |
| COT 0 | 3.03 | 3.03 | 1.03 | – | – | – | – | – |
| COT C 2 | 1.00 | – | – | – | – | – | – | – |
| COT J 4 | – | 1.00 | 3.00 | 3.00 | 2.00 | – | – | – |
| COT K 5 | – | – | – | 1.00 | – | – | – | – |

TABLE 109 (Continued)

| Condition | A | | B | | | C | | D |
|---|---|---|---|---|---|---|---|---|
| Year | 1 | 2 | 3 | 4 | 5 | 6 | 7 | 8 |
| COT I 5 | – | – | – | – | 3.00 | 2.00 | – | – |
| COT M 7 | – | – | – | – | – | 3.00 | 5.40 | 6.70 |
| Total Acreage Cropped | 9.46 | 9.46 | 9.46 | 9.46 | 9.69 | 8.40 | 8.25 | 9.55 |
| Net Cash Income | 748 | 845 | 1029 | 1133 | 1333 | 1555 | 1821 | 2033 |

Notes: Net cash income values are slightly lower than in the sequence of cotton innovations. Costs of inputs on improved food activities have been deducted.

It is assumed that the small plots for food innovations up to year 5 are within the plots of either EML or ECML.

Source: Compiled by the author.

Checking the system through with innovations at these levels shows EML and LML discarded completely, ECML down to .50 acres, LCML to .90 acres, and OC down to .50 acres. Thirty-nine days of casual hired labor are required to run the system, concentrated mainly in December and January and absorbing 142.50 shillings of the available 150 shillings working capital. The level of innovations in year 6 has released fourteen instead of the twelve extra days of family labor required for a smooth progression between the position of years 5 and 7. In year 7, of course, the final solution levels of the food activities are reached and the remaining intercrops ECML and LCML are discarded. In year 8 the scale of COTM is increased to the level of the optimal solution allowed by the increased capital outlay under conditions at point D.

The development of the system through the period and within the conditions of the model is summarized in Table 109.

Extension activity will be organized around the content identified by the planning sequence. The implications for extensions are summarised in Chapter 20.

# CHAPTER 20
# FARM MANAGEMENT ECONOMICS IN TRADITIONAL AFRICAN AGRICULTURE AND EXTENSION ORGANIZATION

Farm management economics makes its contribution through the existing extension services by identifying appropriate extension content. In the first section of this study the structure of the extension services and government policy on agricultural development strategy were identified as important conditions molding the approach which the discipline must adopt. The case was made for a development strategy aimed at improvement within the existing structure of the agricultural sector, and the investigational and planning approach outlined are relevant to such a strategy. However, even with an improvement policy allowing the fuller utilization of lower-grade manpower, staff/farmer ratios will still be high and 1:1,000 will be common for some time. This, combined with the conviction that visits to individual farms are necessary to supervise the implementation of advice, will limit extension coverage to the category defined as innovators in diffusion and adoption research—perhaps less than 2 percent of the total farmer population. Innovators will identify themselves when a viable opportunity for change is offered and will form focal points for the diffusion of improved practices throughout the community. Communication will be served best, in terms of rapport and morale, by prolonged association with this nucleus of adopters, providing they retain their identity with the community. Control of extension content and credit facilities will ensure that these do not expand until they become relevant to the position of the average local farmer.

Within this context, farm management economics makes two major contributions based on the analysis of the existing farming system and evaluation of the improved technology available to the system. It identifies an innovation sequence, initially attractive to smallholders yet with the potential for sustained development of the system. The sequence is a balance among three factors: the priorities and resource position of the farmer, his changing attitudes to change, and government's priority for increased agricultural productivity. The progression of changes provides the content for extension effort on the farms of adopters identified by their willingness to try out the initial innovation.

Selection of content is based on criteria of profitability (qualified by reliability), congruity, complexity, and acceptability. Early in the development period, when initial attractiveness is important and the scale of adoption on the farm is small, criteria of congruity, complexity, and profitability (reflecting national objectives) dominate the evaluation. As scale and incongruity increase, acceptability, used to describe the degree of compatibility with the nonmarket priorities of the farmer—particularly his home-produced food supplies—gains importance in balancing profitability. There was considerable flexibility in the order in which the various criteria were brought to bear on the possible changes available, and the arithmetical approach to scoring was crude.

Working from the conviction that there is no inherent barrier to change in traditional agriculture, it has been stressed that the farmer must make the pace. If the planning is right and the organization effective, changes will be absorbed. It cannot be overemphasized that to be right, planning must be consistent with the objectives of the farmers. So must implementation, since there can be no automatic sequence of adoption. Although planning is directed to selecting the sequence most likely to appeal to the majority of farmers, in the final analysis the speed of implementation is dependent on the communication between farmer and extension agent; and communication will founder unless the pace of change is flexible to variations in the results of innovations and in farmer preferences. The model incorporates the expected pace of development, and individuals will be dispersed around this average. Implementation must adjust to the pace of the individual, a factor which complicates its administration.

## THE ORGANIZATION OF EXTENSION

The remainder of this chapter deals with extension organization, for the investigational and planning sequences raise guidelines for mounting extension programs in the field, and outlines the responsibilities of the usual three levels of the extension hierarchy in the field: the contact staff, the supervisors, and the local organizer.

The need for visits to the farms to supervise the changes recommended limits the coverage of field contact staff. This will be important in planning the deployment of extension staff, both within an area and between areas of different potential. The potential of an area vis-à-vis extension programs will involve estimating the coverage of a given set of manpower. This is governed, for an area, by three factors: the ease of innovator identification; the density of the farm population, the settlement pattern, and the mobility of the adviser; and the complexity of the innovation packages forming the content of the program at any one time.

## Ease of Innovator Identification

Where the innovation package consists solely of husbandry changes, would-be innovators must come forward to the extension service to ask for advice. Once the program requires purchased inputs, those availing themselves of facilities can be identified. The channels available for input distribution influence the ease of identification. Where inputs can be forced through an institutional bottleneck in direct contact with the farming population, the process of identification is easy. Inputs available at small retail outlets widely dispersed throughout the area present some problems. In the case of Sukumaland, inputs are distributed through the cooperative primary societies, which purchase farmers' produce and are a natural focus for initial extension contact with would-be innovators.

Identification is a problem at the start of a program. Difficulties due to no purchased input being included or to widely dispersed distribution outlets may prolong the buildup to the required complement for each contact worker.

## Settlement Patterns and Population Density

The contact staff of the extension services typically uses bicycle transport and therefore has limited mobility, though in very densely populated areas even this form of transport may be unnecessary. In certain terrains the settlement pattern may limit coverage; for example, in a country with a high ridge/valley conformation, movement between valleys is very difficult. Usually population density and the type of community organization—nucleated or farmstead settlement—will be a more important factor.

Sukumaland, with an area of about 30,000 square miles, has a farmstead settlement pattern, with the majority of the population living at a density of between 60 and 120 per square mile. Throughout the area are 530 cooperative primary societies, each a focal point in rural life. The average society has some 600 members, covering 60 percent of the farmers of the area. Each contact worker must cover about 1.5 primary societies, or perhaps 100 square miles, an area having a radius of five or six miles.

In an organized routine, using a bicycle, he can visit three farmers a day. Allowing him 3.5 days in the field gives ten or eleven farm visits each week. Deliberate clustering of cooperating farmers may increase his coverage but also may reduce the diffusion effect. In the example area, then, the physical coverage capacity of the contact worker is ten or eleven farm visits each week. The final factor, the complexity of the innovation package, decides how many different farmers he can cope with.

## Complexity of the Innovation Package

Farm visiting will be concentrated at the time when changes are being implemented. Just as an analysis of necessary timeliness in labor use is vital to a description of bottlenecks in the existing system, a similar analysis aids the organization of extension coverage. A good example is the use of insecticides as an introduction to farmers of purchased inputs in the planned sequence. It is a complex innovation with a variety of points requiring supervision: assembly of the pump, maintenance of the pump, mixing the insecticide, application of the insecticide, intervals between applications, protection from adverse effects of the insecticide, and sanitary arrangements for its storage.

Of these seven points, three—mixing the insecticide, its application, and the interval between applications—are specifically timely. First application must be made eleven weeks after germination, the subsequent five at ten-day intervals. The insecticide must be mixed just prior to use in order to maintain its active life and to preclude mishaps within the farm family. The remaining points are nontimely; for example, pump assembly may be demonstrated by the seller or to a group of purchasers by the extension officer. Indeed, all other points may be demonstrated but require checking in each individual case on the farm. Assembly must be done before time for spraying, and storage of pesticides must be checked when the inputs are taken home. Other points can be demonstrated or checked once the regime is complete. The controlling factors on coverage will be the need to visit all farmers around the eleventh week after their respective plantings and again after ten days. A spread in planting times between farmers is to be expected, and investigational data will show the latitude this gives; but the need to revisit in ten days suggests a feasible coverage of about fifteen farmers, within the physical limitation of ten or eleven visits each week. Other innovations, such as the introduction of maize in year 3, or other parts of the same package, such as correct spacing and early weeding of the sprayed cotton, lend themselves to similar analyses. The visit routine for each farmer can be built into a calendar for the season, as a timetable for the contact worker and a means of control for his supervisor. Changes in the scale of adoption will not affect the visit routine. Over time, the intensity of visits required on established innovations will decline and, unless new innovations occurring in the sequence require visits to compensate for these, a worker's coverage can be increased.

E. M. Kulp has recently demonstrated the contribution that systems analysis can make in designing an infrastructure for planning the aggregated requirements and consequences of increased input uptake and production on the farm. The relationships at this grass-roots level, between contact worker and farmer, are similarly amenable to analysis in O and R terms.[1]

The coverage potential of extension workers is an important additional consideration in planning programme content. Just as later in the adoption sequence less timely supervision will be required and coverage can be increased, so also initial package design can be weighted toward being more or less necessarily timely to allow greater or less coverage. The balance between more visits on the farm and visiting more farms will depend on the shape of the marginal revenue curve for additional innovations on the farm unit. Where the curve is downward-sloping, greater coverage will improve returns to extension resources, whereas if marginal returns are still rising, the better alternative will be more intensive visiting.

The alternatives are further complicated by the balance already being struck in the package between acceptability to the farmer and profitability, the national priority. Good congruity and acceptability and low complexity imply less than potential profitability, and thus the intensification of visits to supervise innovations with better congruity acceptability and complexity scores is never worthwhile if it involves a reduction in coverage of farms. Losses occur at both the extensive and the intensive margins. The corollary is that there can be no case for increasing the acceptability elements in a package unless it increases the coverage capacity of the extension service. To do so without increasing capacity widens a market which cannot be supplied. The exception to this is where the benefits of faster diffusion outweigh the losses in profitability on the farms covered directly.

In planning the progression of innovations, it will be important to know the physical capacity of the contact worker in the area concerned, and to evaluate innovations with an eye to the relative timeliness of supervision they require. Important differences will weight the selection. Innovations which increase the frequency of visiting in a busy period, to the extent that they reduce the coverage of the contact worker, should be shifted to later in the planning sequence, when the intensity of supervision is relaxed, and are to be avoided in the earlier stages of the adoption period. A quantification of the comparative profitability of innovations under various planning conditions would allow an estimate of the opportunity cost of increasing visit frequency at the expense of coverage. However, both the estimation of the necessary visit frequency and the weighing of the other planning criteria with profitability require well-informed judgment.

A second contribution to extension communication is the liaison material revealed by both investigation and planning. A major objective in the analysis of investigational data is a description of the resource/production/consumption relationships of the farming system. This gives extension personnel the opportunity to understand what the farmer is doing with his existing farming practice. Understanding may be modest at contact worker level but thorough at immediate supervisory levels. To be able to demonstrate an awareness of both the sources of farmer satisfaction, such as preferred food combinations, and farmer problems, such as the need to insure against variability in food crop yields or the bottlenecks in labor requirements and the crops and operations causing them, contributes greatly to trust between farmer and adviser. Similarly, the planning procedure identifies those changes of management routine which are associated with innovations. It allows the worker to advise the farmer to hire his casual labor a bit earlier than usual. It pinpoints likely clashes in farmer priorities—for example, between expanding an innovation and weeding his food crops to ensure adequate food. Advance warning of such problems helps the worker resolve uncertainties in the farmer's mind about the effect of changes, and to prepare the ground for other changes needed to prevent the clash from ever occurring. Both types of liaison material allow the extension worker to communicate with the farmer on his own terms. This is very different from the "holier than thou" attitude which has sometimes prevailed, often hiding a lack of confidence in the "advice" being offered. Many extension services have had to oblige their own junior staff to use "improved" methods on their domestic plots, a sad comment on both the advisory content and the understanding of the more senior workers.

## RESPONSIBILITIES OF STAFF

Finally, we have noted that the need for individual innovators to follow their own pace of development will complicate the administration of programs in the field. While this is true, and implies individual attention to adopting farmers, it will affect only the pace, not the content, of the program. It is the decision as to content which creates problems for staff of the level of qualification usual in contact workers in developing agriculture. The farmer sets the pace in deciding his future involvement with the key innovation. As his confidence is won, the extension services seek to interpolate new changes which will allow his involvement to grow. Decisions on the timing of these new changes must be taken on an individual basis by the extension staff but can readily be passed up the hierarchy for the local organizer, since they are annual decisions. The responsibilities vis-à-vis this type of program of the three field levels of the service are set out below.

### The Local Organizer

In most former British areas of Africa, the local organizer would be the district agricultural officer, usually a graduate. He would be responsible for absorbing the program content from the farm management economist and research team. This would include the specification of the innovation progression and its expected timing for each area. There may be several such areas under his control. With the farm management economist he would prepare liaison material, from the results of investigation and planning, for each package of content in the sequence, to be channeled down through his field supervisors to the contact staff. He would limit the coverage of this material to the variety of steps expected to be handled concurrently over the area, further material being prepared as the sequence progresses. The organizer would plan deployment of his staff in relation to local input supply outlets and would have six main recurrent tasks each season:

1. Deciding the new changes to be encouraged for each innovator.

2. Planning the extension requirement, in terms of visit frequency, for the new changes and adjusting worker coverage if required.

3. Liaison with suppliers for timely availability of inputs at the outlets.

4. Supervising the routines of his field supervisors, which will involve visits with the supervisor and the contact officers to participating farmers.

5. Devising instructions and procedures, and procuring equipment and supplies to be channeled down to field level.

6. Channeling the results of program evaluation procedures back to the planning authority.

## The Field Supervisor

The field supervisor, usually a diplomat or experienced and effective contact worker, will be limited to a single type of farming area. His initial task will be to absorb the innovation sequence and liaison material for his area, and to discuss the liaison material with his contact workers. He will allocate his contact workers between the input-supply outlets in his area and instruct them on procedure and the selection of potential innovators for the program. He would have six main recurrent functions:

1. To check input outlets in his area for availability of stocks at the appropriate time in the season, and for adequate buyer documentation.

2. To channel instructions on steps to be encouraged on each innovator's farm to the contact workers , and to discuss with them the steps and the relevant liaison material.

3. To supervise the visiting routines of the contact workers, and to check on the level of cooperation each has established with the farmers by farm visits.

4. From the diaries of the contact workers, to compile regular reports to the local organizer on the progress of each innovator.

5. To keep his own diary covering supervisory visits to contact staff and to include a summary in his report.

6. Before the critical decision on the level of involvement for the season must be made, to visit all farms with the respective contact

workers and to channel farmer decisions, as they are made, back to the local organizer.

### The Contact Worker

Typically the contact worker in extension services following an improvement strategy will have had between eight and ten years' schooling, with two years past school training often leading to a certificate. His main role in establishing the program will be to contact would-be innovators at the input outlets and select participants for the program in consultation with his field supervisor. His five main recurrent tasks would be the following:

1. To visit each cooperator, as indicated by the extension schedule drawn up by the local organizer.

2. To maintain a field diary recording his visits to cooperators and their progress, in particular the purchase of inputs, their application, and the timing of cultural practices on the crops in the program.

3. To take sample counts or weighings on improved and traditional plots on each farm as the basis for program evaluation.

4. To report the decision of the innovator on the level of future participation and to specify new volunteers wanting to be included.

5. To maintain contact with the innovator by occasional visits during the off-season.

It is appreciated that this is not a comprehensive listing of the duties of extension staff, which frequently performs a variety of functions, including implementation of the provisions of agricultural ordinances and, for the senior levels, the administration of personal welfare of their staff. In addition, all levels will be concerned to promote the diffusion of innovations through the community through group meetings and through sponsoring visits to selected innovators as demonstrations in the fields of farmers recognized as members of the community. However, the lists demonstrate that the administration of this type of program can be spread, without undue burden, between the levels of hierarchy usually found in extension services in African agriculture.

## CONCLUSIONS

This completes the primary task of the study. It has described in detail an investigation and planning sequence for farm economics to guide farmers' resource use in traditional peasant agriculture within the conditions of many rurally dominated African economies. Properly organized, farm economics offers the opportunity to break a circle which seems to grip efforts to develop peasant agriculture: a lack of political enthusiasm, lack of funds for research, poor research orientation, inappropriate extension content, extension failures, poor advisory service morale, and again a lack of political enthusiasm. It is hoped that efforts along the lines described in the study will allow a closer identity between national and individual priorities and greater harmony in working to meet those priorities, and will contribute to effective mobilization of the mass of small farmers for economic growth.

Several initial assumptions were made in the study: a fairly strict definition of traditional agriculture, a government policy of speeding the evolution of small farms within the existing structure of the agricultural sector, and a limitation to a micro role for farm economics in guiding farmers' resource use through extension, ignoring its potential contribution to policy formulation. Each of these assumptions requires further comment in the light of conclusions reached on the approach suitable for farm economics in traditional African agriculture.

Throughout the study we have been concerned with traditional agriculture in which tribal affiliation dominates the community and family labor the farm organization. Most economies in Africa, and many agricultural sectors, are dualist in character. Subsectors are structured in a particular way as a result of plantation or estate alienation, and even indigenous agriculture may be at different stages of development, as a result of local market penetration or pockets of dense population where the emphasis has moved from labor to land as the factor limiting the scale of activity. The agricultural developers need a variety of approaches to meet the different needs of this variety of structural characteristics. As long as production units are small, there will be a need for shortcuts to allow cost-effective farm planning. Where market opportunities and technical alternatives are limited, the criteria described in the study will form a useful basis for grouping. Even with large commercial farms, the need to make the best use of the manpower qualified to use farm economic techniques gives importance to the use of representative models. Where differences are of size, without significant economies of scale, planning results can easily be adapted to the particular farm unit. With wholly commercial units, the scope of investigation and the number of planning criteria can be narrowed to reflect the preoccupation with market production. A single farm economic planning unit may need all these alternatives at its fingertips, and a single rigid set of procedures would merely repeat the mistakes made at the technical level in previous years.

Chapter 5 noted that transformation of the structure of the traditional agricultural sector will be a prerequisite for development under particular circumstances. Beyond this, it may continue to be favored by governments unconvinced of the potential of improvement strategies. Under a transformation policy, planning of farm units will be no less important; and many of the aspects of the approach described, particularly in investigation and data preparation, will be just as useful.

The relevance of the approach as a whole, which stresses the needs of the farmer as a vital consideration in effective planning, with voluntary adoption as the basis for cooperation, depends on two aspects of transformation policy.

1. The degree to which government shifts the scheme out of the economic environment facing producers. Where government creates an island of infrastructure around the scheme—new market opportunities, a food retailing system, alien technology—it will become an economic enclave, and planning will be guided by the revision in conditions. Where government cannot afford to do this and the schemes remain dependent on existing opportunities and technology, a good deal of the know-how of the existing farming system will be carried over into the planning. Much of the investigational phase and the representative model and optimizing stage in the planning sequence will remain relevant.

2. The degree to which management can be imposed on the agricultural livelihood of would-be participants. This is dependent on strong local pressures for structural change. Where transformation is adopted as a general strategy for agricultural development, perhaps without the local pressures, farmers must be attracted into schemes. As we have seen, attraction will be dependent on prior analysis of farmer objectives and priorities, so that broader planning criteria and wider investigation of the way in which existing systems meet farmers' needs is a prerequisite. Clearly, under fully centralized management, with members virtually laborers on the schemes, an adoption sequence and criteria centered on acceptability, with the associated investigational aspects, will be superfluous.

The second role for farm economics in contributing to more effective policy formulation, and thus more efficient investment of public funds, has particularly complex methodological aspects, themselves the subject of a great deal of professional effort. The brief comments here are intended only to highlight areas of policy which might be aided in the course of applying the approach presented here.

Particularly directly affected are the strategy and organization of agricultural research. Both the investigation and planning phases help to describe the social and economic environment within which the farmer operates. Superimposed on the ecological environment, which is still the preoccupation of most research effort, this is a much more directly relevant guide to the design of experimental programs which meet the needs of the farmer.

The investigational phase identifies complementarity between crops in the system, particularly subsistence crops and the various roles for each crop. Where foods are grown intercropped, for example, it is unlikely that an extension effort to encourage pure stands of one of the constituents will succeed unless it also covers improved methods of growing the admixtures, and unless it ensures that changes conform to the roles of the existing crops. The Sukuma farmer typically mixes five or six other crops with his maize; if he is advised to plant his maize in a pure stand, what is he expected to do with these? Does he cultivate extra acreage for them? If so, what will be the labor demand for cultivation over the peak period? Typically he plants four or five plots of mixtures, including maize at various times and in various locations. Sometimes one or two may fail. He is advised to plant all his maize at one "optimal" time. What if the whole planting fails for lack of rainfall at germination or later during the period of maximum growth? Bearing in mind the importance of maize as the staple grain and the security of food supply as the dominant motivation for peasant farmers, their apathy toward such recommendations can be understood. Improvement of maize in the farmer's eyes might be higher yields from existing practices, a reduction in vulnerability to drought, or a shorter-term crop to give even greater flexibility in planting time.

The planning phase identifies activities with low productivity at critically busy periods of the season. If productivity on these items can be significantly improved, whether by higher yields or by shifting the timing of the crop to avoid labor intensity in the critical period, a large release of labor allows an expansion of market production. It is this type of analysis which can indicate productive lines of research. This applies equally to mechanization, a field in which answers tend to be provided before problems are diagnosed; and the result is often an enthusiastic sales campaign for an inappropriate item of capital equipment. Both investigation and planning identify bottlenecks in which the constituent operations lend themselves to mechanization. Cotton and groundnut harvests frequently clash in seasonal systems, and both crops spoil if left on the plants. Groundnuts can be dug up and stooked on poles above the ground; and rain runs off the stooks, which rapidly dry out, avoiding sprouting. In this way the subsequent labor-intensive operations of pulling off the nuts can be postponed until after the cotton harvest. Simple techniques of this kind can be interpolated to meet a problem diagnosed by analysis of the system.

The study raised a number of ways in which experimental design could be brought closer to the farmer. Many researchers have noted the need to base recommendations on extensive trials, with replicates located at several sites to allow observation of the effects of microclimatic variation on the treatments. This allows the calculation of expected reliability of the results, a criterion as important to the farmer as the average level of the results. Also important is the description of farm practice from the investigation, which serves as a base to control the selection of experimental variables. Results are more easily communicated the closer the congruity between existing and proposed practices.

More general is the link between planning extension program content at the area level and regional or national planning. The type of model set out in comparing improvement and transformation, its parameters directly dependent on the investigation and planning sequence, shows the expected rates of expansion in both production and the demand for requisites. This gives a useful base for decisions on the development, marketing, transport, and finance facilities within the region. The planning sequence itself gives useful indications for area planning. In our example the immediate provision of credit facilities would have reduced the adoption period. At the same time, the extension progression gave a better basis for deciding the content of credit programs, the timing of the introduction of different lines of credit: for cotton insecticides, in the second year; for maize fertilizer and insecticides, in the third year; and for cotton fertilizer, in the fifth year.

Just as the investigational and planning sequence provided pointers to productive lines of research, it can equally well provide pointers to useful independent infrastructural developments. Say, for example, cowpeas are a favored legume but there is no improved technology for the crop, which is labor-intensive and hence constitutes a bottleneck to changes in practice on other crops. Promoting a flow of cowpeas through new or established retail outlets in the area, perhaps even at a subsidized price, would be an alternative to a research program on the crop. Where an adjacent area was a producer of surpluses of cowpeas, the promotion of trade between the two would be a step toward internal market differentiation and specialization in production. With purchase as an alternative source, the barrier to innovation on complementary crops is lowered and the system can expand marketed output with the labor released from the production of cowpeas.

Finally, the aggregation of expected producer reactions to changed market opportunities can be estimated from the model used in the planning sequence. If the conditions surrounding the model are reexpressed, particularly by the use of social pricing techniques for both factors and products, useful conclusions can be drawn on area potential and the necessary shifts in prices to realize it. The required diversion of productive resources could be directly stimulated by both price changes and a revision of extension content.

Clearly the discipline has a great deal to offer at the macro level. However, to do these aspects justice would require a number of studies as detailed as this one has tried to be in probing the role of farm economics in planning extension content in African peasant agriculture.

## NOTE

1. E. M. Kulp, Rural Development Planning (New York: Praeger, 1970).

# BIBLIOGRAPHY

Adegboye, Basu, and Olatunbosun. "The Impact of Western Nigerian Farm Settlements on Surrounding Farmers," Nigerian Journal of Economic and Social Studies, XI, 2 (1969).

Ashton, J., and R. F. Lord, eds. Research, Education and Extension in Agriculture. Iowa City: Iowa State University Press, 1969.

Barker, A. S. "The Example Farm in Management Investigation," Journal of Agricultural Economics, XI, 4 (1956).

Barlow, C. "Estimation of Crop Production Relations and Their Use in Farm Planning," Experimental Agriculture, II, 4.

Barnard, C. S. "Farm Models, Market Objectives and the Bounded Planning Environment," Journal of the Agricultural Economics Society, XV (1963).

Bauer, P. T. The Rubber Industry. London, 1958.

Baum, E. "Land Use in the Kilombero," in Smallholder Agriculture and Development in Tanzania. Edited by H. Ruthenburg. "Africa Studies," XXIV. Munich: IFO, 1968.

Beck, R. S. "Coffee Farming on Kilimanjaro." Dar es Salaam: Tanzania Ministry of Agriculture, 1963. (Mimeographed.)

Belshaw, D. G. R., and M. Hall. "The Analysis and Use of Agricultural Experimental Data." East African Agricultural Economists Conference paper. 1970.

Berry, L. Economic Zones of Tanzania. University of Dar es Salaam, Bureau of Resource and Land Use Planning.

Bessel, J. E., J. A. Roberts, and N. Vanzetti. "Survey Field Work." Agricultural Labour Productivity Investigation Report no. 1. Universities of Nottingham and Zambia, 1968.

________. "Some Determinants of Agricultural Labour Productivity in Zambia." Agricultural Labour Productivity Investigation Report no. 3. Universities of Nottingham and Zambia.

Bowden, E., and J. Moris. "Social Characteristics of Progressive Baganda Farmers," East African Journal of Rural Development, II, 1 (1969).

Brander and Kearl. "Evaluation for Congruence as a Factor in the Adoption Rate of Innovations," Rural Sociology, XXIX (1964).

Brock, B. "The Sociology of the Innovator." East African Agricultural Economists Conference paper. 1969.

Brown, K. J. East African Agricultural and Forestry Journal, XXVI (1960).

________. "Rainfall, Tie-Ridging and Crop Yields," Empire Cotton Growing Review (January, 1963).

________, and J. E. Peat. "Yield Responses of Raingrown Cotton in Lake Province, Tanganyika," Empire Journal of Experimental Science (1962).

Carter, H. O. "Representative Farms as Guides for Decision Making," Journal of Farm Economics, XLV, 5 (1963).

Catt, D. C. "Surveying Peasant Farmers—Some Experiences," Journal of Agricultural Economics, XVII, 1 (1966).

Chambers, Robert. Settlement Schemes in Tropical Africa. New York: Praeger, 1969.

Clayton, E. "Note on Research Methodology in Peasant Agriculture," Farm Economist, XVIII, 6 (1957).

________. Economic Planning in Peasant Agriculture. "Department of Economics Monographs." University of London, Wye College, 1963.

________. "Opportunity Costs and Decision Making in Peasant Agriculture," Netherlands Journal of Agricultural Science, XVI (1968).

Cleave, J. "The Collection, Analysis and Use of Farm Management Data in Uganda." East African Agricultural Economists Conference paper. 1965.

Clough, R. H. "Recent Experience with Farm Management Surveys in Kenya." East African Agricultural Economists Conference paper. 1970.

Collinson, M. P. "Bukumbi Area." Farm Economic Survey no. 1. Dar es Salaam: Tanzania Department of Agriculture, 1961. (Mimeographed.)

________. "Usmao Area." Farm Economic Survey no. 2. Dar es Salaam: Tanzania Department of Agriculture, 1962. (Mimeographed.)

________. "Maswa Area." Farm Economic Survey no. 3. Dar es Salaam: Tanzania Department of Agriculture, 1963. (Mimeographed.)

________. Three Farm Surveys, Nyumba ya Mungu Dam Appraisal. Dar es Salaam: Tanzania Department of Agriculture, 1963.

________. "Lwenge District." Farm Economic Survey no. 4. Dar es Salaam: Tanzania Department of Agriculture, 1964. (Mimeographed.)

________. "Experience with a Trial Management Farm in Tanzania, 1962-65," East African Journal of Rural Development, II, 2 (1969).

________. "The Evaluation of Innovations in Peasant Agriculture," East African Journal of Rural Development, I. 2 (1968).

________. "Aromatic Tobacco and Virginia Tobacco: A Comparative Survey of Two Tobaccos on Family Farms in the Tabora Region of Western Tanzania." East African Agricultural Economists Conference paper. 1970.

________. "A Survey of Innovations in Traditional Agriculture in Tanzania." East African Agricultural Economists Conference paper. 1970.

Conklin, H. C. "Hanunóo Agriculture: A Report on An Integral System of Shifting Cultivation in the Phillipines". Rome: FAO, 1957

Cory, H. Sukuma Law and Custom. London: Oxford University Press, 1954

Cotton Research Corporation. Progress Reports from Experimental Stations: Western Research Centre, Tanzania.

Coughenor, C. M. "Towards a Theory of the Diffusion of Technology," in First Interamerican Research Symposium on the Role of Communications in Agricultural Development. 1964.

Davidson H. A. and B. R. Martin. "Relationships Between Field and Experimental Yields," Australian Journal of Agricultural Economics, IX (1965).

Day, L. M. "The Use of Representative Farms in Studies of Interregional Competition and Production Response," Journal of Farm Economics, XLV, 5 (1963).

Day, R. H. "On Aggregating Linear Programming Models of Production," Journal of Farm Economics, XLV, 4 (1963).

________. "More on the Aggregation Problem," Journal of Farm Economics, LI, 3 (1969).

De Schlippe, P. Agricultural Statistics for Developing Countries. 1969.

________, and B. L. Batwell. "Preliminary Study of the Nyangwara System of Agriculture in the Southern Sudan," Africa, XXV, 4 (1955).

De Wilde, J. C. Experiences with Agricultural Development in Tropical Africa. Baltimore: Johns Hopkins Press, 1967.

Dexter, K. "Report on Discussion Group D," in Proceedings of the Conference of the International Association of Agricultural Economists. 1967.

Doggett, H. "Bird Damage in Sorghum," East African Agricultural and Forestry Journal, XVI, 1.

East African Statistical Unit. Investigation into the Measurement of Acreage Statistics. Kampala, 1959.

Eisgruber, L. M. "Micro and Macro Potential of Agricultural Information Systems," Journal of Farm Economics, XLIX (1967).

El Adeemy, A. S. and J. D. MacArthur. "The Identification of Modal Type Farm Situation in North Wales," Farm Economist, XI (1970).

Elliot, F. F. "The Representative Firm Idea Applied to Research and Extension in Agricultural Economics," Journal of Farm Economics, X (1928).

Emery, F. E. and O. A. Oeser. Information, Decision and Action: A Study of the Psychological Determinates of Changes in Farming Technique. Carleton, Victoria: Melbourne University Press, 1958.

Evans, A. C. "Studies of Intercropping. I," East African Agricultural and Forestry Journal, XXVI (1960).

________. "Time of Planting Studies in Tanzania," East African Agricultural and Forestry Journal, XXVI (1964-65).

________, and Sheedran. "Intercropping Studies. II," East African Agricultural and Forestry Journal (1961).

Fallers, L. A. The King's Men. London: Oxford University Press, 1964.

Fortes, M., and F. L. Fortes. "Food in the Domestic Economy of the Tallensi," Africa, IX.

Fox, R. H. "Studies of the Energy Intake and Expenditure Balance Among African Farmers in the Gambia." Ph.D thesis. London: Medical Research Council, 1953.

Frick, G. E. and R. H. Andrews. "Aggregation Bias and Four Methods of Summing Farm Supply Functions," Journal of Farm Economics, XLVII, 3 (1965).

Geertz, C. Agricultural Involution. Berkeley and Los Angeles: University of California Press, 1968.

Ghosh, A. "Accuracy of Family Budget Data with Reference to the Period of Recall," Bulletin of the Calcutta Statistical Association, V (1953).

Goodfellow, D. M. Principles of Economic Sociology. London, 1939.

Grimble. Kampala: Uganda Department of Agriculture. (Mimeographed.)

Groeneveld, S. Problems of Agricultural Development in the Coastal Region of East Africa, "Africa Studies," XIX Munich: IFO, 1967.

Grundy, Healy, and Rees. "The Economic Choice of the Amount of Experimentation," Journal of the Royal Statistical Society, series B, VIII (1956).

Hall, M. "A Review of Farm Management Research in East Africa," East African Agricultural Economists Conference Paper (1970).

Handbook of the Natural Resources of East Africa.

Hartley, H. O. "Experimental Designs for Estimating the Characteristics of Production Functions." OECD Agricultural Documentation no. 65.

Heady, E. O. "Objectives of Farm Management Research," Journal of Farm Economics, XXX (1948).

________. "The Agricultural Economist and His Tools," in Proceedings of the Conference of the International Association of Agricultural Economists. 1961.

Herskovits and Horowitz, eds. Estimates made by the UN Economic Commission for Africa.

Heyer, J. "Linear Programming Models for Peasant Agriculture." Ph.D. thesis. London.

________. "Seasonal Labour Inputs in Peasant Agriculture." East African Agricultural Economists Conference paper. 1965.

Hill, P. "The Myth of the Amorphous Peasantry—a Northern Nigerian Case Study," Nigerian Journal of Social and Economic Studies, X, 2 (July, 1968).

Hirsch, G. P. "Peasant Experimental Farms as a Research Technique," Farm Economist, VIII, 9 (1957).

Houseman, E. E. and H. F. Huddleston. "Forecasting and Estimating Crop Yields from Plant Measurements," FAO Monthly Bulletin of Agricultural Economics and Statistics, XV, 10 (1966).

Hunt, K. E. Agricultural Statistics for Developing Countries. Oxford: Oxford University Press, 1970 (revised).

Hurley, R. "Problems Relating to the Criteria for Farm Classification," Journal of Farm Economics, XLVII, 5 (1965).

Hutchinson, J. "The Objectives of Research in Tropical Agriculture," Empire Cotton Growing Review, XXXV (1958).

IBRD. The Economic Development of Tanganyika. 1960.

Johnson, R. W. M. "The African Village Economy, an Analytical Model," Farm Economist, XI, 9 (1968).

________. "The Labour Economy of the Reserves," Dcc. paper no. 4. Salisbury: University College of Rhodesia and Nyasaland, 1964.

Jolly, A. L. Report on Unit Farms to the Caribbean Secretariat. Trinidad: Imperial College of Tropical Agriculture, 1953. (Mimeographed.)

Joosten, J. H. L. Wirtschaftliche und Agrarpolitische Aspecte Tropischer Landbau Systeme. 1962.

Jowett, D. "Use of Rank Correlation Techniques to Determine Food Preferences," Experimental Agriculture, II, 3 (1966).

Joy, J. L. "The Economics of Food Production," African Affairs, XXVI.

Kennedy, T. J. "Cotton Farmers' Motivations in Kisumu," East African Economic Review, new series II.

Kenya Farm Economic Survey Unit. Series of farm surveys.

Kiehl, E. R. "A Critical Appraisal of the State of Agricultural Economics in the Mid Sixties," Journal of Farm Economics, XLIII (1961).

Kiray, M., and J. Hinderink. "Interdependencies Between Agroeconomic Development and Social Change," Journal of Development Studies, IV, 4 (1968).

Kreinin, M. E. "The Introduction of Israel's Land Settlement Plan to Nigeria," Journal of Farm Economics, XLV, 3 (1963).

Kulp, E. M. Rural Development Planning. New York: Praeger, 1970.

Larsen, A. "A Choice of Sampling Population in Rural Sukumaland." East African Agricultural Economists Conference paper. 1970.

Lewis, A. W. "Education and Economic Development," International Social Science Journal, XIV, 4 (1962).

Ludwig, H. D., Ukara. A Special Case of Land Use in the Tropics. Edited by H. Ruthenburg. "Africa Studies," XXIV. Munich: IFO, 1967. (In German.)

Luning, H. A. An Agro-Economic Survey of Katsina Province. Northern Nigeria Government Printer, 1963.

________. "Patterns of Choice Behaviour on Peasant Farms in Northern Nigeria," Netherlands Journal of Agricultural Science, XV (1967).

________. "Farm Economic Survey and Planning, Rungwe, Tanzania." "Africa Studies," Technical Paper 3. University of Leiden, 1969.

MacArthur, J. D. "The Economic Study of African Small Farms—Some Kenya Experiences," Journal of Agricultural Economics, XIX, 2 (May, 1968).

________. "Labour Costs and Utilisation in Rice Production of the Mwea/Tebere Irrigation Scheme," East African Agricultural and Forestry Journal, XXXIII, 4 (1968).

Mahalanobis, P. C., and S. B. Sen. "Some Aspects of the Indian National Sample Survey," Bulletin of the International Statistical Institute, XXXIV, 2 (1954).

McMeekan, C. P. "Co-ordinating Economic and Technical Research," in Proceedings of the Conference of the International Association of Agricultural Economists. 1964.

Mellor, J. "Family Labour in Agricultural Development," Journal of Farm Economics, XLV (1963).

Moody, A. "A Report on a Farm Economic Survey of Tea Smallholdings in Bukoba District, Tanzania." East African Agricultural Economists Conference paper. 1970.

Moris, J. "The Application of Adoption Theory to the Study of Agricultural Development in East Africa." East African Agricultural Economists Conference paper. 1969.

Morrow. The Study of Labour Use on Aromatic Tobacco Growing on Large Scale Farms. Salisbury. (Mimeographed.)

Mynt, H. The Economics of Developing Countries. London, 1964.

________. "Dualism and the Internal Integration of the Underdeveloped Economies." (Mimeographed.)

Newiger, N. "The Village Settlement Schemes in Tanzania," in Smallholder Agriculture and Development in Tanzania. Edited by H. Ruthenburg. "Africa Studies," XXIV. Munich: IFO.

Panse. Various papers on yield surveys.

Parker, R. N. "Intercropping." B.Sc. dissertation. University of Reading, 1969.

Peart, B. "Farm Management Advisory Work," Agriculture, LXXVI, 9. 1969.

Pudsey, D. "A Pilot Study of 12 Farms in Toro." Kampala: Uganda Department of Agriculture, 1966. (Mimeographed.)

________. "An Economic Study of the Farming of the Wet Long Grass Area of Toro." Kampala: Uganda Department of Agriculture, 1967. (Mimeographed.)

________. "The Economics of Outgrower Tea Production in Toro, Western Uganda." Kampala: Uganda Department of Agriculture, 1967. (Mimeographed.)

Raeburn, J. in Proceedings of the Conference of the International Association of Agricultural Economists. 1955.

Read, M. "Native Standards of Living and African Cultural Change," supplement to Africa, XI, 3 (1938).

Reh, E. A Manual on Household Consumption Surveys. "Nutritional Studies," 18. Rome: FAO, 1962.

Reid, I. G. Discussion of Barker paper, Journal of Agricultural Economics, XI, 4 (1956).

Renbourg, U. Studies on the Planning Environment of the Agricultural Firm. Uppsala, Sweden: University of Uppsala, 1962.

________. Discussion of Weinschenk paper, in Proceedings of the Conference of the International Association of Agricultural Economists. 1964.

Richards, A. T. Land, Labour and Diet in Northern Rhodesia. Oxford: Oxford University Press, 1939.

Robertson, and Stoner. Shell Symposium on New Possibilities and Techniques for Land Use Surveys with Special Reference to Developing Countries.

Robinson, J. D. "Extensive Trials with Perennial Crops on Smallholders' Fields in Africa," Experimental Agriculture, II, 4.

Rogers, E. M. "Categorising Adopters of Agricultural Practices," Rural Sociology, XXIII (1958).

Rounce, N. V. "Crop Acreage Survey of Sukumaland." 1945. (Manuscript.)

________. The Agriculture of the Cultivation Steppe. Longmans, 1948.

Roy, P., F. C. Fliegel, J. E. Kilvin, and L. K. Sen. Agricultural Innovation Among Indian Farmers. 1968.

Ruthenburg, H., ed. Smallholder Agriculture and Development in Tanzania. "Africa Studies," XXIV. Munich: IFO, 1968

Ruttan, V. W. "Issues in the Evolution of Production Economics," Journal of Farm Economics, XLIX (1967).

Scaife, A. M. "Maize Fertiliser Experiments in Western Tanzania," Journal of Agricultural Science, LXX (1968).

Scheffler, W. Smallholder Production Under Close Supervision: Tobacco Growing in Tanzania, "Africa Studies," XXVII. Munich: IFO, 1968. (In German.)

Schroder, W. R. "The Application of Systems Models to the Analysis of Swine Management Problems," Dissertation Abstracts (Ann Arbor), XXIX, 3 (1968).

Schultz, T. "The Theory of the Firm and Farm Management Research," Journal of Farm Economics, XXI (1939).

Sheehy, S. J. and R. H. McAlexander. "The Selection of Benchmark Farms," Journal of Farm Economics, XLVII (1965).

Shepherd, D. H. "Advisors' Opinions on Advisory Methods," NAAS Quarterly Review, 69 (1965).

Simmons, N. S. "Plant Interactions and Crop Yields," New Scientist, XVIII (1963).

Southworth, H. M., and B. F. Johnston, eds. Agricultural Development and Economic Growth. Ithaca, New York: Cornell University Press, 1967.

Tanzania Central Statistical Bureau. Kilosa Acreage Survey. Dar es Salaam, 1963

________. 1967 Population Census. Dar es Salaam, 1968.

________. Tanzania Population Changes 1948-67. Dar es Salaam, 1968.

Tanzania Government Printer. First Five-Year Plan. Dar es Salaam, 1964.

________. Second Five-Year Plan, I and IV. Dar es Salaam, 1969.

Tanzania Ministry of Agriculture. Annual reports.

Thompson, J. F. "Defining Typical Resource Situations," U.S. Southern Cooperative, Farm Management Bulletin, 56 (1958).

Thompson, S. C. Monte Carlo Programming Techniques. University of Reading, Department of Agriculture, 1970.

UNESCO. World Education Statistics.

Upton, M. Agriculture in South-Western Nigeria. "Development Studies," 3. University of Reading, 1967.

U.S. Southern Cooperative. Farm Management Bulletin, 56 (1958).

Von Rotenham, D. Land Use and Animal Husbandry in Sukumaland, "Africa Studies," XI. Munich: IFO, 1966. (In German.)

Weinschenk G. "Quantitative Research in Agricultural Economics," in Proceedings of the 1964 Conference of the International Association of Agricultural Economists. London: Oxford University Press, 1964.

Wells, J. C. "Nigerian Government Spending on Agricultural Development, 1962-67," Nigerian Journal of Economic and Social Studies, IX, 3.

Wilcox, W. "Types of Farming Research and Farm Management," Journal of Farm Economics, XX (1938).

Wragg, S. R. "Co-operative Research in Agriculture and the Provision of Input/Output Coefficients," Journal of Agricultural Economics, XXI (1970).

Yates, F. Sampling for Census and Survey. London, 1949.

________. "Principles Governing the Amount of Experimentation Needed in Development Work," Nature, CLXX (1952).

Zarkovich, S. S. The Quality of Sample Statistics. Rome: FAO, 1964.

# POSTSCRIPT

Interest in a viable approach for applying the principles of farm management to peasant agriculture has accelerated since the late 1960s and early 1970s. Several of the small number of farm management economists working in Africa at that time were preoccupied with three aspects of the problem:

1. The economic logic of the practices of small farmers, given their circumstances. (David Norman's work in northern Nigeria broke ground in this area; see, for example, Norman, 1974.)
2. The need to evaluate new technologies within the context of the system being operated by the farmer, a theme central to this book.
3. A growing feeling that technologies emerging from traditional agricultural research were often not relevant to the needs of small farmers (see, for example, Belshaw and Hall, 1972).

Work in these areas has resulted in farming systems research—the link between farm management principles and agricultural research. Using the investigative concepts and methods of farm management to understand, and sometimes to model, existing farm systems, the farm systems economist identifies the key constraints on the expansion of the system. The economist works alongside technical scientists, who suggest alternatives or additions to present management practices that will relax or avoid those constraints. Materials and practices that may solve farmers' problems, and that can be handled with their limited resources, become the elements of a program of adaptive experiments to be carried out on local fields. Recommendations on specific treatments are made to the extension services if they receive a favorable assessment from farmers. The interaction between economists and technical scientists allows the selection and testing of those technologies closely relevant to farmers' needs and capabilities; it also focuses the programs of disciplinary and commodity researchers by referring back to research centers the unsolved technical problems identified as important to the development of small farms.

There is no consensus in the profession that farming systems research is *the* way to apply farm management principles to African peasant agriculture. Support, however, is growing for the approach as a cost-effective way of increasing the relevance of technology extended to small farmers.

## REFERENCES


Belshaw, D. G. R., and M. Hall. "The Analysis and Use of Experimental Data in Tropical Africa," *Eastern African Journal of Rural Development* V, 39–72 (1972).

Norman, D. W. "Rationalising Mixed Cropping Under Indigenous Conditions: The Example of Northern Nigeria," *Journal of Development Studies* XI, 3–21 (1974).

## BIBLIOGRAPHY

For those interested in additional information on farming systems and the farming systems research approach, I suggest:

Collinson, M. P. "Microlevel Accomplishments and Challenges for the Less Developed World," in G. L. Johnson and A. Maunder, eds., *Rural Change: The Challenge for Agricultural Economists.* Proceedings of the Seventeenth International Association of Agricultural Economists Conference (1979).

Norman, D. W., E. B. Simmons, and H. M. Hays. *Farming Systems in the Nigerian Savanna: Research and Strategies for Development.* Westview Press (1982).

Ruthenberg, H. *Farming Systems in the Tropics,* 3rd edition. Oxford University Press (1980).

Shaner, W. W., P. F. Philipp, and W. R. Schmehl, eds. *Farming Systems Research and Development: Guidelines for Developing Countries.* Westview Press (1982).

Shaner, W. W., P. F. Philipp, and W. R. Schmehl, eds. *Readings in Farming Systems Research and Development.* Westview Press (1982).

# INDEX

# About the Book and Author

## *Farm Management in Peasant Agriculture* by Michael Collinson

First published in 1972, *Farm Management in Peasant Agriculture* remains the only detailed discussion of on-site research techniques for economists working on the development of small-holder agriculture in Africa. Part 1 describes the conditions of the agricultural sector within which the African peasant farmer must operate, and then outlines an approach to farm management tailored to those conditions. Part 2 sets out the research planning and investigation tasks implied by the approach. Survey techniques, as well as the value of a pre-survey for understanding general attributes of a farm system, are reviewed, and alternative data-collection methods are elaborated. Part 3 shows how research data can be used in planning content for extension programs. Dr. Collinson concludes with the details of a planning method that interpolates changes in farm practice into a model of the existing farm system and that projects a sequence of changes, representing a sequence of extension content, on the basis of farmer acceptability.

Dr. Collinson is regional economist for eastern, central, and southern Africa with the International Maize and Wheat Improvement Center (CIMMYT).

For Product Safety Concerns and Information please contact our EU representative GPSR@taylorandfrancis.com Taylor & Francis Verlag GmbH, Kaufingerstraße 24, 80331 München, Germany

Batch number: 10725913

Printed by Printforce, the Netherlands